One–dimensional and Two–dimensional NMR Spectra by Modern Pulse Techniques

One-dimensional and Two-dimensional NMR Spectra by Modern Pulse Techniques

Edited by

Koji NAKANISHI

Professor, Department of Chemistry, Columbia University, New York 10027, U.S.A.
and
Director, Suntory Institute for Bioorganic Research, Osaka 618, Japan

KODANSHA
Tokyo

1990

UNIVERSITY SCIENCE BOOKS
Mill Valley, California

Copublished by
Kodansha Ltd., Tokyo
and
University Science Books, Mill Valley, California

Distributed exclusively in Japan by Kodansha Ltd.
Distributed exclusively in North/South America,
Continental Europe, Australia and New Zealand by
University Science Books.

ISBN 4-06-203665-7 (Japan)
ISBN 0-935702-63-6
LIBRARY OF CONGRESS CATALOG CARD NUMBER: 90-070367

Printed in Japan

Contributors

Parts in parentheses indicate the parts to which the authors contributed.

Takenori KUSUMI (Parts I & II)
Department of Chemistry, University of Tsukuba, Tsukuba 305, Japan

Takashi IWASHITA (Parts I, II & III)
Suntory Instutute for Bioorganic Research, Osaka 618, Japan

Hideo NAOKI (A part of parts I & II)
Suntory Institute for Bioorganic Research, Osaka 618, Japan

The initials in brackets indicate the operator of each spectrum.

[TK]	Takenori Kusumi, Ph. D. (The University of Tsukuba)
[TI]	Takashi Iwashita, Ph. D. (Sunbor)
[HN]	Hideo Naoki, Ph. D. (Sunbor)
[KM]	Kosei Mizukawa, Ph. D. (Suntory Co., Ltd)
[LJ]	LeRoy F. Johnson, Ph. D. (GE NMR Instruments)
[MW]	Markus R. Wälchli, Ph. D. (Bruker Japan Co. Ltd.)
[MZ]	Michael G. Zagorski, Ph. D. (Columbia University)
[JEOL]	JEOL Ltd.

Preface

Of the various physicochemical measurements, none is more popular than nuclear magnetic resonance. Its applications are boundless. It can handle samples ranging from conventional organic compounds to biopolymers, plant fruits to human bodies, most elements listed in the periodic table, dynamic and static problems, etc., while solid state NMR enables one to measure membrane proteins and pursue enzyme reactions. The strength of the magnetic field was 60 MHz (for protons) several years ago, but is now 500 MHz, and 600 MHz instruments are not unusual.

A major problem for most users, however, is that new techniques and new pulse sequences are being developed so rapidly that unless one is a professional NMR spectroscopist, it is not easy to select the best practical technique. The difficulty is greater for scientists and students who lack direct access to NMR facilities. Even for researchers who deal daily with NMR spectra, it is time-consuming to retrieve from journals or monographs suitable spectra which demonstrate the advantages of a specific NMR technique.

The first 360 MHz instrument was installed at the Suntory Institute for Bioorganic Research (Sunbor) in 1980, and since then additional 300 MHz and 500 MHz instruments have been added. Over the years, the Sunbor NMR lab has measured many samples from numerous institutes. To facilitate practical use we decided to publish selected spectra in a picture book format with minimal theory. Each example is accompanied by brief comments, and each starts on a new page to facilitate the reading. Part I depicts 1D NMR, Part II shows examples of 2D NMR, and Part III gives brief explanations of basic FT NMR principles.

We have attempted to illustrate a wide variety of NMR applications and, in some cases, have selected simple compounds to demonstrate the characteristics of a particular technique. There are cases where a technique is omitted because of lack of suitable examples; we welcome input from readers for appropriate examples.

The current volume is based on a Japanese edition published in 1986, which has been revised and updated. We are grateful to Professor Kunio Hikiji, Hokkaido University, and Professor Hiroshi Kakisawa, the University of Tsukuba for suggestions, to various authors for permission to quote spectra, to Professor Dinshaw J. Patel, Columbia University, Bruker Japan, JEOL, and Dr. LeRoy F. Johnson of GE for supplying spectra. To Dr. Ritsuko N. Arison and Dr. Peter T. M. Kenny (Sunbor) for translating the manuscript into English, To Dr. Yoko Naya and the Sunbor staff for helpful discussions and to Messrs. Shizuo Sawada, Ippei Ohta and Masanori Sato of Kodansha Scientific for their patience with the authors.

Koji Nakanishi
December 22, 1989

Glossary

Data Table

[1D-NMR]	
ONE-PULSE SEQUENCE	Name of pulse Sequence
P2= 4.00 USEC	Pulse width ($\theta°$: μsec)
D5= 2.00 SEC	Recovery interval for spin-lattice relaxation
NA= 16	No. of acquisitions
SIZE= 32768	No. of data points
AT= 3.28 SEC	Data acquisition time
QPD ON=1 ABC ON	QPD : Phase of the pulse changes 0°, 90°, 180°, and 270° with ABC
BUTTERWORTH FILTER ON DB ATT.=3	Filter for removing noise
ADC= 12 BITS	No. of bits in analog-digital converter
AI= 2	Absolute intensity
SW=+/-2500.00	Spectral width (Hz)
DW= 200	Dwell time per point for data (μsec) (reciprocal of spectral width)
RG= 10 USEC	Width of trigger pulse for start of FID acquisition
DE= 200 USEC	Delay before start of FID acquisition
TL HIGH POWER ON	Switch on transmitter output
F2= 360.053338	Decoupler frequency (MHz)
BB MODULATION OFF	Turn on when noise modulation is applied to decoupler (^{13}C-NMR)
OF= 1690.57	Offset frequency from TMS signal to the carrier radiofrequency (Hz)
SF= 360.053338	Spectrometer frequency (MHz) (Carrier radiofrequency)
EM= 0.20	Exponential filter for FT (Line broadening factor is 0.2 Hz in this case.)
PA= 280.9 PB= 77.8	Phase correction parameter
SCALE 105.90 HZ/CM = .2941 PPM/CM	Scale (horizontal) for printout

(The abbreviations in this table are used in NT and GN spectrometers which are made by Nicolet Instrument Corparation or General Electric Company. Therefore, other instruments may use different abbreviations. The notation, e.g. P1, D3, D4, D6 etc., has meanings different from this table in some spectra.)

[2D-NMR]	
H, X-COSY HA=8★, QP=+0, AB=-I8= DW/2(H)	Name of pulse sequence and comment
P1= 46.00 USEC	180° pulse (^{13}C)
P2= 23.00 USEC	90° pulse (^{13}C)
D3= 3.30 MSEC	Δ_1 (see Part III section2.4)
D4= 2.00 MSEC	Δ_2 (see Part III section 2.4)
D5= 1.00 SEC	Delay time (T)
D6= 36.00 USEC	90° pulse (^{1}H)
D8= 180.00 USEC	$t_1/2$: corresponds to spectral width of ^{1}H side (see Part III section 2.4)
I8= 180.00 USEC	
NA= 256	
SIZE= 4096	
AT= 126.98 MSEC	
QPD ON= 0 ABC OFF	Phase of the pulse is controlled in pulse sequence
BUTTERWORTH FILTER ON	
DB ATT.=3	
ADC= 12 BITS	
AI= 4	
SW=+/-8064.51	
DW= 62	
RG= 10 USEC	
DE= 62 USEC	
TL HIGH POWER ON	
F2= 360.055380	Decoupler frequency (^{1}H ; MHz)
BB MODULATION ON	
OF= 8876.74	
SF= 90.544405	Spectrometer frequency of X nucleus (^{13}C in this instance ; MHz)
SCALE 845.16 HZ/CM = 7.1253 PPM/CM	
MATRIX (F1 X F2)=256 X 4K	Size of data matrix
F1, F2 : DM=3	Digital filter for data processing
CONTOURLEVEL=6 (YS=11371)	Contour level at data readout

Pulse Sequence

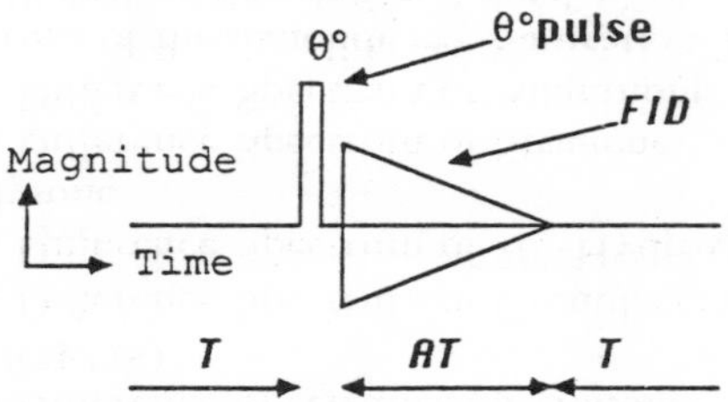

pulse : Radiofrequency pulse

θ° : Pulse rotation angle (flip angle)

ϕ, φ, x, y, , , : The phase change

FID : Free induction decay

AT : Acquisition time

t_2 : The elapsed time of recording the FID

t_1 : The length of evolution period in 2D-NMR

T : Recovery interval for spin-lattice relaxation

 : The irradiation time for decoupling

CW : Continuous wave irradiation

B.B.D. : Broadband decoupling

τ, Δ : Delay time

τ_m : Mixing time

Spectrum

(a) 1D-NMR

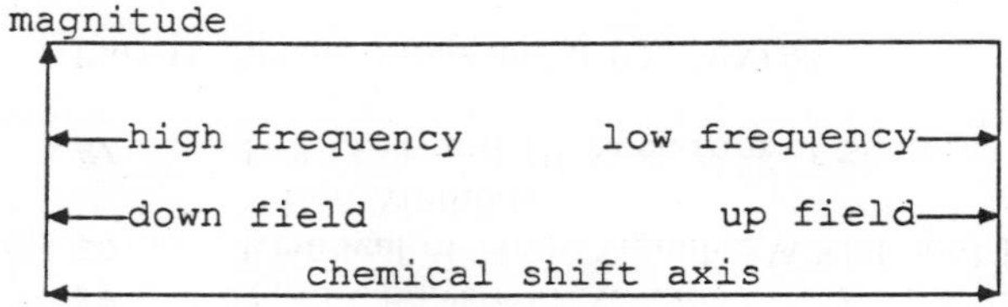

(b) 2D-NMR (Stacked plots)

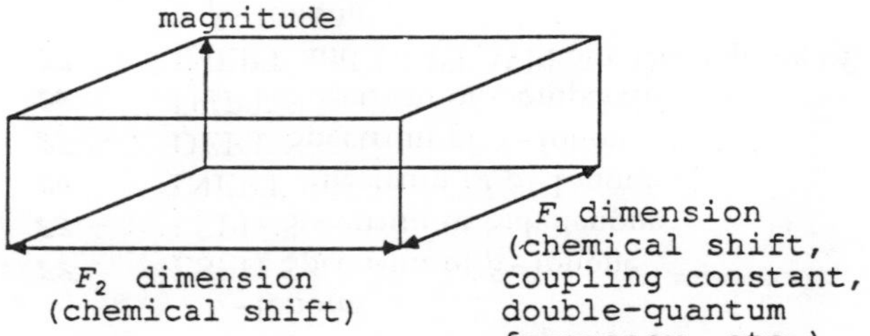

(c) 2D-NMR (Contour plot)

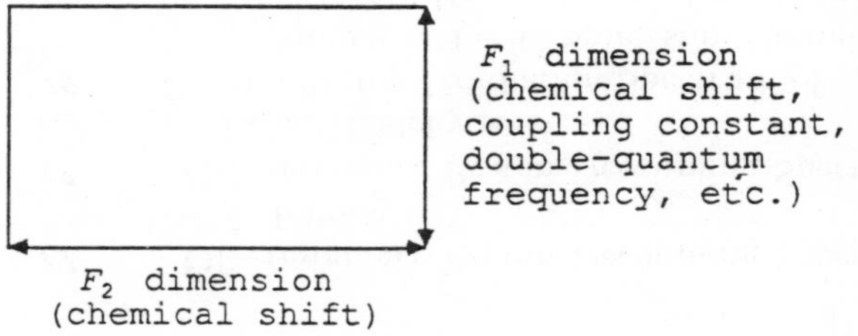

Contents

Preface vii
Glossary viii

Part I One-dimensional FT-NMR

1 ^{1}H-NMR Spectra of α-Santonin at 100 and 500MHz 2
2 ^{1}H-NMR Spectrum of Acetoxypachydiol Dibenzoate at 90 MHz 4
3 ^{1}H-NMR of Methyl 9,12-Octadecadienoate (with Shift Reagent) 6
4 Spin Decoupling Difference Spectrum 8
5 Spin Decoupling Difference Spectrum of β-Ionone 10
6 Spin Decoupling Difference Spectrum of Cholic Acid 12
7 Spin Decoupling Difference Spectrum of Mosesin-2 14
8 NOE Difference Spectrum of β-Ionone (NOEDS) 16
9 NOE Difference Spectrum of Crenulacetal A 18
10 NOE Difference Spectrum of 20-Hydroxyecdysone 20
11 NOE Difference Spectrum of Striatene 22
12 NOE Difference Spectrum of Ailanthone 24
13 Spectrum of Bassianolide — Saturation Transfer 26
14 SPT (Selective Population Transfer) of β-Ionone 28
15 SPT Spectrum of Aspartic Acid 30
16 Broadband Proton Decoupling Spectrum of β-Ionone 32
17 Off-resonance Proton Decoupling Spectrum of β-Ionone 34
18 Off-resonance Proton Decoupling Spectrum of Isobutylaldehyde 36
19 Gated Proton Irradiation Spectrum of β-Ionone (^{1}H-^{13}C Spin-spin Couplings) 38
20 Inverse Gated Proton Decoupling Spectrum of β-Ionone (No NOE) 40
21 Selective Proton Decoupling Spectrum of β-Ionone 42
22 LSPD Spectrum of β-Ionone 44
23 LSPD Spectrum of Salvilenone 46
24 INEPT Spectrum of β-Ionone 48
25 DEPT Spectrum of β-Ionone 50
26 DEPT Spectrum of Compactin 52
27 INEPT and DEPT Without Decoupling of β-Ionone 54
28 T_1 (Longitudinal Relaxation/Spin-lattice Relaxation) of β-Ionone 56
29 T_1 of Cholesteryl Acetate 58
30 Removal of Huge Signals (WEFT and Presaturation Methods) 60
31 1-3-3-1 and JR Spectrum of Sucrose 62

Part II Two-dimensional FT-NMR

32 J-Resolved Spectrum of α-Santonin 66
33 J-Resolved Spectrum of Mugineic Acid 68
34 J-Resolved Spectrum of Semburin 70
35 J-Resolved Spectrum of Trichilin-A 72
36 Heteronuclear J-Resolved Spectrum of β-Ionone 74

37 COSY Spectrum of Ethyl Acetate 76
38 COSY Spectrum of β-Ionone 78
39 COSY Spectrum of Compactin (I) 80
40 COSY Spectrum of Compactin (II) 82
41 COSY Spectrum of Chromazonarol (I) 84
42 COSY Spectrum of Chromazonarol (II) 86
43 COSY Spectrum of Chromazonarol (III) 88
44 COSY Spectrum of Chromazonarol (IV) 90
45 COSY Spectrum of Albocycline 92
46 COSY Spectrum of Adenanthin 94
47 COSY Spectrum of Aphanamol-I 96
48 DCOSY Spectrum Emphasizing Long-range Coupling (Aphanamol-I) 98
49 COSY-45 Spectrum 100
50 COSY-45 Spectrum of Aphanamol-I 102
51 Phase-sensitive DQF-COSY Spectrum of L-Aspartic Acid (I) 104
52 Phase-sensitive DQF-COSY Spectrum of L-Aspartic Acid (II) 106
53 Phase-sensitive DQF-COSY Spectrum of Chromazonarol (I) 108
54 Phase-sensitive DQF-COSY Spectrum of Chromazonarol (II) 110
55 HOHAHA Spectrum of Gramicidin S 112
56 HOHAHA Spectrum of Brevetoxin A (I) 114
57 HOHAHA Spectrum of Brevetoxin A (II) 116
58 NOESY Spectrum of β-Ionone 118
59 NOESY Spectrum of Ailanthone 120
60 NOESY Spectrum of Aphanamol-I 122
61 NOESY Spectrum of Chromazonarol 124
62 Phase-sensitive NOESY Spectrum Ailanthone (I) 126
63 Phase-sensitive NOESY Spectrum of Ailanthone (II) 128
64 Phase-sensitive NOESY Spectrum of Gramicidin S 130
65 Phase-sensitive NOESY Spectrum of DNA 12-mer 132
66 Phase-sensitive NOESY Spectrum of Bassanolide 134
67 ROESY (CAMELSPIN) Spectrum of Gramicidin S 136
68 H,C-COSY Spectrum of β-Ionone 138
69 H,C-COSY Spectrum of L-Menthol 140
70 H,C-COSY Spectrum of Compactin (I) 142
71 H,C-COSY Spectrum of Compactin (II) 144
72 H,C-COSY Spectrum of Aphanamol-I 146
73 H,C-COSY Spectrum of Metasequoic Acid A (I) 148
74 H,C-COSY Spectrum of Metasequoic Acid A (II) 150
75 H,C-COSY Spectrum of Methyl 2,3,5-tri-*O*-acetyl-β-D-fucofuranoside (I) 152
76 H,C-COSY Spectrum of Methyl 2,3,5-tri-*O*-acetyl-β-D-fucofuranoside (II) 154
77 H,P-COSY Spectrum of Lipid A 156
78 Phase-sensitive H,C-COSY Spectrum of Strychnine 158
79 HMQC Spectrum of Strychnine 160
80 Long-range H,C-COSY Spectrum of β-Ionone 162
81 COLOC Spectrum of Dictyotalide B 164
82 Long-range H,C-COSY Spectrum of Compactin 166
83 HMBC Spectrum of Strychnine 168
84 HMBC Spectrum of Cyanoviridin RR (I) 170
85 HMBC Spectrum of Cyanoviridin RR (II) 172
86 Relayed H,C-COSY of α-Santonin 174
87 Relayed H,C-COSY/HOHAHA of α-Santonin 176
88 2D-INADEQUATE Spectrum of L-Menthol 178
89 2D-INADEQUATE Spectrum of Cholesteryl Acetate 180
90 2D-INADEQUATE Spectrum of Rotenone 182

91	2D-INADEQUATE Spectrum of Compound X'	184
92	2D-INADEQUATE Spectrum of Brevetoxin B	186

Part III Principles of FT-NMR

1	FT-NMR	190
1. 1	1 Pulse Experiment	192
1. 2	Spin Decoupling	194
1. 3	Spin Decoupling Difference Spectrum (SDDS)	195
1. 4	Long-range Selective Proton Decoupling (LSPD)	195
1. 5	Gated Decoupling and Inverse Gated Decoupling	196
1. 6	NOE Difference Spectrum (NOEDS), Selective Population Transfer (SPT), and Saturation Transfer	196
1. 7	Inversion Recovery (T_1 Measurement), PRFT,WEFT	199
1. 8	Spin Echo (T_2 Measurement)	200
1. 9	INEPT	202
1.10	DEPT	203
1.11	Selective Excitation : DANTE, Redfield 214, Jump and Return (JR), 1-3-3-1	204
1.12	INADEQUATE	205
2	Two-dimensional FT-NMR	207
2. 1	*J*-Resolved Spectroscopy (^{1}H)	208
2. 2	Heteronuclear *J*-Resolved Spectroscopy	209
2. 3	COSY,COSY-45, PCOSY,DCOSY	211
2. 4	H,X-COSY	214
2. 5	NOESY	218
2. 6	Phase-Sensitive NOESY	219
2. 7	ROESY (or CAMELSPIN) and HOHAHA	222
2. 8	RELAY	224
2. 9	HMQC and HMBC	226
2.10	2D-INADEQUATE	227
2.11	Multiple-quantum Filter	228

Bibliography	230
Index	231

Part I

One-dimensional FT-NMR

1 ^{1}H-NMR Spectra of α-Santonin at 100 and 500 MHz

With increased availability of superconducting magnets, high field NMR spectroscopy has become a reality. High magnetic field provides the following advantages: 1) signal sensitivity is increased; 2) signal separation is improved; and 3) the pattern of splitting approaches first-order (increased symmetry of pattern), which facilitates interpretation. Signal separation increases because, in addition to increased separation of resonances at higher fields, overlapping is further reduced by the fact that spin-spin coupling constants (J) are independent of magnetic field strength[1)], and hence a multiplet will take up relatively less spectral width at the higher fields. This should be apparent from the illustration (right). In a 60 MHz spectrum, doublet A ($J = 12$ Hz) with a chemical shift of 1.70 ppm overlaps with quartet ($J = 6$ Hz), which has a chemical shift of 1.55 ppm. In a 360 MHz spectrum, on the other hand, with the chemical shift axis on the same scale, the spin-spin coupling in ppm is 1/6 of the above and the signals are completely separated.

Spectrum (A) is a ^{1}H-NMR of α-santonin obtained at 100 MHz, and spectrum (B) is that of the same sample at 500 MHz. It is obvious that signal separation is greatly superior in (B). In (A), signals in the 1.5-2.7 ppm region completely overlap and analysis is impossible, whereas in (B) the overlap is eliminated and the signals are all separated. The power of high field NMR is appreciated when an expansion of this region is studied.

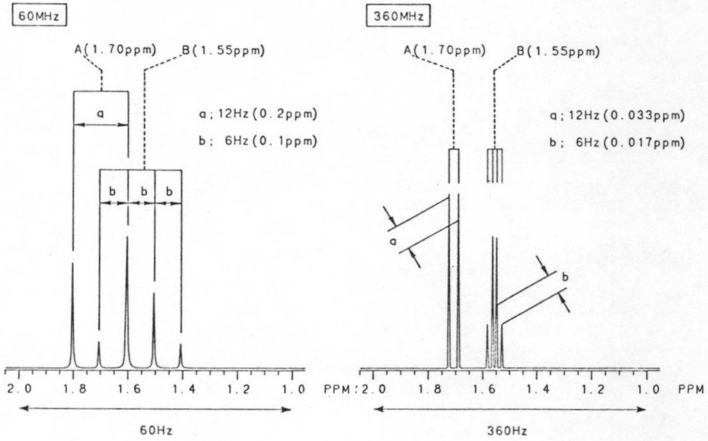

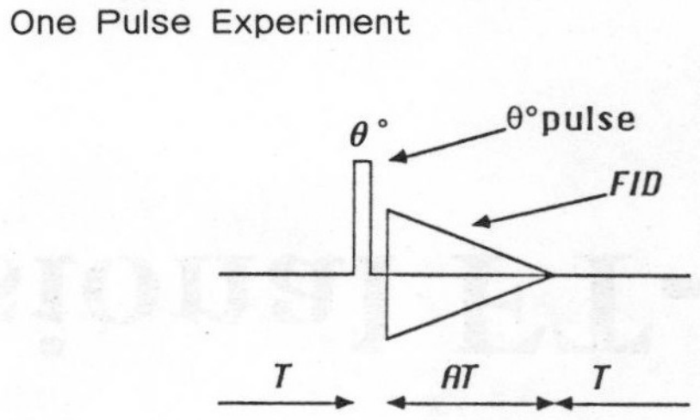

pulse : Radiofrequency pulse
$\theta°$: Pulse rotation angle (flip angle)
FID : Free induction decay
T : Recovery interval for spin-lattice relaxation
AT : Acquisition time

1) R.K. Harris, *Nuclear Magnetic Resonance Spectroscopy—A Physicochemical View*, Pitman Books Ltd., London (1983) p.17

(B) 500 MHz ^{1}H-NMR ($CDCl_3$) [TI]

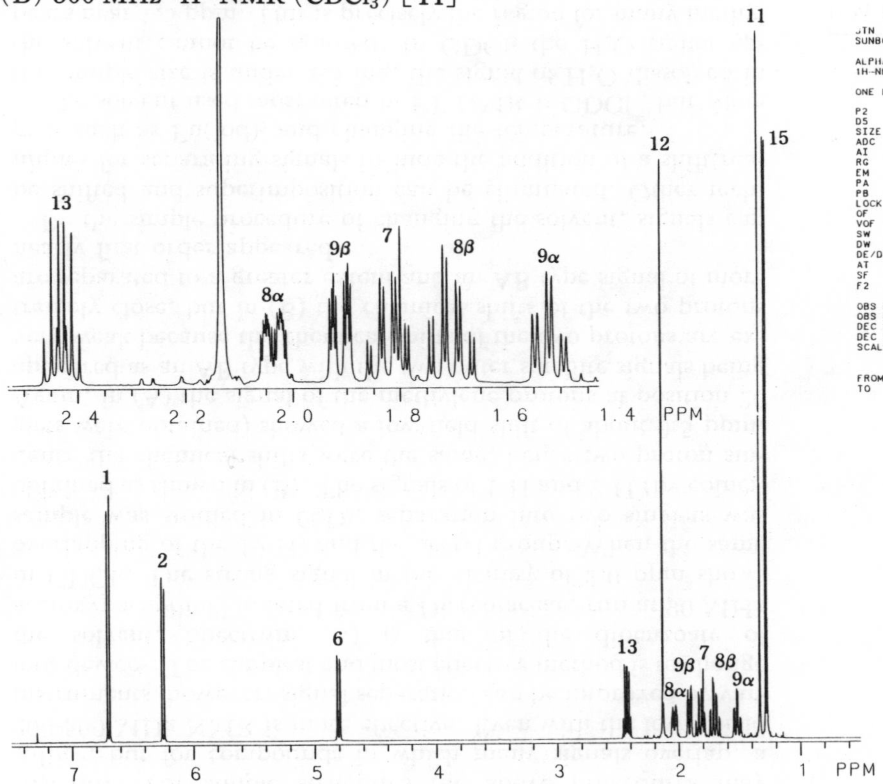

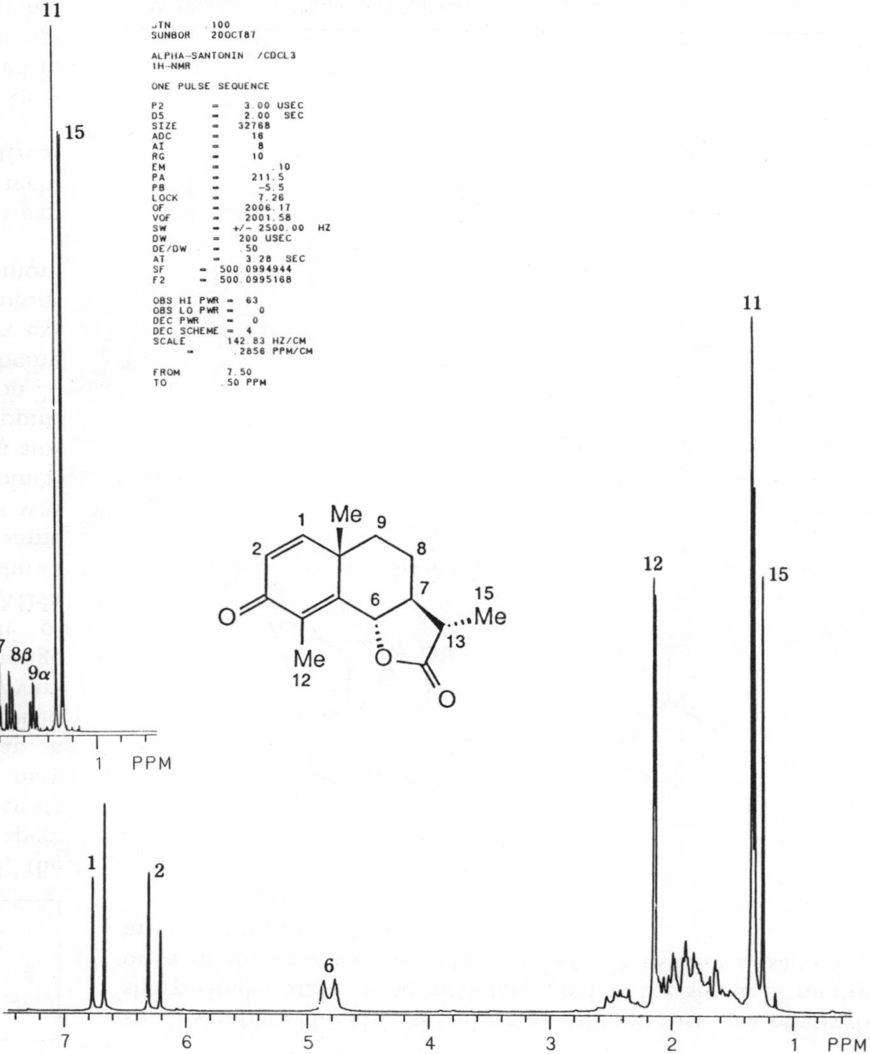

(A) 100 MHz ^{1}H-NMR ($CDCl_3$) [TI]

3

2 ^{1}H-NMR Spectrum of Acetoxypachydiol Dibenzoate at 90 MHz

In determining the structure of an organic compound, the usual practice is to run a spectrum using the 60-100 MHz spectrometers to inspect the purity of the sample and to deduce its structure. For simple structures the above procedures may suffice, but for compounds in which many signals overlap, a 200-500 MHz NMR is more effective. Even with the lower field instruments, however, signal separation can be improved by various devices. The simplest and most effective method is to change the solvent. Spectrum (A) is that of the dibenzoate of acetoxypachydiol[1)] isolated from a Dictyotaceae, run at 90 MHz in $CDCl_3$. The strong signal in the vicinity of 2.0 ppm shows overlapping of the 19-H_3 and the acetyl group. When the same sample was studied in C_6D_6, separation into two singlets was obtained as shown in (B). The signals of 1-H and 2-H (by coincidence the chemical shifts were the same, hence two proton singlets were obtained) showed a low-field shift of about 0.5 ppm. Again, in (A) the signal of the methylene protons at position 20 appeared as an AB type with the two outer satellite signals being very weak because the chemical shifts of the two protons are extremely close, but in (B) the chemical shifts of the two protons are separated to a greater extent and an AB type signal of more nearly first order appeared.

By the simple procedure of changing the solvent, signals can be shifted and superimposition can be eliminated. Other techniques for separating signals include the addition of a shift reagent such as Eu(fod)$_3$ and changing the temperature.

The solvent used most often in FT-NMR is $CDCl_3$, but when the sample size is under 2-3 mg, the signal of H_2O dissolved in the solvent cannot be ignored. In $CDCl_3$ the H_2O signal appears near 1.5 ppm. This is precisely the region for many methyl and methylene protons, thus the problem becomes serious for a small sample. The H_2O signal frequently appears as a relatively sharp singlet and can be misinterpreted as the signal of methyl on a quaternary carbon. H_2O in C_6D_6 is usually observed at around 0.5 ppm.

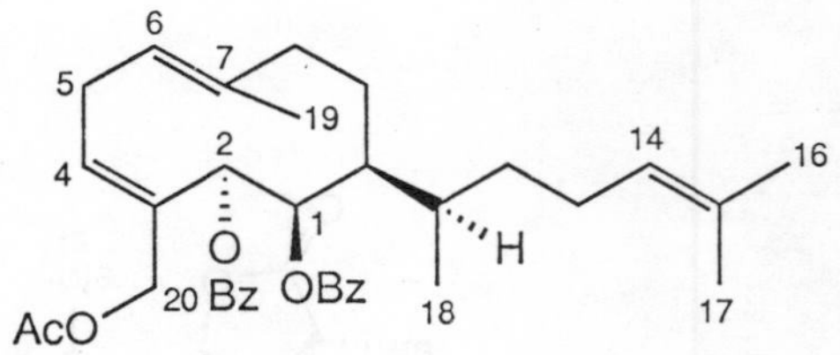

Bz:COPh

1) M.Ishitsuka, T.Kusumi, H.Kakisawa, Y.Kawakami, Y.Nagai and T.Sato, *Tetrahedron Lett.*, **27**, 2639 (1986)

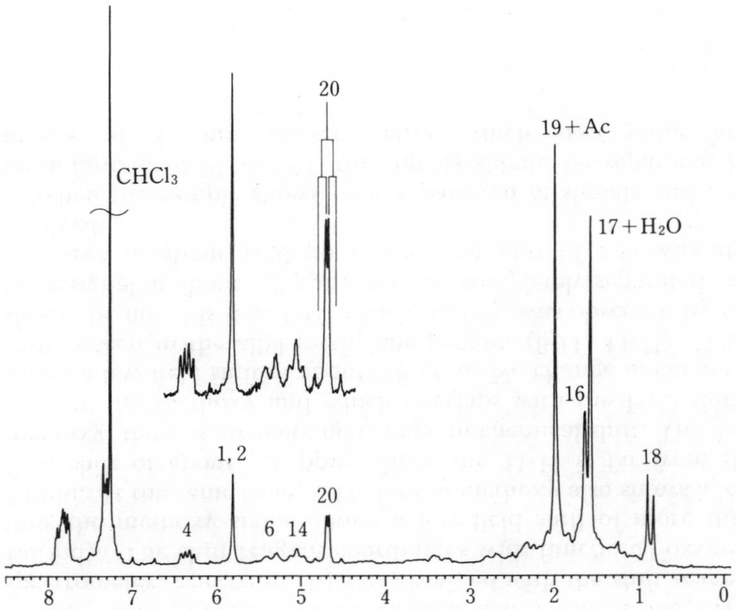

(A) 90MHz ¹H-NMR (CDCl₃) [TK]

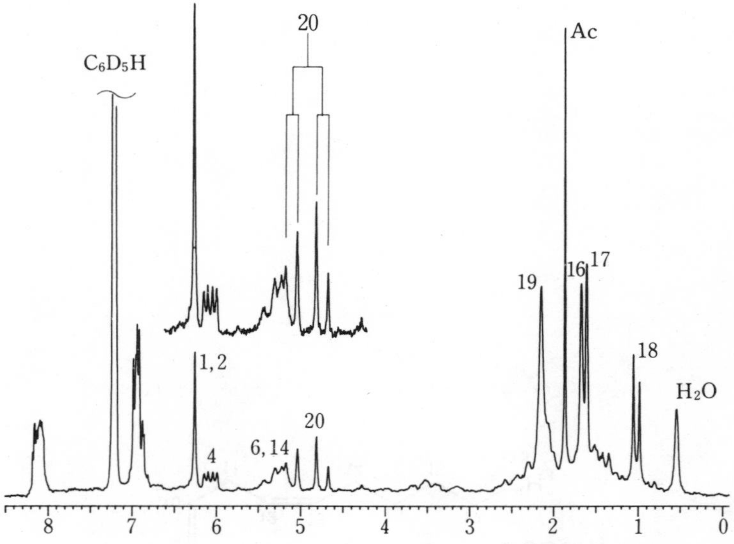

(B) 90MHz ¹H-NMR (C₆D₆) [TK]

3 ^{1}H-NMR of Methyl 9,12-Octadecadienoate (with Shift Reagent)

The use of a shift reagent is an effective technique for overcoming signal overlap.[1, 2]

Spectrum (A) was obtained on a methyl ester of an unsaturated fatty acid (methyl 9,12-octadecadienoate) with a 300 MHz spectrometer. Spectrum (B) was obtained with the shift reagent Eu(fod)$_3$. The shift reagent coordinates with functional oxygen, thus the methoxy signal shows a low field shift of more than 1 ppm. At the same time, 2-H close to methoxy also shows a low field shift of about 1.2 ppm. Since the 11-H is far from the methoxy, there is virtually no change in chemical shift. The 3-H close to the methoxy and which overlaps with the H_2O signal shows a low field shift of about 0.8 ppm. No change in chemical shift is seen in the allyl methylene protons (8-H, 14-H). What should be noted is that 4-H, which, in (A), was obscured by the large signal at about 1.3 ppm, became completely separated and appeared at about 1.70 ppm, and that part of 5-H was also resolved.

When the sample shows poor separation of signals and contains functional N and O, the signals should be separated by means of a shift reagent after which decoupling and 2-dimensional spectroscopy may be attempted. It should be remembered that in 2-dimensional spectroscopy with a shift reagent added, measurement should be done within a short time, because change in chemical shifts may occur.

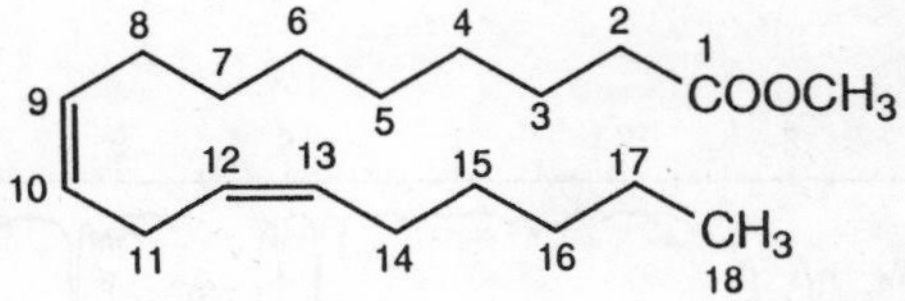

1) A.F.Cockerill, G.L.O.Davies, R.C. Harden and D.M. Rackham, *Chem. Rev.*, **73**, 553 (1973)
2) R.E. Sievers (Ed.), *Nuclear Magnetic Resonance Shift Reagents*, Academic Press, New York and London (1973)
3) Sample provided by Prof. I.Kubo, University of California

(A) Normal

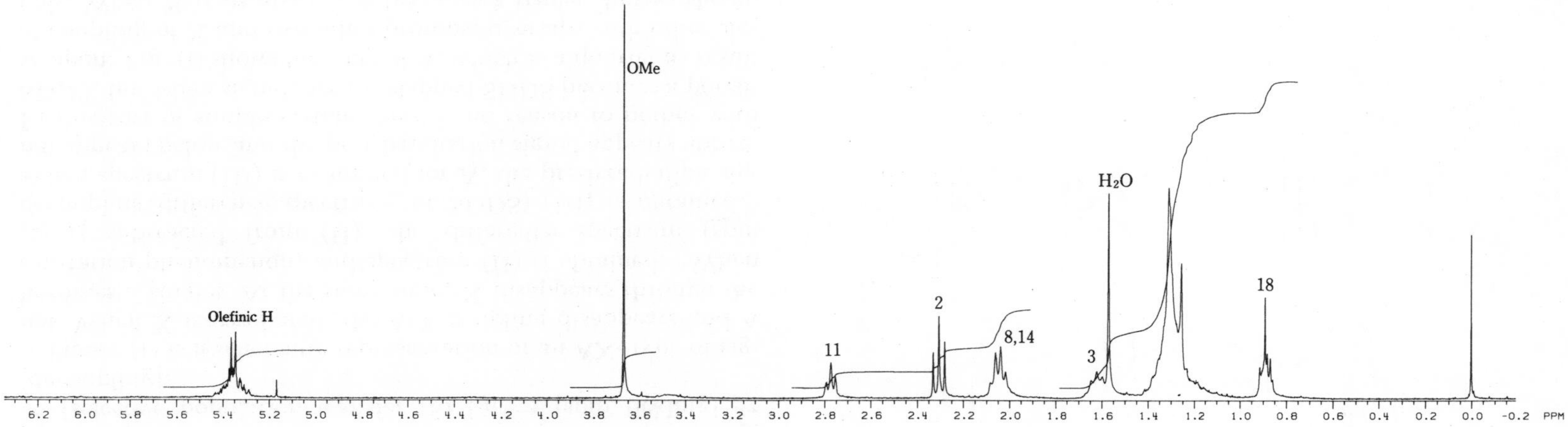

(B) With 0.1 equivalent Eu $(fod)_3$

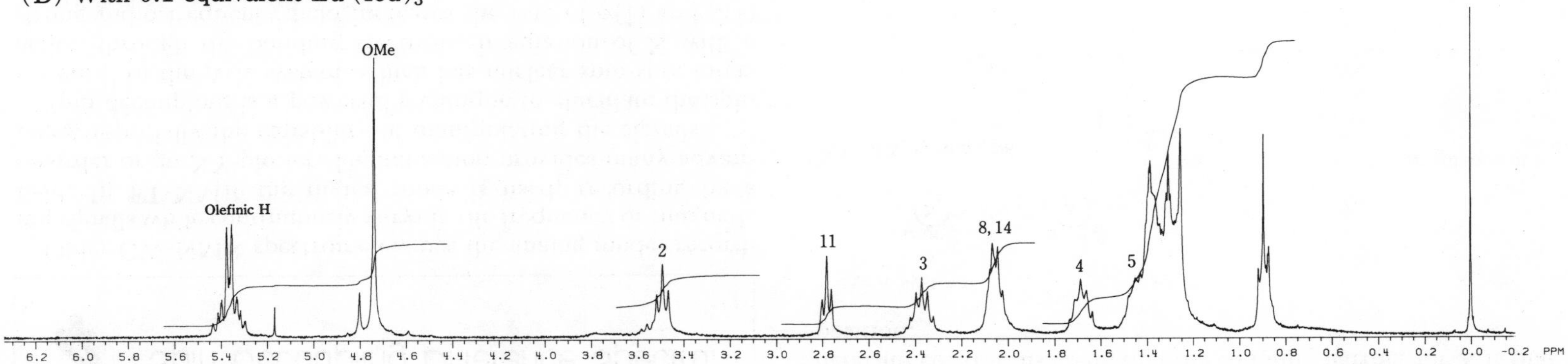

(A), (B) 300 MHz ^{1}H-NMR ($CDCl_3$)[3] [HN]

4 Spin Decoupling Difference Spectrum

Older CW-NMR spectrometers use the analog mode, recording signals while continuously varying the frequency or magnetic field. In FT-NMR the digital mode is used, recording by a recorder or an XY-plotter. Digitalization provides many advantages, especially the capability of manipulating the signals.

Spin decoupling is a powerful technique to elucidate the spin system.[1] In the A-X system, which has nuclear spin-spin interaction through the bonding electrons, irradiation of X with a strong radio frequency field increases the rate of $\alpha(\uparrow)$ and $\beta(\downarrow)$ spin state exchange in X to the extent that A can no longer differentiate the two states. Depending on the strength of the radio frequency field, the following phenomena occur: [coupling] → [selective population transfer (SPT)] → [spin tickling] → [decoupling].

Figure (I) is a schematic representation of an AX_3 type of signal. When X is irradiated, the A-X coupling disappears and A becomes a singlet. At the same time, X disappears through the saturation phenomenon, and spectrum (II) is obtained.[1] When (I) is subtracted from (II), the difference spectrum (spin decoupling difference spectrum, or SDDS) (III) is obtained.[2] When spectrum (III) is examined for A, the pre-irradiation signal appears below and the post-irradiation signal appears above. In this sort of simple system there is no reason to bother with SDDS, but when signals are overlapped SDDS becomes a potent weapon. Fig. (i) shows how signal A, which is a quartet as result of coupling of X and two other protons, overlaps with other signals. When X is irradiated, A becomes a triplet, but as shown in (ii), there is not much change in the shape of the overlapping part. When, however, (i) is subtracted from (ii), the strength of other signals unaffected by irradiation of X becomes zero, and A appears clearly as shown in (iii). By observing the signal in the positive direction, the splitting pattern of A after irradiation is determined. It is also seen that the coupling partner for X is that position.

A–X system

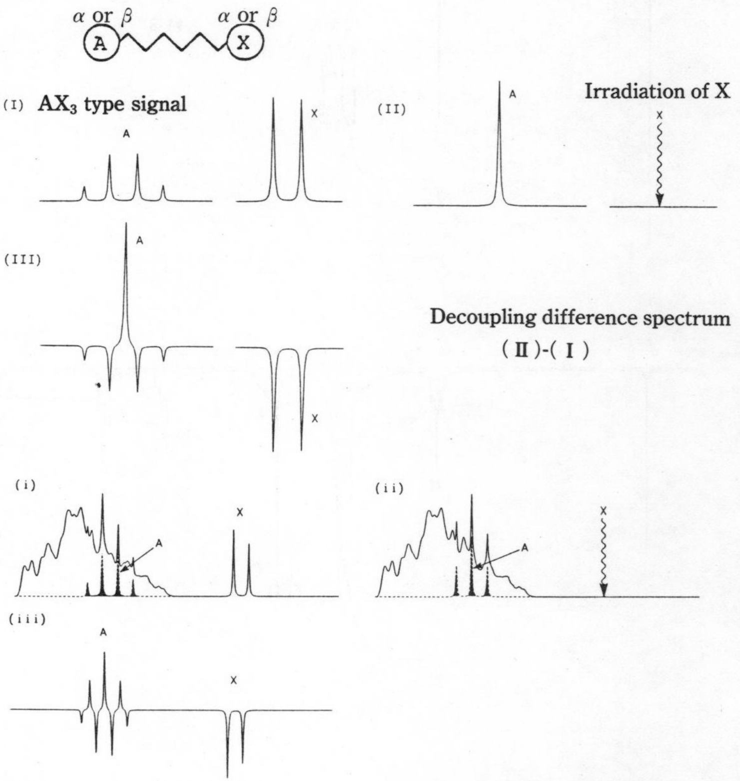

1) H.Günther (translated by R.W.Gleason), *NMR Spectroscopy — An Introduction*, John Wiley & Sons, Ltd., New York (1980), pp.285-292

2) J.K.M.Sanders and J.D.Mersh, *Prog. Nucl. Magn. Reson.*, **15**, 353 (1982), pp.355-361

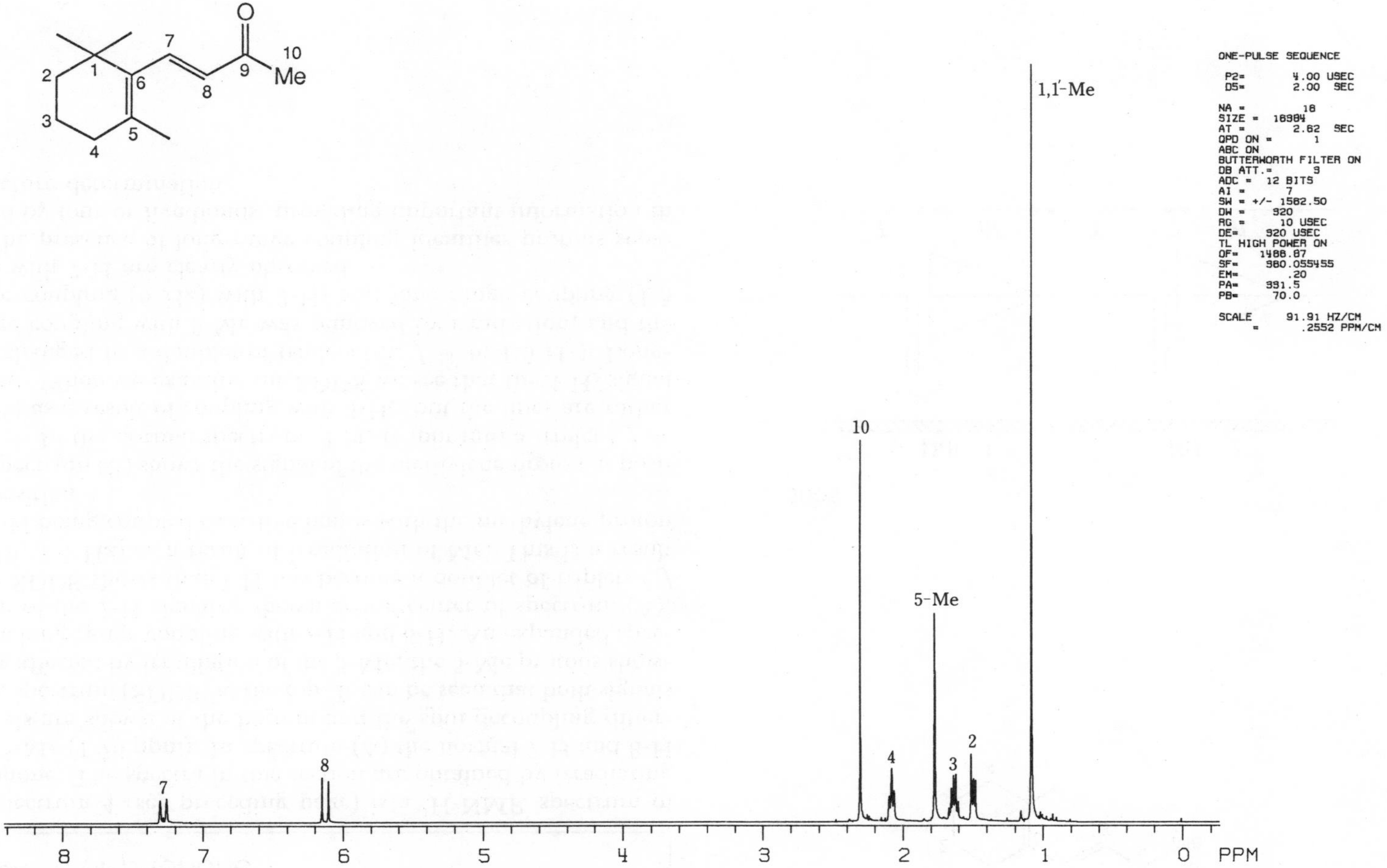

Normal ^{1}H-NMR spectrum of *β*-Ionone (360 MHz, $CDCl_3$) [TI]

5 Spin Decoupling Difference Spectrum of β-Ionone

Spectrum 4 (see preceding page) is a ^{1}H-NMR spectrum of β-ionone. The spectra in this section are obtained by irradiating the 5-Me (1.76 ppm). In spectrum (A) the normal 7-H and 8-H signals are shown at the bottom and the spin decoupling difference spectrum (SDDS) at the top. It can be seen that both signals were affected by irradiation of the 5-Me, the 5-Me protons showing a long-range coupling with 7-H and 8-H. An expanded spectrum of the 7-H signal is shown at the center of spectrum (A). The SDDS shows that 7-H has become a doublet of triplets (J = 18, 1.5 Hz) as a result of irradiation of Me. This is a result of 7-H being coupled over five bonds with the methylene proton at position 4.

Spectrum (B) shows the signal of the methylene proton at position 4. In the normal spectrum, 4-H_2 is split into a triplet (J = 6 Hz) as a result of coupling with 3-H_2, but the lines are rather broad. When we examine the SDDS we see that the 4-H_2 signal has changed to a doublet of triplets (dt; J = 6, 1.5 Hz). Long-range coupling with 5-Me was removed by irradiation, and the large coupling (6 Hz) with 3-H_2 and long-range coupling (1.5 Hz) with 7-H are clearly observed.

The presence of long-range coupling identifies protons separated by four or five bonds, providing important information in structure determination.

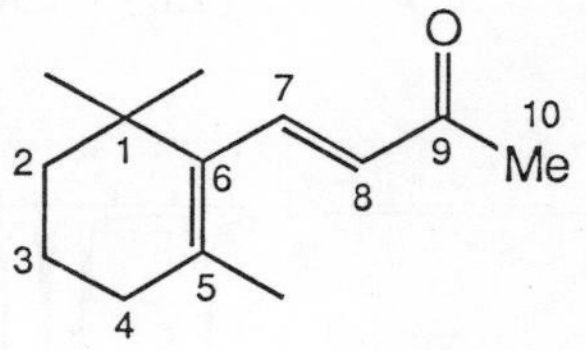

SDDS

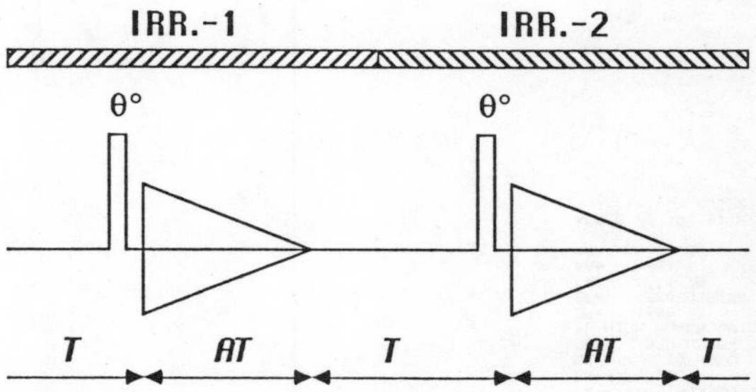

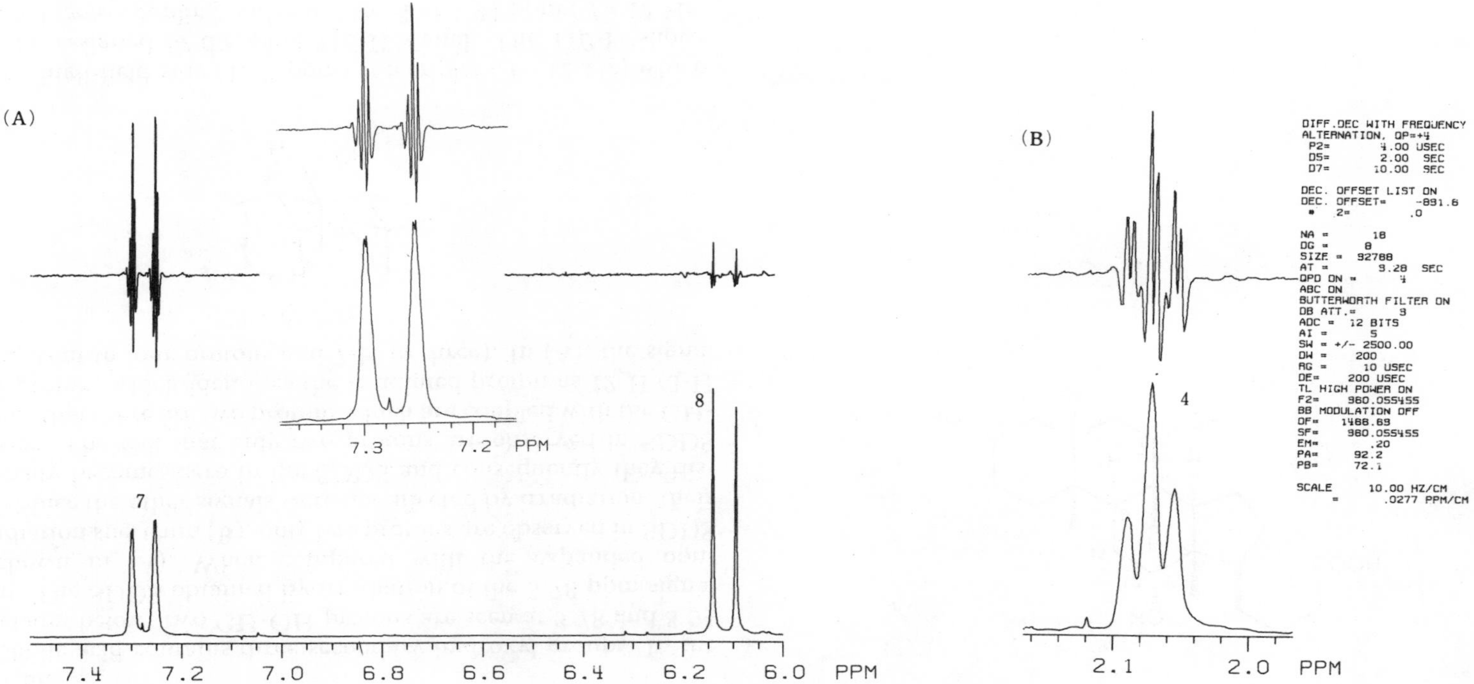

(A), (B) 360 MHz ^{1}H-NMR ($CDCl_3$), 5-Me irradiated in both spectra. [TI]

6 Spin Decoupling Difference Spectrum of Cholic Acid

Signals of aliphatic protons (1.0-3.0 ppm) in compounds having many sp^3 protons, e.g., cholic acid, show complex spin-spin coupling and the signals overlap making analysis extremely difficult.

Cholic acid contains three secondary hydroxyl groups. In the spectrum below, two C**H**-OH protons are seen at 3.78 and 3.94 ppm. The SDDS obtained by irradiation of the 3.78 ppm signal is shown in (A). When compared with the expanded non-irradiation spectrum (B), only two protons are observed in SDDS (A). Since the other signals were not affected by irradiation, their intensity becomes zero in the SDDS and consequently they disappear. The fact that only two protons are observed in SDDS means that there are two protons which are coupled with the C**H**-OH proton, which identifies the irradiated proton as 12-H (3-H is adjacent to four protons and 7-H to three). In (A), the signal

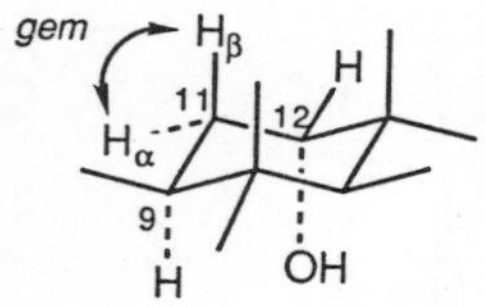

on the high-field side (1.53 ppm) is a triplet (J = 12 Hz) which may be assigned to the axial 11β-H signal. The 11β-H shows geminal (*gem*) coupling with the 11α-H at 1.94 ppm (J = 12 Hz) and also with the axial 9-H, resulting in a triplet. In addition to the 12 Hz *gem* splitting, the equatorial 11α-H shows a 5 Hz coupling with 9-H, typical of equatorial-axial interaction, and hence appears as a doublet of doublets (see spectrum (A)).

It should be noted that these spectra became well defined only as difference spectra. It would not be possible to obtain a detailed pattern of the splitting of the coupling partners of 12-H from ordinary decoupling spectra.

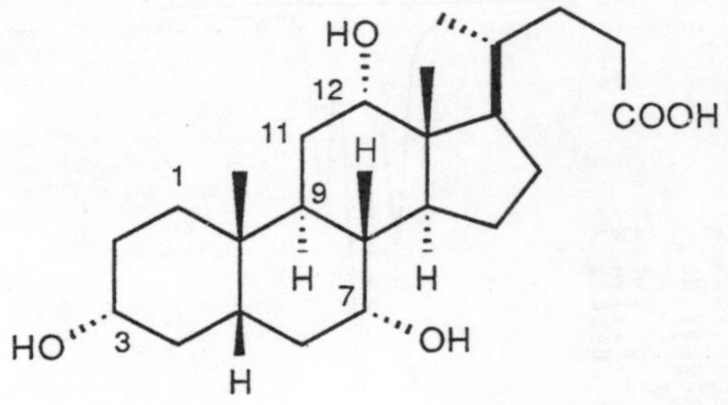

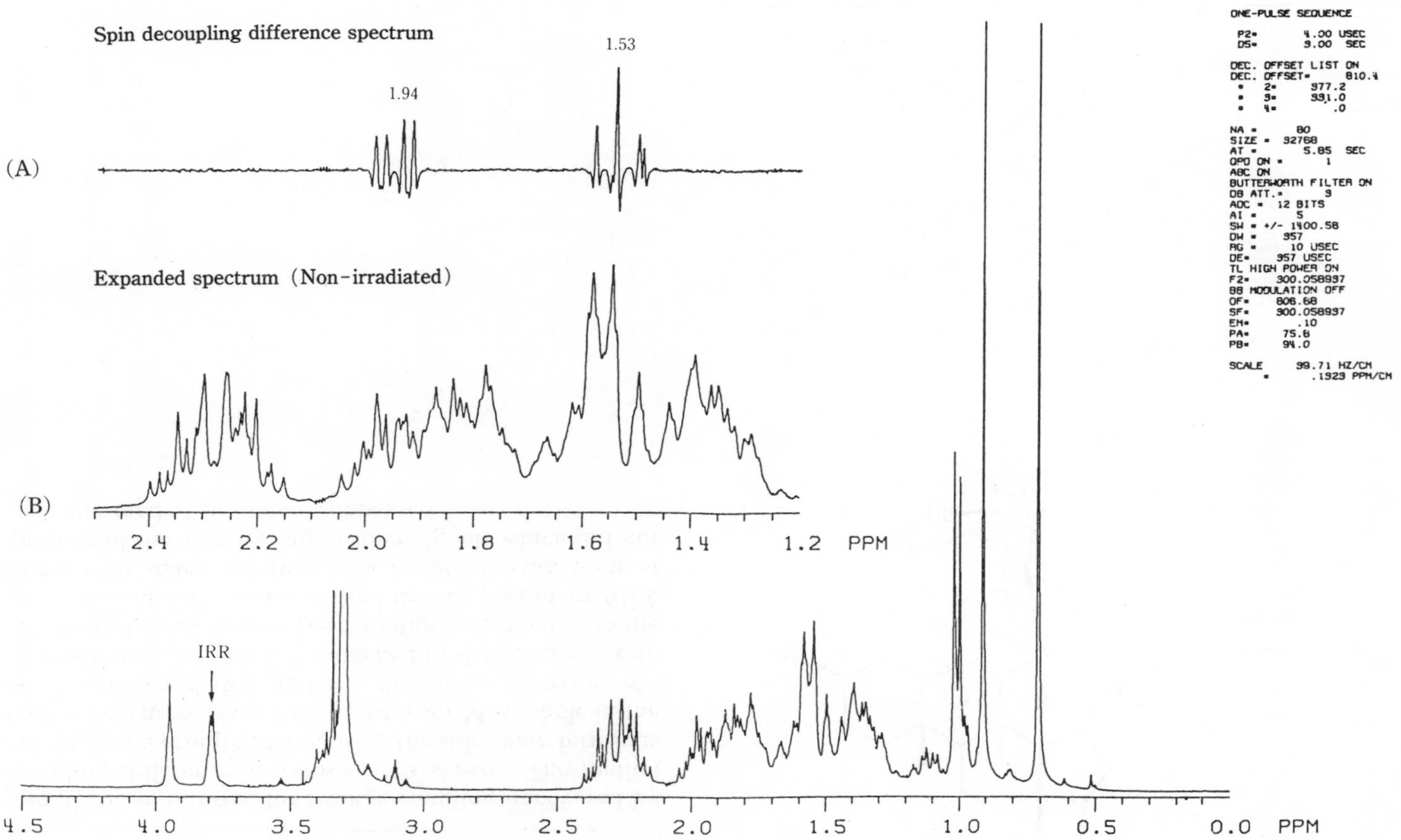

300MHz ^{1}H-NMR (CD_3OD) [HN]

7 Spin Decoupling Difference Spectrum of Mosesin-2

An example of an unusual long-range coupling uncovered by spin decoupling difference spectroscopy is shown. The methyl protons of the acetyl group (2.03 ppm) at the side chain terminus of the shark repellent mosesin-2 obtained from Moses Sole in the Red Sea[1, 2] were irradiated and the difference spectrum observed. It was found that the CH_2 attached to the acetoxy group (3.82-3.96 ppm) showed long-range coupling over 5 bonds to the methyl protons. The CH_2 signal showed the AB part of an ABX type pattern with sharp splitting as a result of irradiation of methyl protons of the acetyl group. Other signals which did not couple with the methyl protons disappeared.

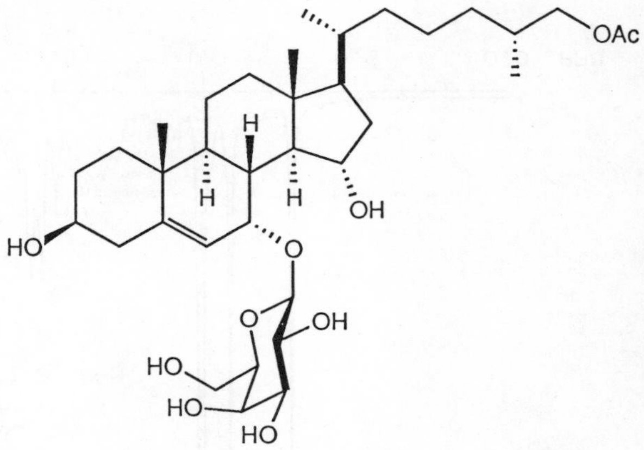

1) K.Tachibana, K.Nakanishi and S.H.Gruber, Abstracts of the 27th Symposium on Natural Organic Compounds, Hiroshima, 1985, p.545
2) K. Tachibana and S.H.Gruber, *Toxicon*, **26**, 839 (1988)

The methyl protons of acetyl group are irradiated.

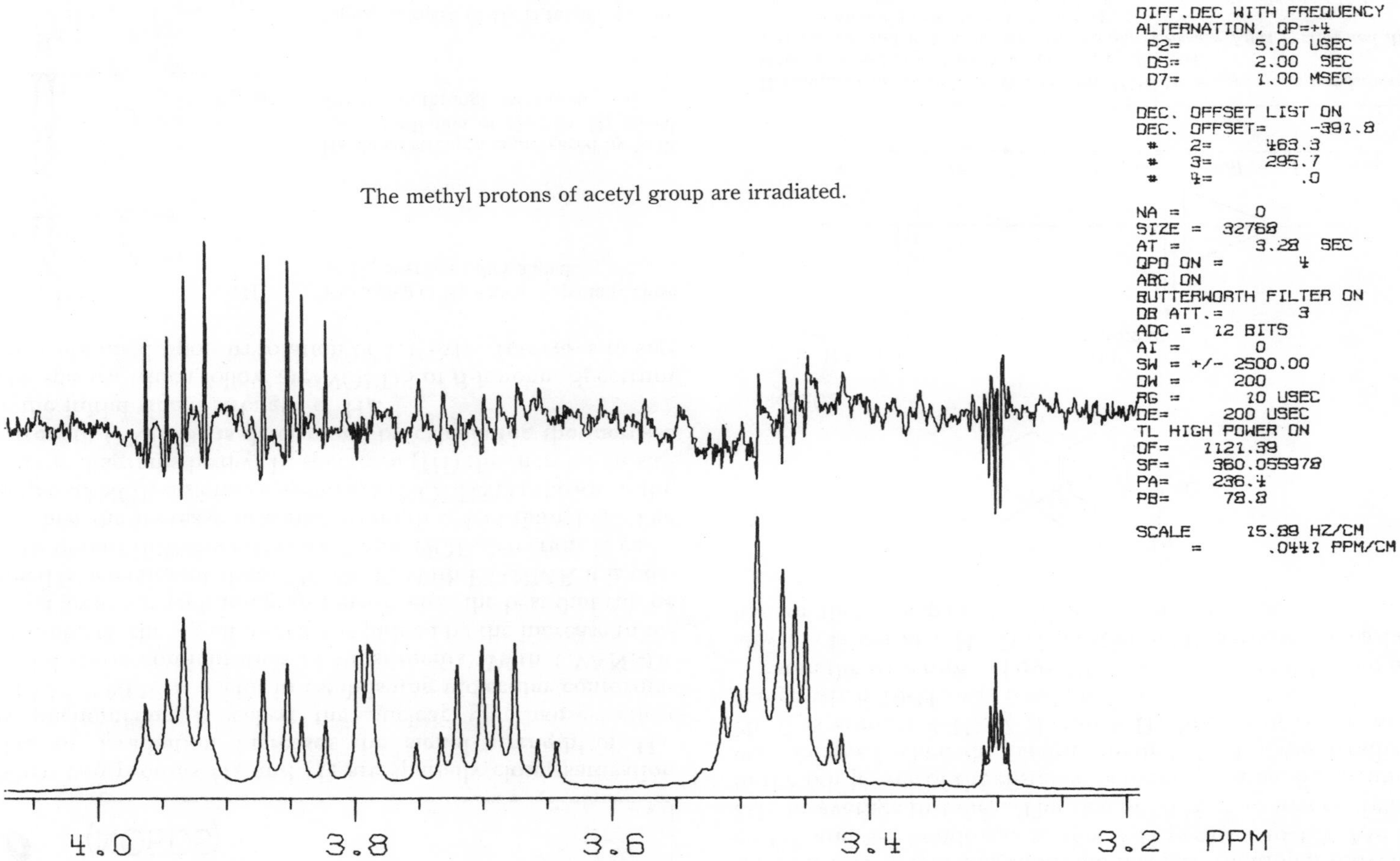

360MHz ^{1}H-NMR (CD_3OD) [TI]

8 NOE Difference Spectrum of β-Ionone (NOEDS)

When two protons H_A and H_B are spatially close, saturation of H_A by irradiation increases the signal strength of H_B. This phenomenon is called the nuclear Overhauser effect (NOE).[1, 2, 3)] NOE is useful in establishing molecular conformation and steric configuration of substituents. With CW-NMR spectrometers, the signal increase is judged by the increase in integrated area, but with integrator error, etc., the best that can be attained is detection of about 5% NOE. With FT-NMR it is possible to obtain difference spectra,[4)] and NOE detection is easy, even when the increase in signal strength is less than 1%. The principle of NOE difference spectrum (NOEDS) is shown in the schematic diagrams below. In spectrum (III) the increase in signal strength (NOE%) is determined by calculating the increase over the initial signal strength of H_B.

The spectra which follow are NOEDS of β-ionone. Spectrum (A) was obtained upon irradiation of 1,1′-Me. Increases in signal strengths of 5.5%, 9.8% and 12.9% are seen at 2-H, 8-H and 7-H, respectively. The reason why NOE on 7-H is larger than on 8-H is that this molecule has free rotational movement on 6-7 and 8-9 bondings, so that 7-H is closer to 1,1′-Me than 8-H in average distance. The size of NOE is in inverse relation to the 6th power of the distance between protons. Spectrum (B) was obtained when the olefinic methyl (5-Me) was irradiated. NOE is seen at 4-H_2, 7-H and 8-H. Spectrum (C) is an instance when 10-H_3 was irradiated.

From the structure, a large NOE is expected at 8-H, but actually it is larger at 7-H. This is presumably because the carbonyl has for the most part the following conformation.

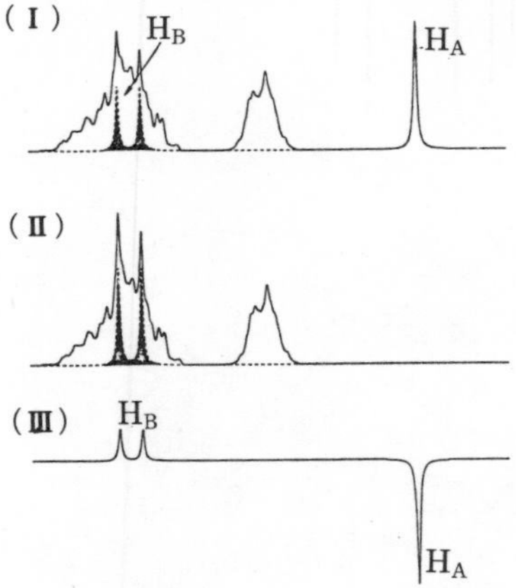

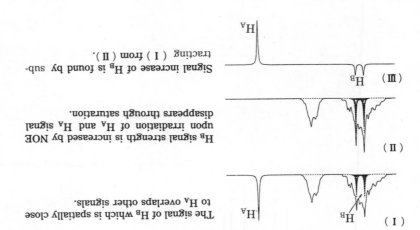

(I) The signal of H_B which is spatially close to H_A overlaps other signals.

(II) H_B signal strength is increased by NOE upon irradiation of H_A and H_A signal disappears through saturation.

(III) Signal increase of H_B is found by subtracting (I) from (II).

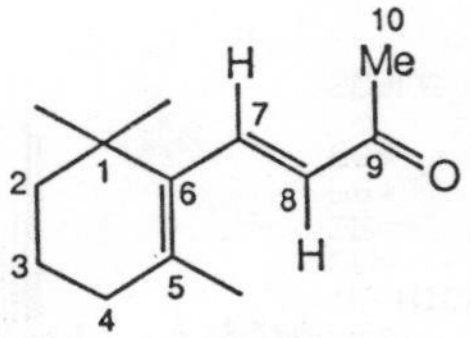

NOEDS

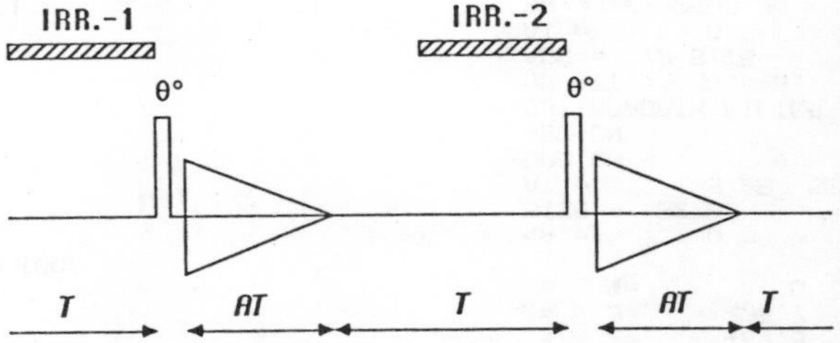

1) H.Günther (translated by R.W.Gleason), *NMR Spectroscopy — An Introduction*, John Wiley & Sons, Ltd., New York (1980), pp.299-304

2) J.H.Noggle and R.E.Schirmer, *The Nuclear Overhauser Effect — Chemical Applications*, Academic Press, New York (1971)

3) The signal enhancement is observed in the case of small molecules. If the motion of molecules (large-size) is slow, the NOE becomes negative.

4) J.K.M.Sanders and J.D.Mersh, *Prog. Nucl. Magn. Reson.*, **15**, 353 (1982), pp.361-380

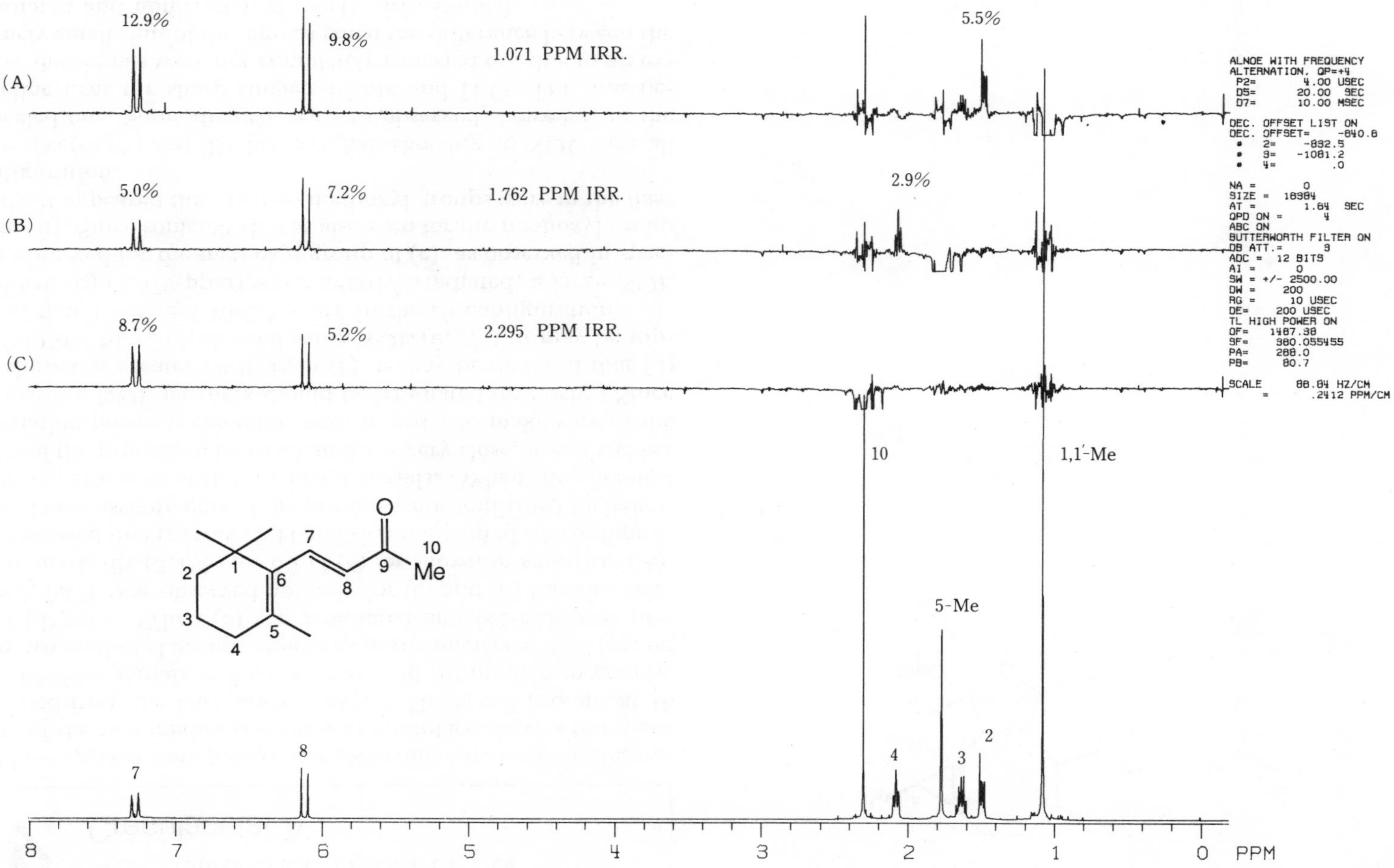

360MHz ^{1}H-NMR ($CDCl_3$)　　[TI]

9 NOE Difference Spectrum of Crenulacetal A

These spectra were prepared to determine the steric configurations of the two methoxyl groups of crenulacetal A, a diterpene obtained from the Dictyotaceae alga.[1] The acetal protons at 18 and 19 show signals at 5.63 (a) and 5.58 (b) ppm, respectively. The two methoxyl groups appear as sharp singlets at 3.52 (c) and 3.35 (d) ppm. When (a) was irradiated and NOEDS was prepared, NOE was observed not only for (c) and (d) but also relatively markedly (3.1%) for 8β-H (e), as shown in spectrum (A). This showed that (a) was 19-H and that it assumed a β configuration. These assignments of the protons were confirmed by irradiation experiments at the methoxyl signals. (When the chemical shifts of the protons to be irradiated are very close, use of weaker irradiation power is essential, and, in order to make sure of the results, the NOE partners should be irradiated inversely.) Since (d) showed a greater NOE than (c), it may be deduced that (d) is 19-OMe. Since (c) showed some NOE (0.5%), it may be supposed that 19-H and 18-OMe are in the *cis* configuration.

When (b) (5.579 ppm) was similarly irradiated, a large NOE was observed for the methoxyl group of (c), as observed in spectrum (B). Since some NOE was also seen for the methoxyl group of (d), it appeared that the two methoxyl groups were in the *trans* configuration.

In spectra (A) and (B), other signals showing no NOE were all canceled out. Some disturbance was observed, however, in the baseline near the sharp singlets of Me and H_2O. This was because the signals were not completely canceled out due to an extremely small shift of the signals when the difference between the irradiated and nonirradiated spectra was studied.

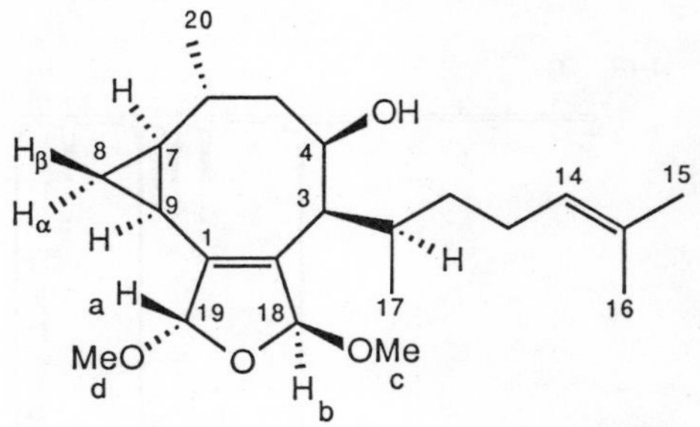

1) T. Kusumi, D. Muanza-Nkongolo, M. Goya, M. Ishitsuka, T. Iwashita and H. Kakisawa, *J. Org. Chem.*, **51**, 384 (1986)

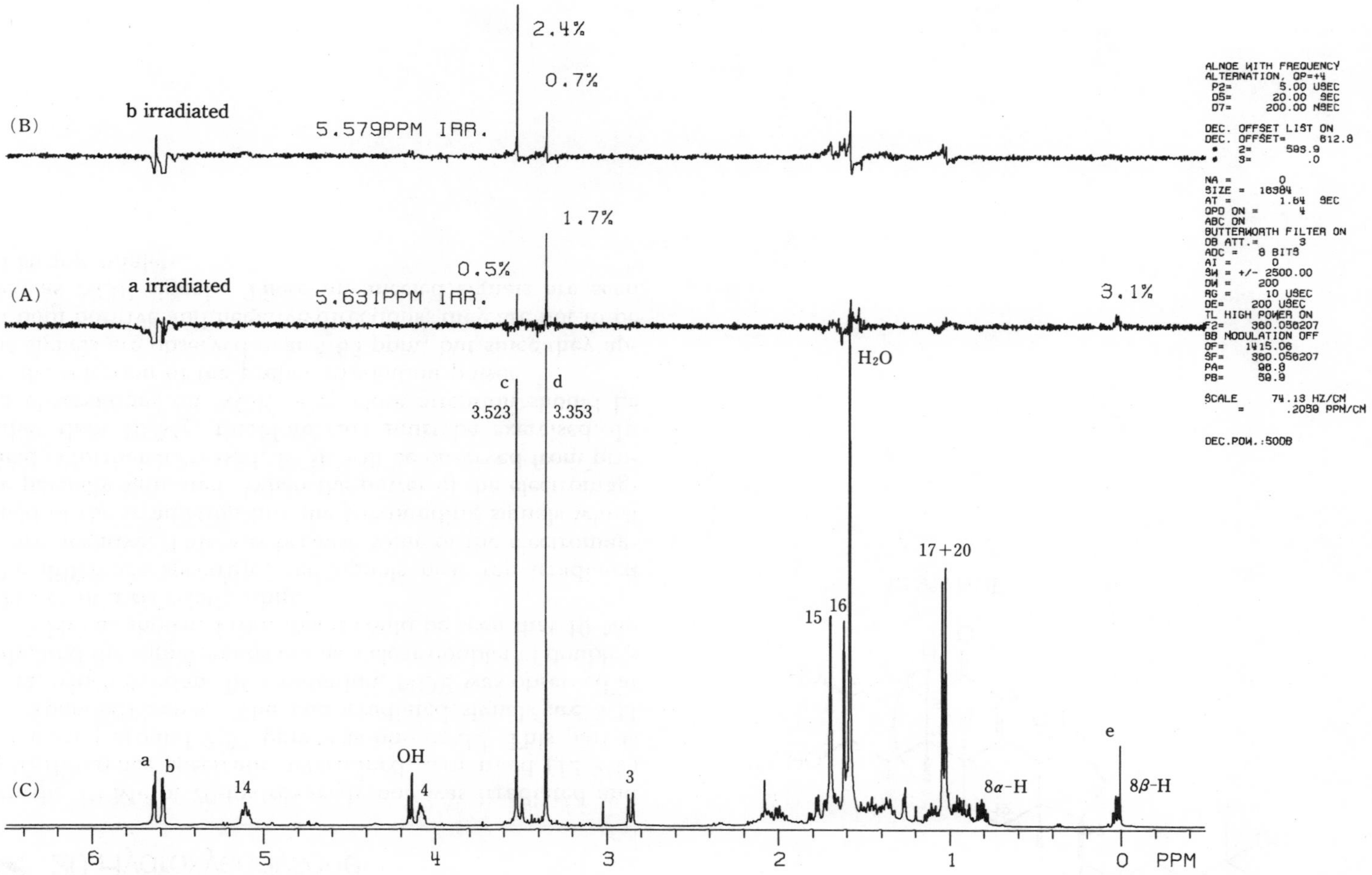

360MHz ^{1}H-NMR ($CDCl_3$) [TI]

10 NOE Difference Spectrum of 20-Hydroxyecdysone

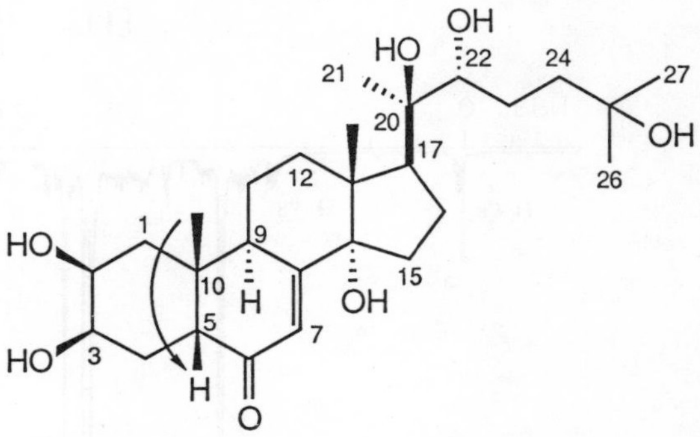

12.9% NOE

When the 10-Me of 20-hydroxyecdysone was irradiated and the NOE difference spectrum determined, a marked (12.9%) NOE centering around 2.37 ppm was observed.[1)] This part is shown (expanded) above. The non-irradiated signals are 5-H and 17-H, which overlap. By irradiation, NOE was observed at 5-H only, and the signal stands out as a clear doublet of doublets (J = 15, 5 Hz) as shown. From this it could be seen that 10-Me and 5-H are in a *cis* relationship.

In the difference spectrum, the signals near the irradiated 10-Me are negative. This was because some of the electromagnetic field of the irradiation hits the surrounding signals which become partially saturated. When the power of the electromagnetic field is further increased, NOE will be observed from protons other than 10-Me, therefore care must be exercised. In making observations on NOE, very close attention should be paid to the selection of the proper irradiation power.

Some signals are observed near 4.85 ppm, but since they appear in both positive and negative directions, they are not to be regarded as NOE signals. These uncanceled signals are seen around strong singlets.

1) I.Kubo, A.Matsumoto and F.J. Hanke, *Agri. Biol. Chem.*, **49**, 243 (1985)

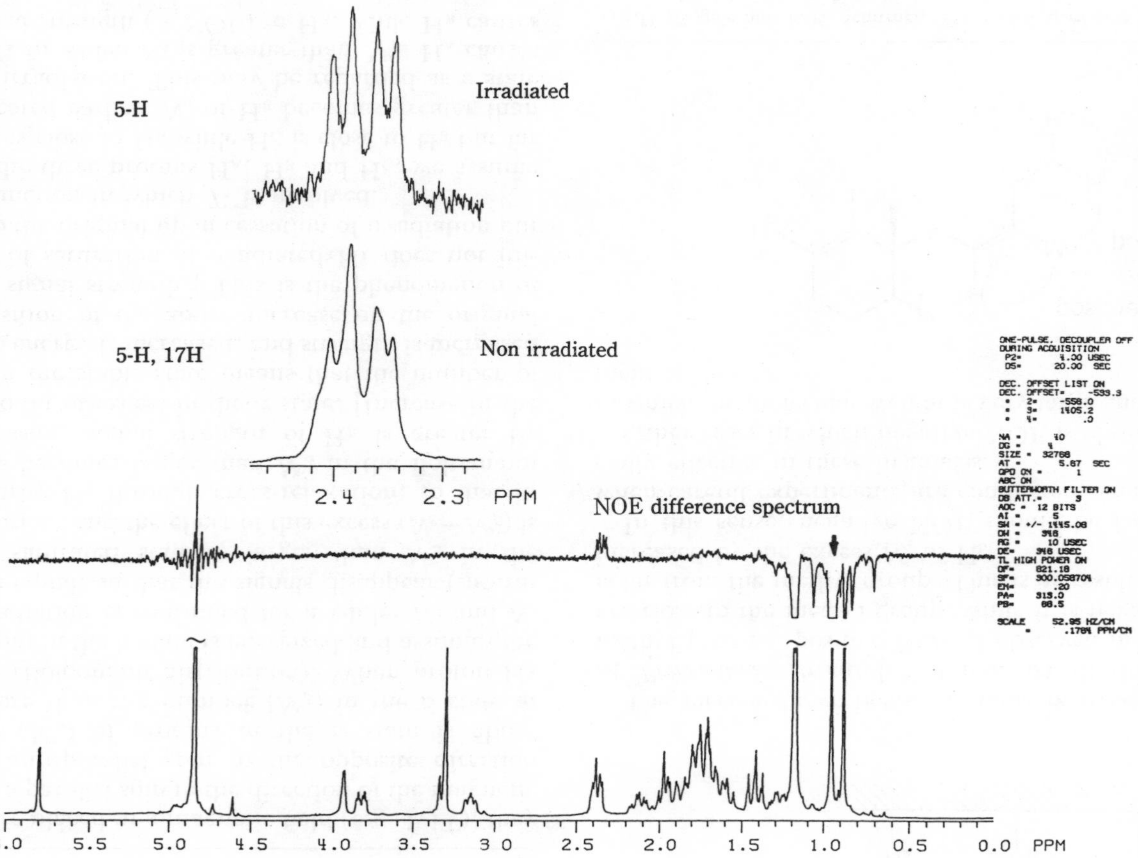

300MHz ^{1}H-NMR (CD_3OD) [HN]

11 NOE Difference Spectrum of Striatene

When protons are placed in a magnetic field (e.g. 1.4T), they assume two states: a parallel spin in the direction of the magnetic field (α) and an anti-parallel spin in the opposite direction (β). The number (N°_{α}) of protons in the α state is about (1 + 1/100,000) more than the number (N°_{β}) in the β state at room temperature (Boltzmann distribution). When proton H_A is irradiated, the spin in the α state is energized and assumes the β state. When irradiation is continued for a while, N_α and N_β eventually become equal, so that the signals disappear (saturation). N_β in the saturated state is larger than N°_{β} in the Boltzmann distribution, and the effect of this excess ($N_\beta - N^{\circ}_{\beta}$) is transmitted to nearby H_B through cross-relaxation, so that in the end, N_α of H_B becomes larger than N°_{α} in the Boltzmann state. For this reason, signal strength of H_B is greater by ($N_\alpha - N^{\circ}_{\alpha}$), the amount of excess in the α state. [Increase in the number of spins in the stable state means that the number of spins which absorb energy is increased, and strength is increased by the superimposition of the above increase on the original (Boltzmann state) signal strength.] This is the phenomenon of NOE.[1)] The state of saturation of irradiated H_A does not immediately revert to the original upon cessation of irradiation but follows the time function in which T_1 is involved.

Meanwhile, in the three protons H_A, H_B and H_C, we assume that irradiated H_A is close to H_B while H_C is close to H_B but far from H_A. As indicated earlier, N_α of H_B becomes greater than N°_{α} as a result of irradiation. This may be regarded as a state opposite that of H_A in which N_β is greater than N°_{β}. H_A causes an increase in signal strength (+NOE) in H_B, while H_B causes a decrease (−NOE) in H_C which is close by.[2, 3)]

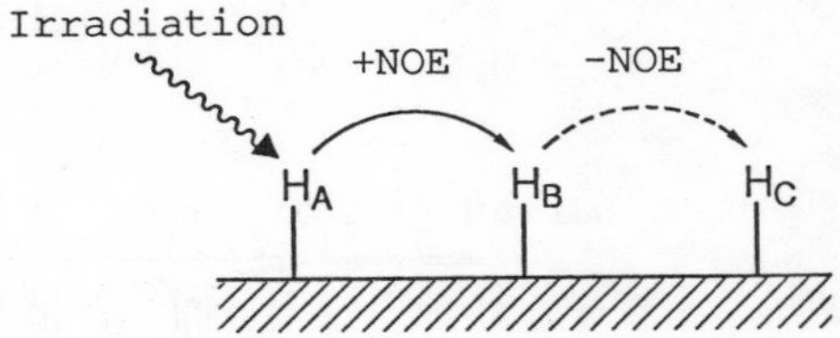

The spectra shown below are those of striatene, a component of *Ptychanthus striatitals*.[1)] Upon irradiation of the allylic methyl protons, positive NOE is observed in H_b and H_c which are close to the methyl group, while it is negative for H_d which is far from the methyl group. This is as result of N_β of H_d being increased by the excess N_α of H_b.

In this sense, negative NOE should be frequently observed when careful experiments are conducted, and NOEDS is especially effective in these instances.

Other cases in which negative NOE is observed include those in which the molecular weight is very large and molecular movement is slow.

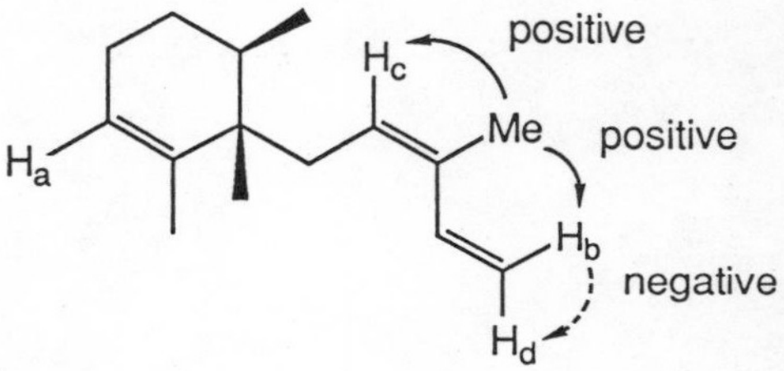

1) J.H. Noggle and R.E. Schirmer, *The Nuclear Overhauser Effect—Chemical Applications*, Academic Press, New York (1971) pp.15-17
2) *Ibid.*, p.57-64
3) R.A. Bell and J.K. Saunders, *Can. J. Chem.*, **46**, 3421 (1968)
4) R.Takeda, R. Mori, Y. Hirose, *Chem. Lett.*, **1982**, 1625

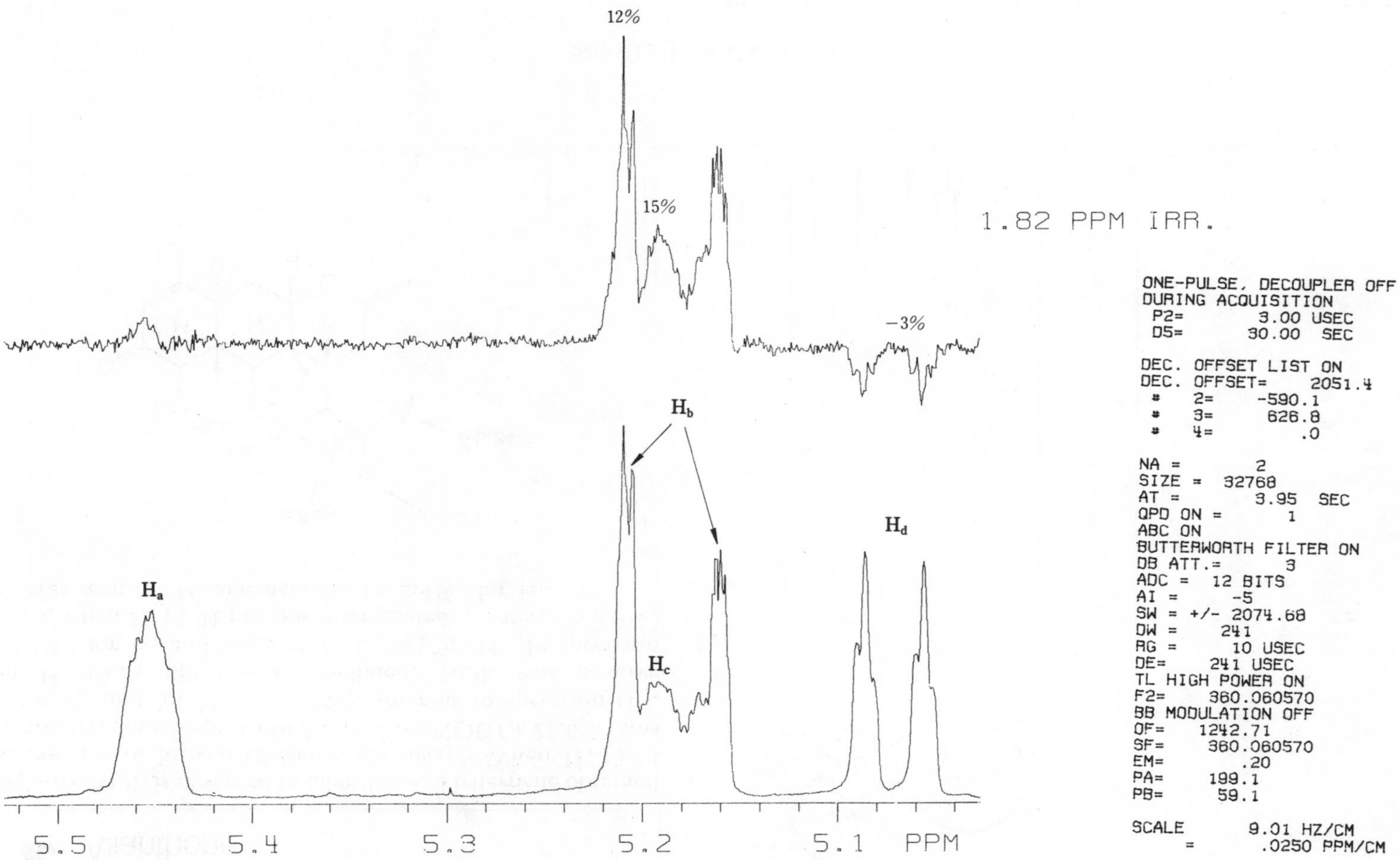

360MHz ^{1}H-NMR ($CDCl_3$) [TI]

12 NOE Difference Spectrum of Ailanthone

Negative NOE is observed in ailanthone, a triterpene obtained from the tree of heaven (*Ailanthus altissima*).[1] When H_a (5.33 ppm) was irradiated (spectrum A), positive NOE (+21.8%) was seen for H_b and for H_c (+11.2%), whereas in spectrum (B), when H_b (5.24 ppm) was irradiated, NOE was positive (+22.1%) for H_a and negative (−2.2%) for H_c. In spectrum (C) also, when H_c (4.04 ppm) was irradiated, positive (+6.9%) NOE was seen for H_a and negative (−2.4%) for H_b.

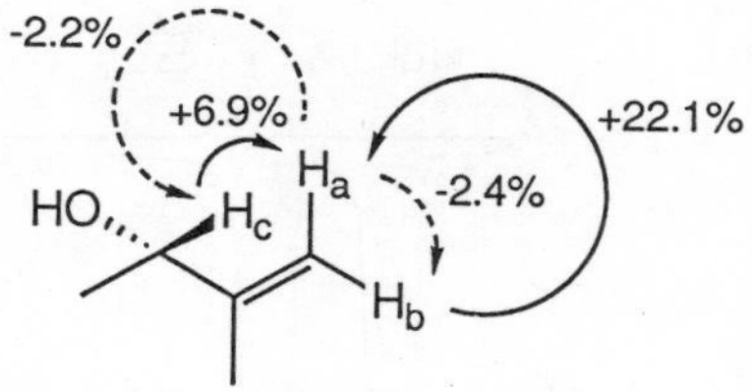

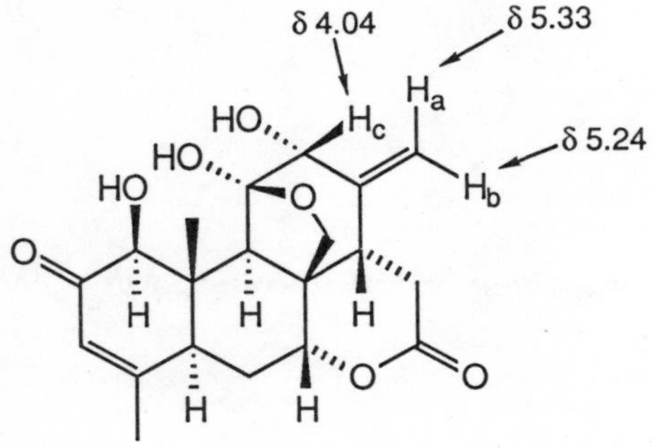

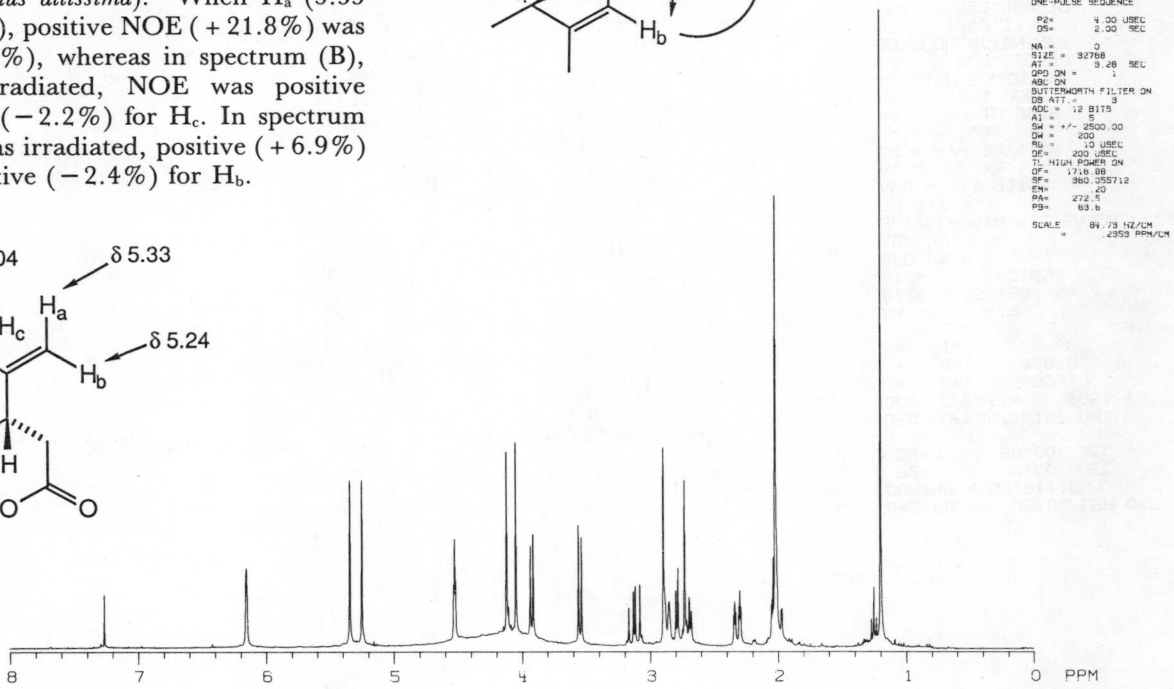

360MHz ¹H-NMR ($CDCl_3$)

1) H.Naora, T.Furuno, M.Ishibashi, T.Tsuyuki, T.Takahashi, A.Itai, Y.Iitaka and J.Polosnky, *Chem. Lett.*, **1982**, 661

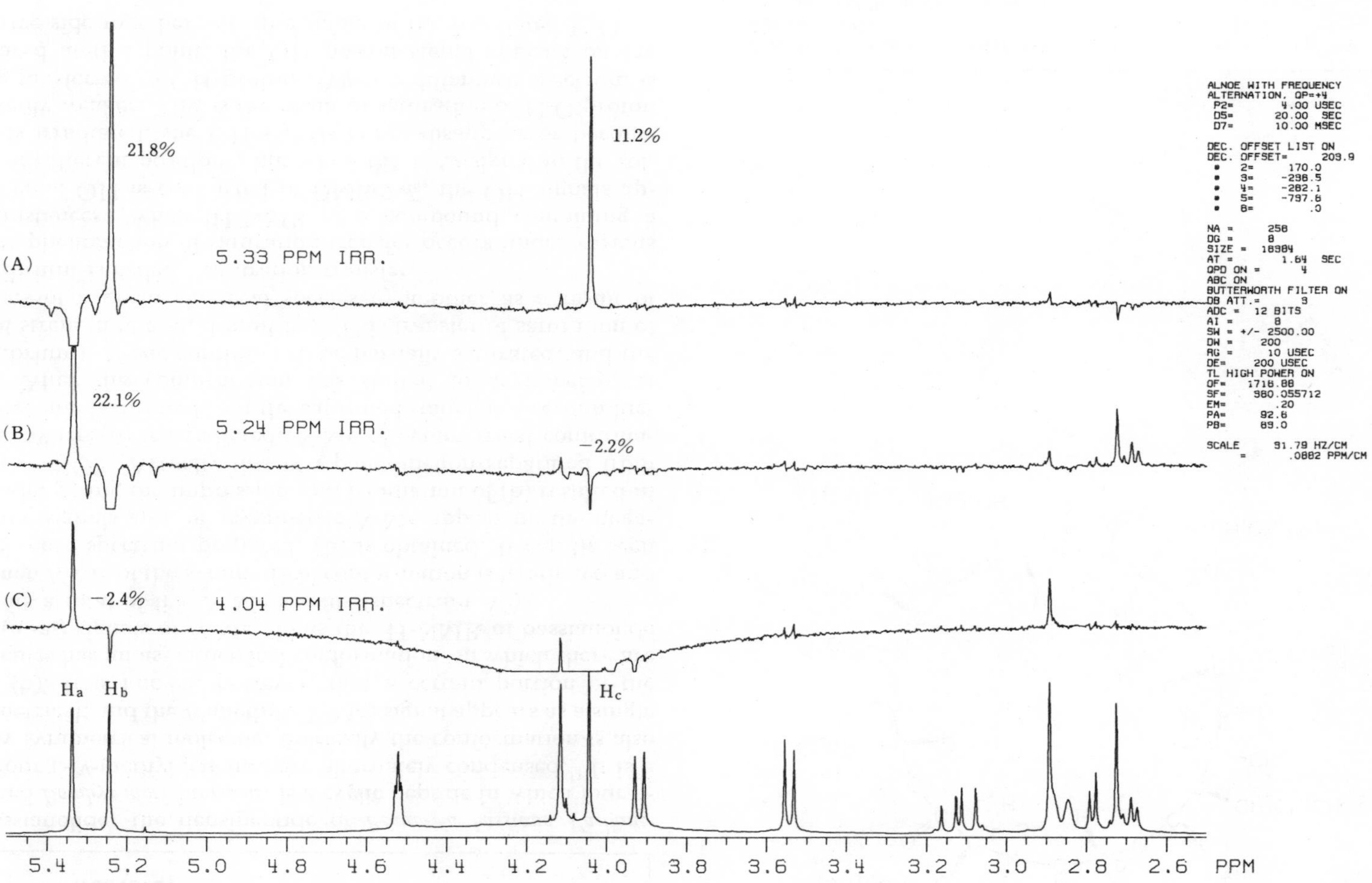

360MHz ^{1}H-NMR ($CDCl_3$) [TI]

13 Spectrum of Bassianolide—Saturation Transfer

Bassianolide, the depsipeptide of *Beauveria bassiana Vaillant*-infected *Bombyx mori Linnaeus*, is a cyclic peptide in which four D- and four L-*N*-methyl leucines are alternately condensed.[1)] It is a highly symmetrical molecule. Sterically the conformation is also symmetrical, and the *N*-methyl (*N*-Me) signal appears as a single peak (b). It is known, however, that a certain portion of the molecules has an asymmetrical conformation, in which there are four (a_1-a_4) signals of *N*-Me. Thus the ^{1}H-NMR of bassianolide exhibits a total of five *N*-Me signals (spectrum A[2)]).

When *N*-Me of the symmetrical conformation is irradiated and a difference spectrum prepared, (B) is obtained. It can be seen that the signals a_1-a_4 of asymmetric *N*-Me appear on the negative side, giving the impression that irradiation of (b) resulted in negative NOE. Actually this is a phenomenon separated from NOE. When (b) is irradiated, *N*-Me of symmetrical conformation becomes saturated, but the saturated state has a certain lifetime. After the conformation has shifted to asymmetry (at equilibrium), *N*-Me continues to be partially saturated, and the signal strength of a_1-a_4 diminishes. This transfer of saturation of protons of a given chemical species to another as a result of equilibrium is called ''saturation transfer.''

The phenomenon of saturation transfer occurs under various circumstances. When ^{1}H-NMR of a compound containing a number of OH is conducted in DMSO-d_6, the OH signals appear at different positions, but when the H_2O signal in the solvent is irradiated, the OH signals either disappear or become markedly weaker. This is the result of saturation of H_2O proton being tansferred to OH proton. When a difference spectrum is prepared at this point, the OH proton signal appears on the negative side together with the signal of the irradiated H_2O.

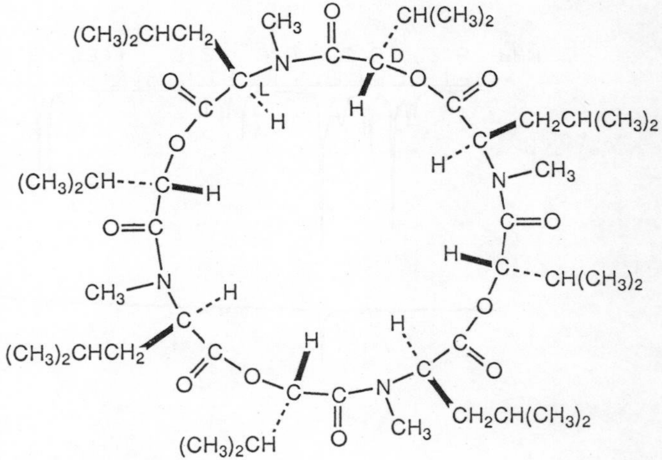

1) A.Suzuki, M.Kanaoka, A.Isogai, S.Murakoshi, M.Ichinoe and S.Tamura, *Tetrahedron Lett.*, **1977**, 2167
2) K.Nomoto and T.Iwashita, unpublished work.

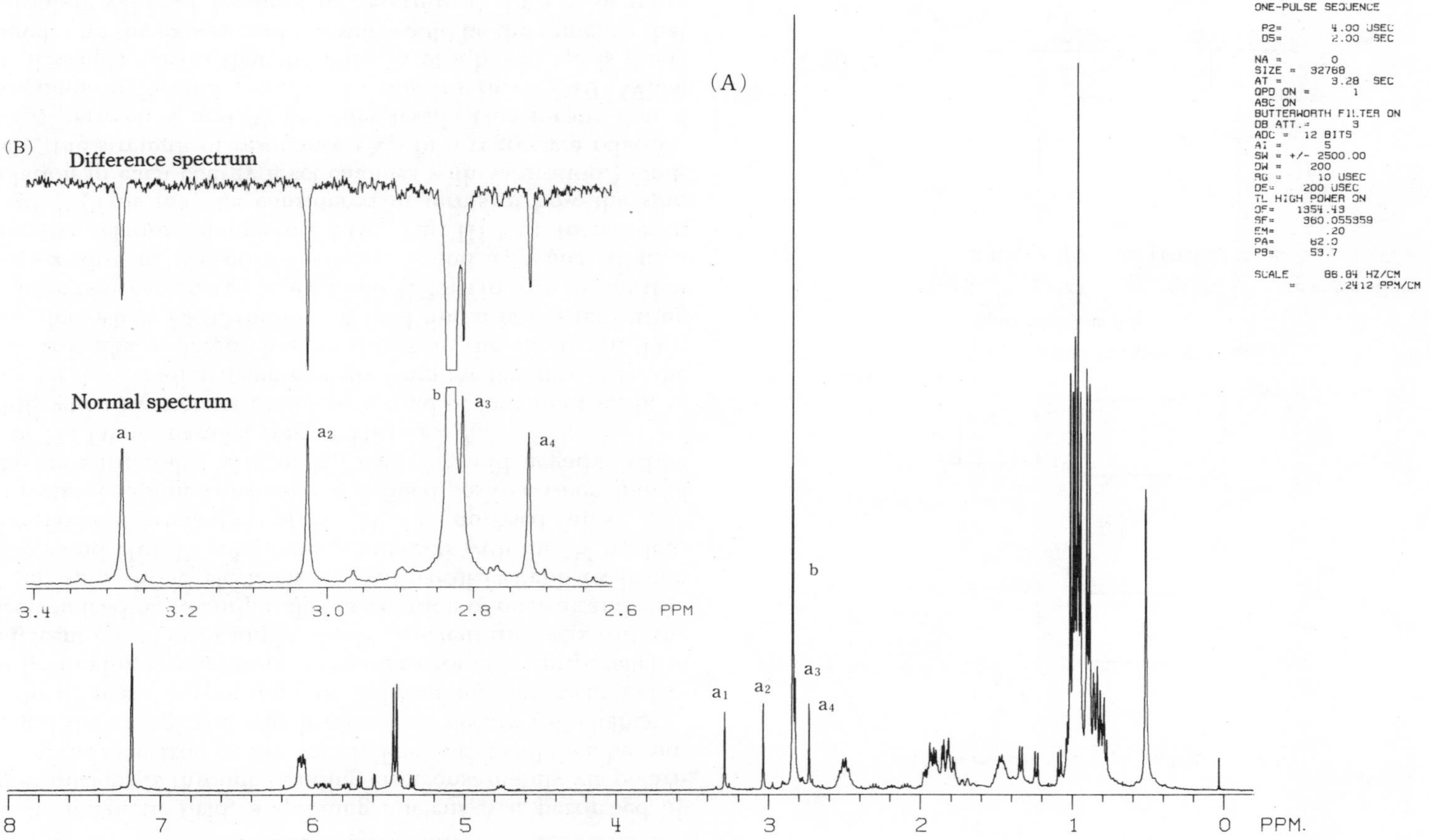

360MHz ^{1}H-NMR (C_6D_6) [TI]

14 SPT (Selective Population Transfer) of β-Ionone

In the magnetic field, a spinning nucleus A is perturbed by another nucleus X through bonding electrons, mainly via polarization of the *s* electron by the nuclei. The nuclei will then be coupled and the energies at which resonance occurs will change.

In the more stable spin state, the spins of both interacting electrons occupying the bonding molecular orbital is antiparallel to an adjacent one. For example, the ^{13}C nucleus interacts with the 2*s* electron leading to antiparallel orientation of nuclear and electron (e_1) spin. Bond electrons are antiparallel (Pauli exclusion principle and Hund's rule), and e_2 interacts with the ^1H nucleus to give favored orientation of ^{13}C-^1H, *i.e.*, opposed spins.

Spin-spin coupling constant J is defined positive when spin of H_A favors antiparallel spin of H_x, *e.g.*, J_{vic}, and negative when spin of H_A favors parallel spin of H_B, *e.g.*, J_{gem}.

Thus when two sets of doublets, a total of four lines (a, b, c, d), are weakly irradiated successively from the low field side, the signal strengths undergo changes shown in the spectrum. This means that when $J>0$, the lowest field line a is A_1, indicating saturation between energy levels 2 and 4. This in turn means that resonance due to transition between 2 and 1, which is in a progressive relation relative to 2-4 (see Part III 1.6), increases in strength.[1)] (This may be considered in terms of how the spin population in each energy level changes with saturation.) Conversely, the strength of resonance (X_1) in a regressive relationship (as between 4 and 3) becomes less.[1)] This means that d corresponds to X_2 and c to X_1. In this instance, $J>0$. When $J<0$, it simply means that the relation of a,b and c,d is interchanged, and the experimental result would be the same, so that the absolute value of J cannot be determined. In a 3- or more spin system such as AMX (discussed next), the relative signs of J can be determined.

The nuclear spin–spin interaction.

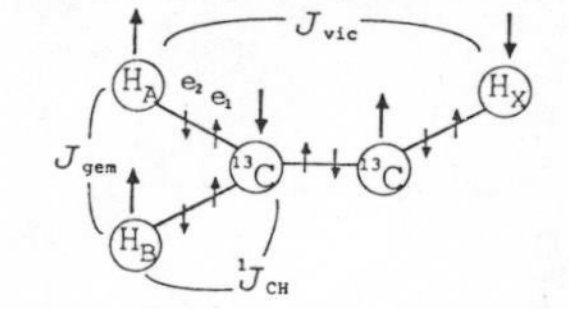

(Pauli exclusion principle and Hund's rule)

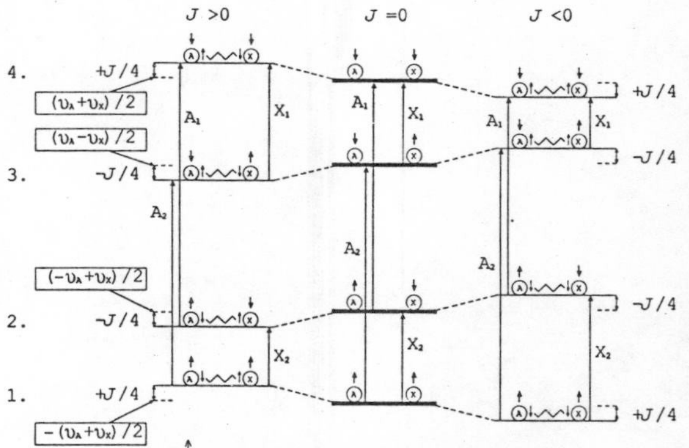

Ⓐ (↑) means that the spin of A is α, oppositely Ⓐ (↓) is β.

$(\nu_A+\nu_X)/2$, $(\nu_A-\nu_X)/2$, ··· are eigenvalues at the no spin–spin coupling state.

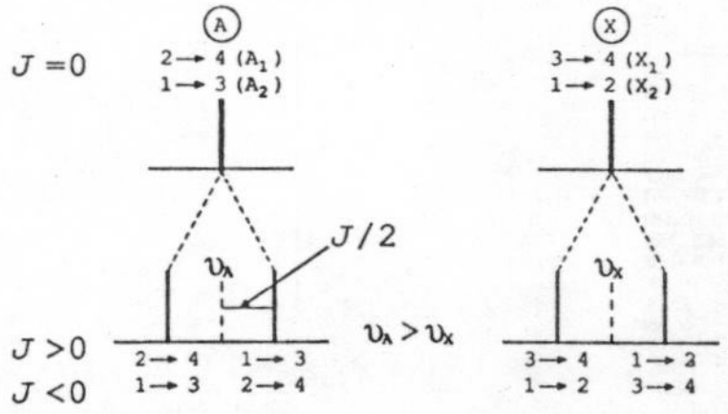

1) R.K.Harris, *Nuclear Magnetic Resonance Spectroscopy — A Physicochemical View*, Pitman Books Ltd., London (1983), p.101

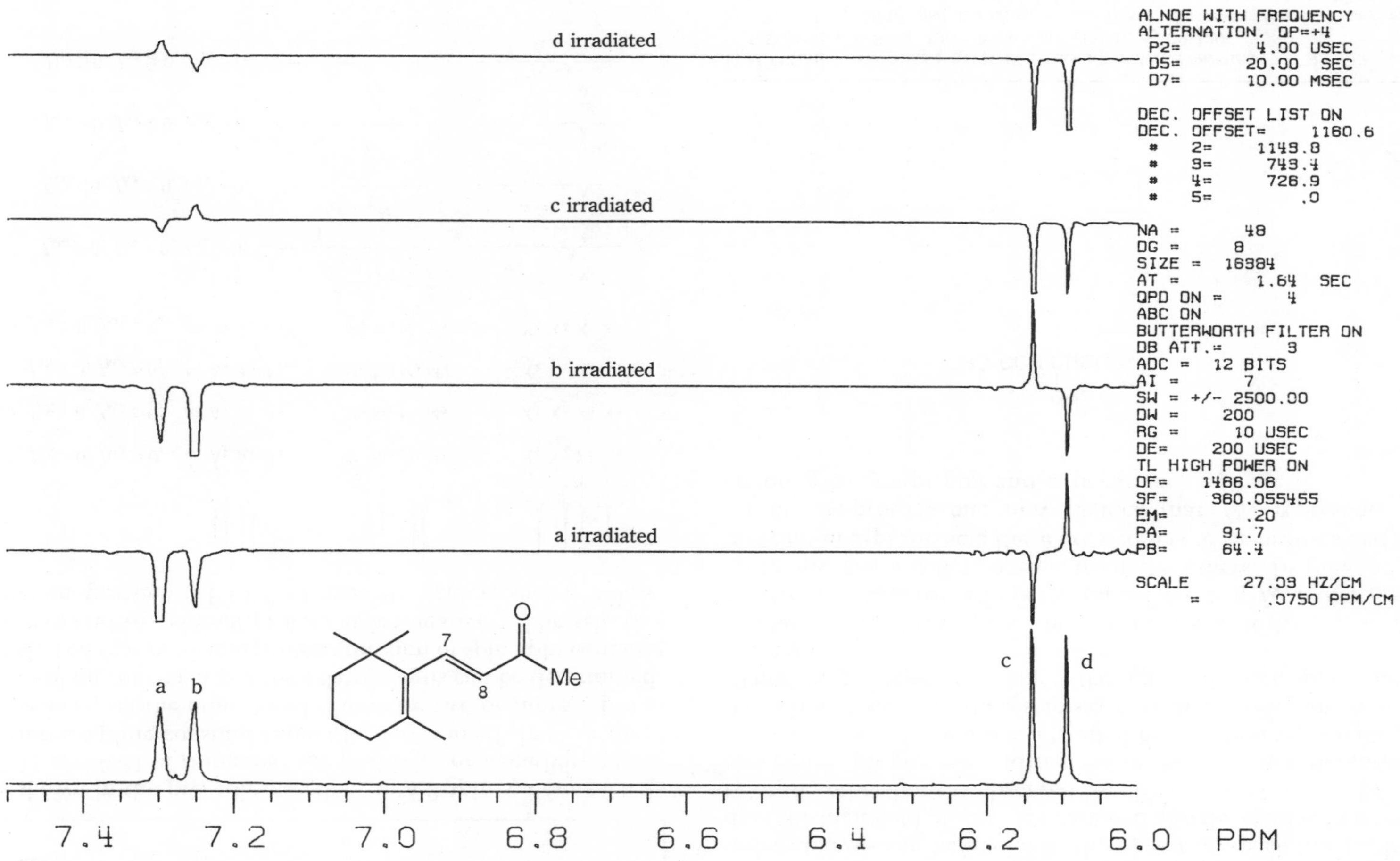

360 MHz ¹H-NMR ($CDCl_3$) [TI]

15 SPT Spectrum of Aspartic Acid

As shown in the preceding section, in a 2-spin system such as the 7-H and 8-H of β-ionone, the positivity or negativity of the spin-spin coupling constant cannot be determined. In 3- or more spin systems, on the other hand, there are two or more types of spin coupling constants and the relative signs can be determined. It is believed that in vicinal protons the sign of spin-spin coupling is positive and in geminal protons it is negative.[1] The selective population transfer (SPT)[2,3] of aspartic acid is shown below.

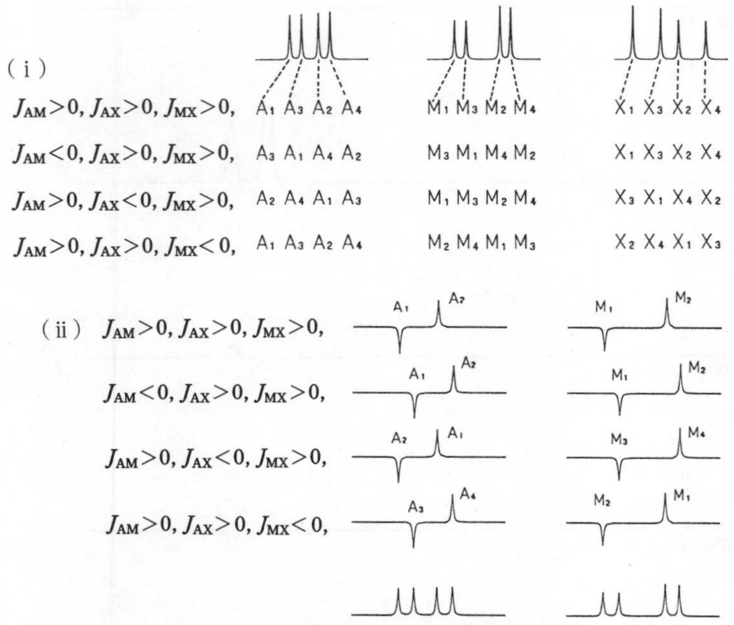

Assignment of lines in AMX system

Aspartic acid has a typical 3-spin system (^{1}H-NMR; AMX from the low field side). The chemical shift difference is large compared with the spin-spin coupling constants, therefore a first order analysis can be made. In this spectrum, each line (a, b, c, d) of the high field protons is irradiated and the change in intensity of the other lines is observed by difference spectrum. By this procedure, progressive and regressive changes are differentiated as in the case of β-ionone. Detailed explanation is provided in part III section 1.6; suffice it to say here that spectra given under ii) may be predicted according to the signs of the spin-spin coupling constants.

The predicted spectrum which coincides with the actual SPT spectrum of aspartic acid is the one of $J_{AM}>0$, $J_{AX}>0$, $J_{MX}<0$. Only J_{MX} has a different sign from the others. In view of the fact that in aspartic acid there are two sets of *vic*-protons and one set of *gem*-proton, one may deduce that J_{MX} represents the coupling by *gem*-proton and is negative.

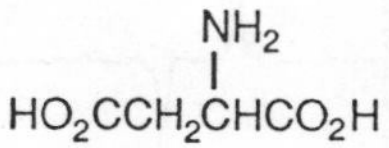

1) H.Günther (translated by R.W.Gleason). *NMR Spectroscopy — An Introduction*, John Wiley & Sons, Ltd., New York (1980), pp.99-120
2) J.W. Akitt, *NMR and Chemistry — An Introduction to the Fourier Transform — multinuclear era*, Chapman and Hall Ltd., New York (1983), p.143-146
3) R.K.Harris, *Nuclear Magnetic Resonance Spectroscopy — A Physicochemical View*, Pitman Books Ltd., London (1983), pp.100-101 and p.167

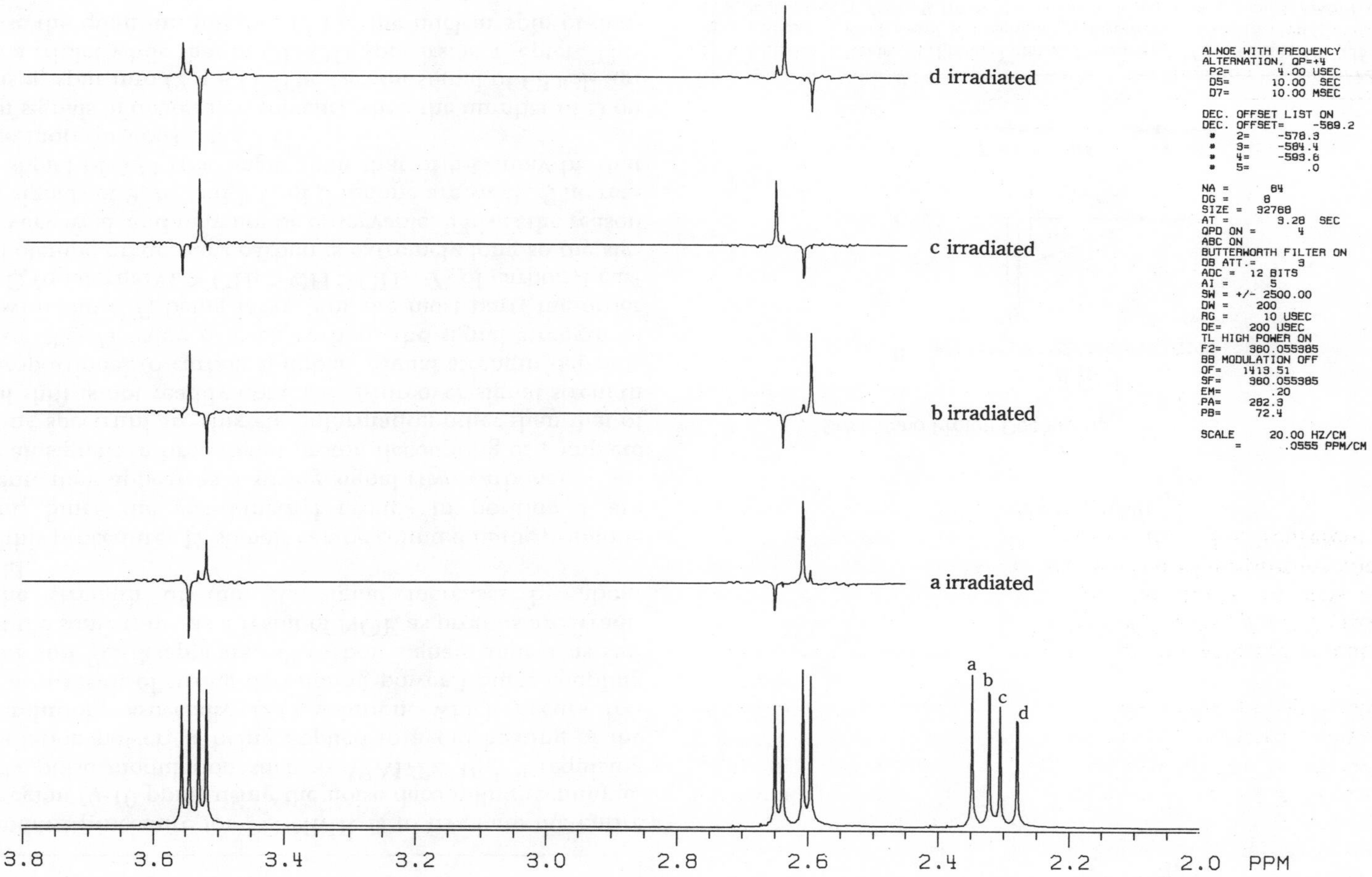

360MHz ^{1}H-NMR (D_2O-NaOD) [TI]

16 Broadband Proton Decoupling Spectrum of β-Ionone

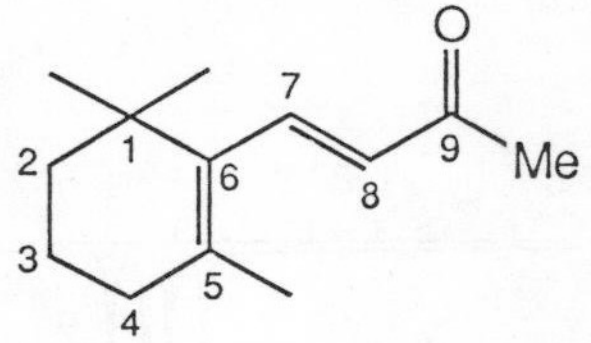

A standard procedure for ^{13}C-NMR is to irradiate the entire proton region (0-10 ppm) using the noise decoupling technique. (Recently pulse modulation such as WALTZ-16,[1,2] requiring less irradiation power, is being applied to avoid heating of the sample solution, especially D_2O solution, which occurs frequently as a result of strong decoupling power.) Since coupling of proton and ^{13}C disappears, all carbon signals appear as singlets. At the same time, as a result of NOE as protons are irradiated, the strength of the ^{13}C signal increases by about three-fold.

With this procedure, 12 signals can be counted in the β-ionone spectrum. Since the *gem*-dimethyl groups in position 1 are equivalent, they appear as a strong signal (two carbons).

Since all signals in broadband proton decoupling or complete decoupling spectrum are singlets, information other than that of chemical shift is not readily obtained. Moreover signal strength is not proportional to carbon number. Signal strength depends mainly on the T_1 value of each carbon, the signal strength of carbon with short T_1 being large. For the most part, the order of T_1 is C (quaternary) $\gg CH_3 > CH > CH_2$. T_1 of carbonyl carbon and olefinic quaternary carbon is extremely long so the signals are very weak and may not be observable. This is the reason why the signals of 9, 6, and 5-C of β-ionone are weak. The reason the signal of 5-C is stronger than that of 6-C may be that there are more protons near 5-C.

Carbon signals of deuterated solvent, when the number of D on carbon is n, split into $(2n+1)$. The carbon signal of $CDCl_3$ appears as a triplet while that of CD_3OD appears as a septet. This is because the quantum number (I) of the nuclear spin of deuterium is 1.

Broadband Proton Decoupling

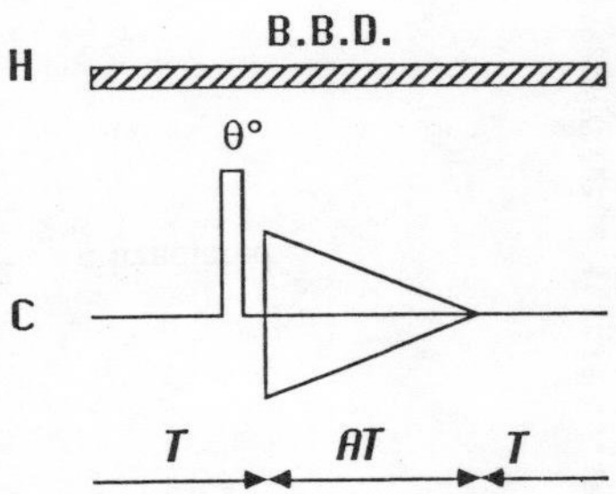

1) A.J.Shaka, J.Keeler, T.Frenkiel and R.Freeman, *J.Magn.Reson*., **52**, 335 (1983)
2) A.J.Shaka, J.Keeler and R.Freeman, *J.Magn.Reson*., **53**, 313 (1983)
3) K.Nakanishi, T.Goto, S.Itô, S.Natori and S. Nozoe (eds.), *Natural Products Chemistry* vol. 3, Kodansha Ltd., Tokyo (co-published by University Science Books, California) (1983), p.2

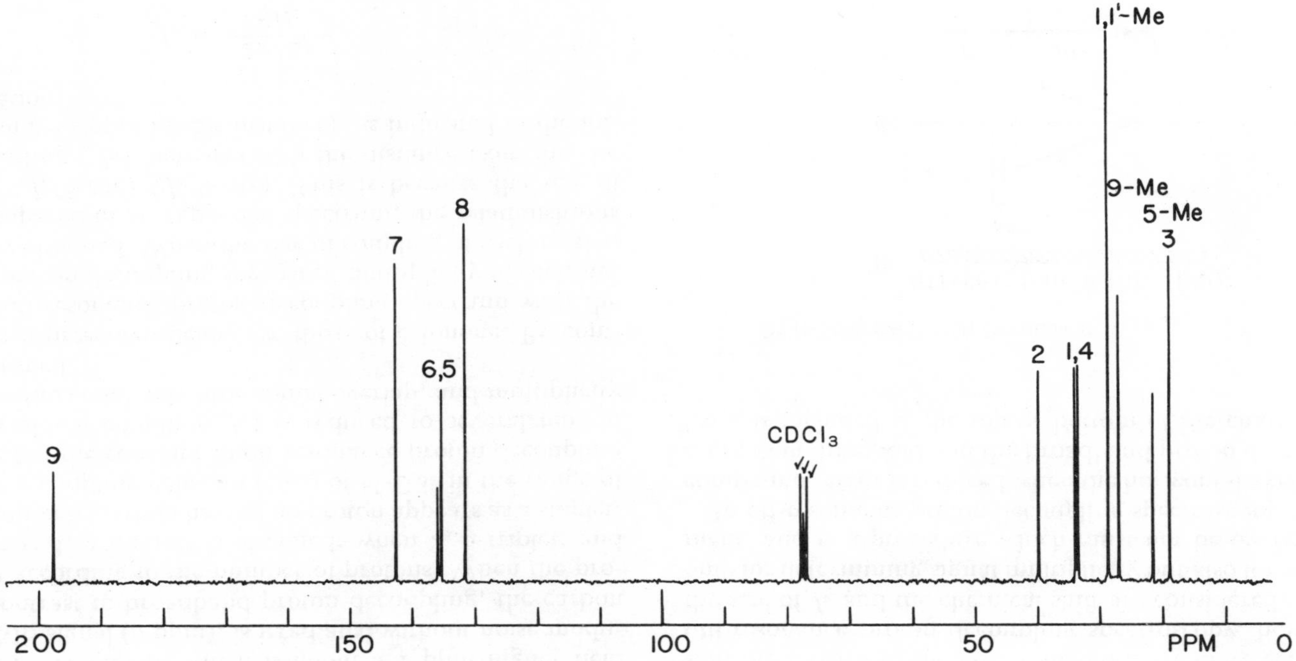

62.5MHz ^{13}C-NMR ($CDCl_3$)[3)]

17 Off-resonance Proton Decoupling Spectrum of β-Ionone

In a broadband proton decoupling spectrum, the multiplicity of each signal cannot be determined. The objective of the off-resonance proton decoupling spectrum is to determine multiplicity. When the ^1H region, which is about a 1 ppm higher field than the TMS signal (0 ppm), is irradiated without noise modulation, in contrast to broadband proton decoupling, the carbon signals split according to the number of protons. When the proton number is 3, a quartet is obtained; when 2, a triplet; and when 1, a doublet. Carbon having no proton appears as a singlet. The spin-spin coupling constant ($^1J_{CH}$) of H-C is in the range of 120-250 Hz but the constant in off-resonance proton decoupling spectrum (residual coupling; J_R) is reduced to several tens of Hz. There is thus relatively little signal overlap, and multiplicity can be discerned.

The spectra presented below are those of β-ionone. By comparing the off-resonance proton decoupling spectrum with the broadband proton decoupling spectrum, multiplicity of each signal is readily obtained. When the size of coupling of each methyl signal is compared in an expanded spectrum, the relationship is $J_R(1,1'\mathrm{Me}) < J_R(5\text{-Me}) < J_R(9\text{-Me})$. This is because the size of residual coupling (J_R) increases with the distance from the site of irradiation (−1 ppm in this instance), as indicated in the following equation.[1)]

$$J_R \simeq \frac{2\pi J \Delta\nu}{\gamma H_2}$$

γ: Magnetogyric ratio
H_2: Decoupler output
J: True size of C-H coupling constant
$\Delta\nu$: Difference between the site of irradiation and chemical shift

When the relationship $\delta_{9\text{-Me}} > \delta_{5\text{-Me}} > \delta_{1.1'\text{-Me}}$ is applied to ^1H-spectrum in the preceding section (see section 4), the above relation is understood.

When the molecule is complex, overlap of signals in off-resonance proton decoupling spectrum increases, and it becomes difficult to precisely determine the multiplicity. Sensitivity also becomes less than that of broadband proton decoupling, and data acquisition time must be increased about three-fold. For this reason, the INEPT or the DEPT methods are more generally used. Off-resonance proton decoupling spectroscopy, however, when the size of J_R and the chemical shift are considered, is useful not only for determining signal multiplicity but also for signal assignment, and is a procedure which must not be overlooked.

In off-resonance proton decoupling spectroscopy of a complex compound, error is reduced when the horizontal axis is expanded more than three-fold and the broadband proton decoupling spectrum is included at the top or bottom of the chart.

Off-Resonance Proton Decoupling

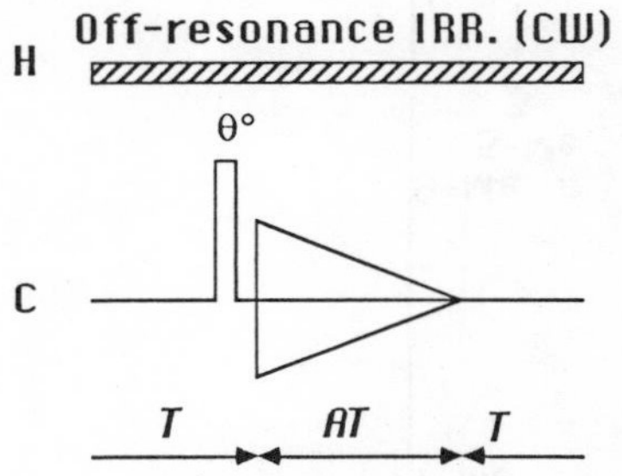

1) H. Günther (translated by R.W.Gleason), *NMR Spectroscopy — An Introduction*, John Wiley & Sons, Ltd., New York (1980), pp.310-312
2) K.Nakanishi, T.Goto, S.Itô, S.Natori and S.Nozoe (eds.), *Natural Products Chemistry*, vol. 3, Kodansha Ltd., Tokyo (co-published by University Science Books, California) (1983), p.2-3

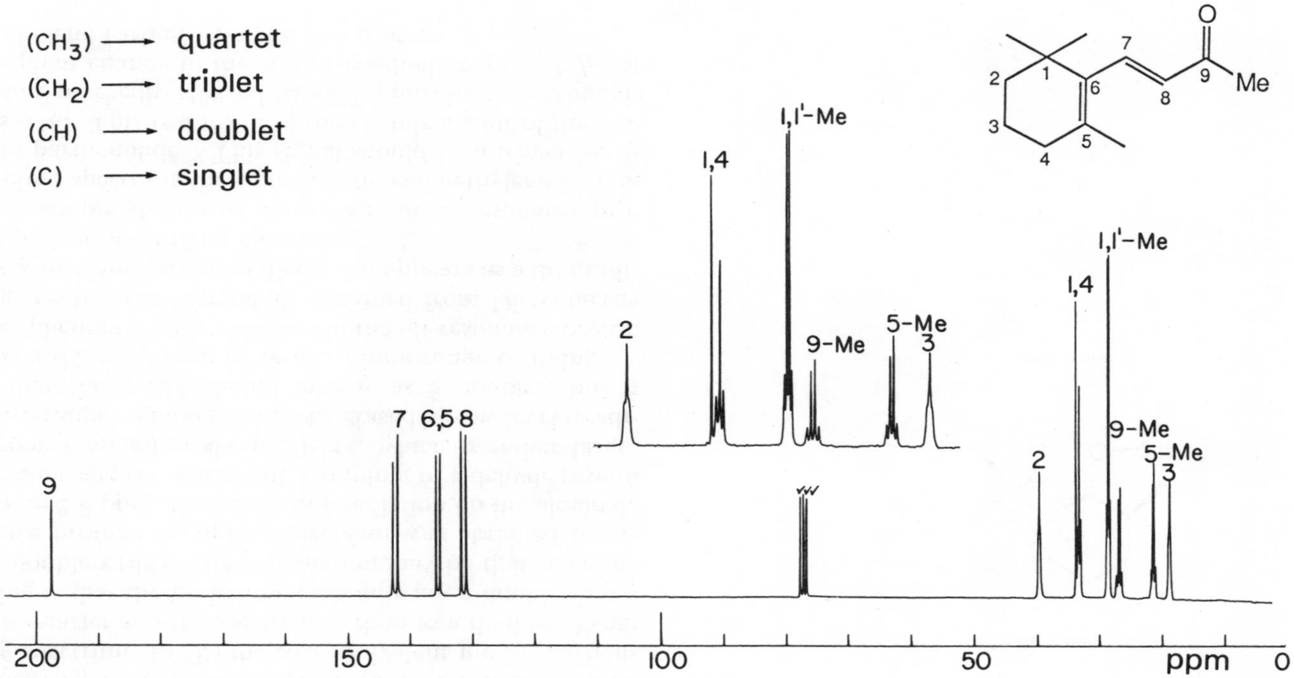

62.5MHz ^{13}C-NMR ($CDCl_3$)[2)]

18 Off-resonance Proton Decoupling Spectrum of Isobutylaldehyde

Spectrum (A) is a broadband proton decoupling spectrum of isobutylaldehyde, while spectrum (B) is its off-resonance proton decoupling spectrum. In (B) the two equivalent methyl carbons appear as a quartet and the aldehyde carbon as a doublet. What is interesting is that the 2-methine carbon is not a doublet but a doublet of doublets (dd). The explanation may be that since the off-resonance proton decoupling spectrum was obtained by irradiating at −2.8 ppm, the power of irradiation on the aldehyde proton at about 10 ppm was small. Coupling of aldehyde proton with 2-carbon ($^2J_{C\text{-}CHO}$) is about 30 Hz, which is rather large, and this long-range coupling cannot be erased by low level irradiation, so that the 2-CH should appear as a doublet, but it emerged as a dd as a result of residual long-range coupling.

A similar phenomenon is observed in the off-resonance proton decoupling spectrum of dictyodial[1] obtained from Dictyotaceae algae. The 2-methine carbon of dictyodial appears as a dd in off-resonance proton decoupling spectrum.[2]

Another example of unusual multiplicity in off-resonance proton decoupling spectra is the signal of the exomethylene carbon ($=CH_2$) on parthenolide.[3] This signal should be a triplet but it appears as a dd. This is a result of the chemical shift of the two protons being markedly affected ($\Delta = 0.7$ ppm) by the carbonyl, with consequent change in the size of residual coupling (J_R) of the exomethylene carbon with the two protons.

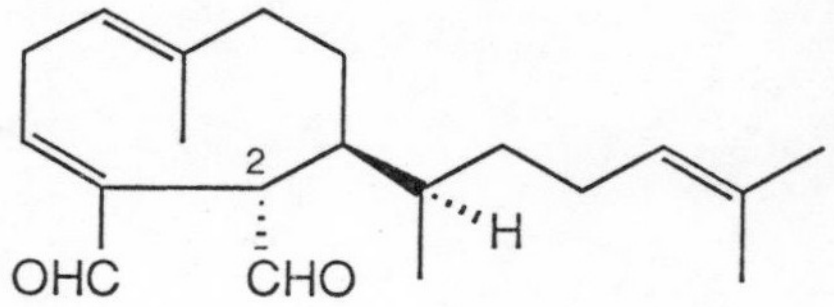

dictyodial

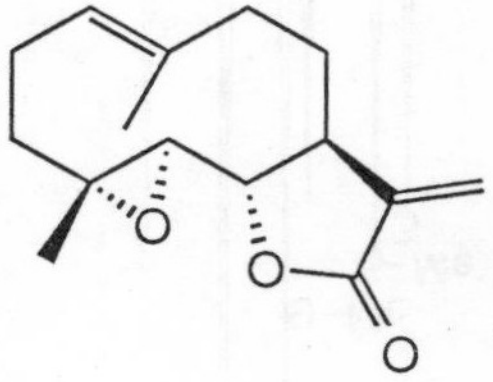

parthenolide

1) J.Finer, J.Clardy, W.Fenical, L.Minale, R.Riccio, J.Battaile, M.Kirkup and R.E.Moore, *J. Org. Chem.*, **44**, #12, 1979, p.2047
2) T.Kusumi, unpublished result.
3) M.Soucek, V.Herout and F.Sorm, *Collect. Czech. Chem. Comm.*, **26**, 803 (1961)

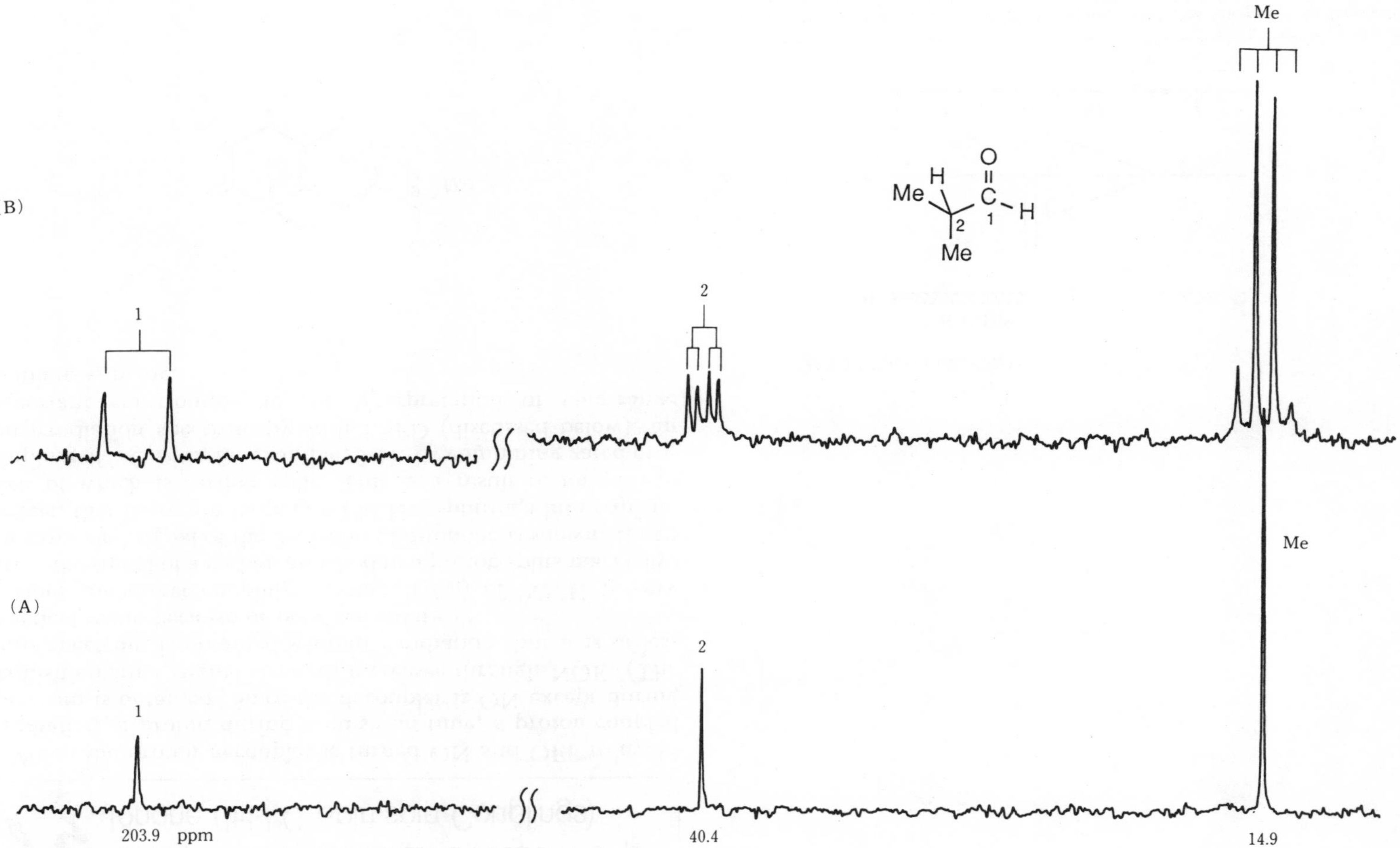

22.5MHz ^{13}C-NMR ($CDCl_3$) [TK]

19 Gated Proton Irradiation Spectrum of β-Ionone (^1H-^{13}C Spin-spin Couplings)

When the proton decoupler is turned ON and OFF to avoid irradiation of proton during acquisition time, a proton coupled spectrum is obtained. Since the decoupler is ON except during acquisition time, signal strength increases through NOE. (The same spectrum is obtained without irradiation, but it is of less practical value because of poor sensitivity.)[1]

Since the direct coupling constant ($^1J_{CH}$) of ^{13}C-^1H is very large, the signal of a carbon attached to a proton splits markedly. An expanded signal of the 2-carbon of β-ionone is shown. It can be seen that there are large ($J \simeq 130$ Hz) splittings into triplets, each of which is further split. This is a result of long-range coupling of 2-C with, for example, 3-H. By combining gated proton irradiation spectroscopy with LSPD (discussed below), an important contribution to the interpretation of long-range coupling is made.

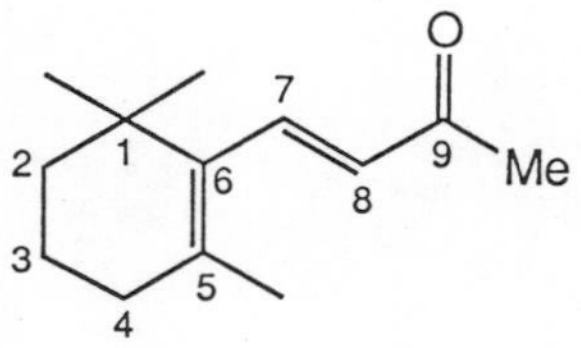

Gated Proton Irradiation

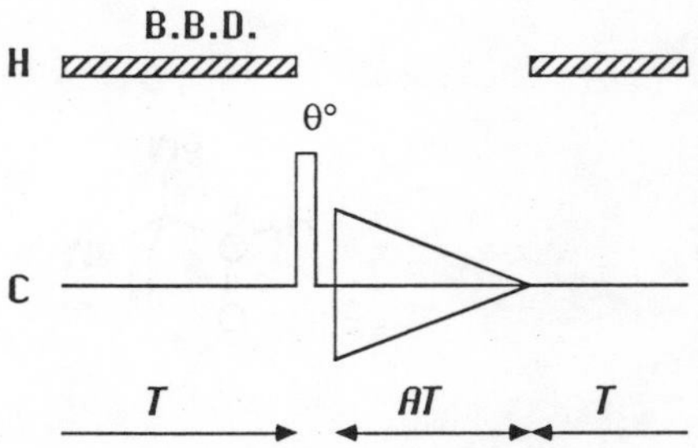

1) H. Günther (translated by R.W. Gleason), *NMR Spectroscopy — An Introduction*, John Wiley & Sons, Ltd., New York (1980), pp.359-360
2) K. Nakanishi, T. Goto, S. Itô, S. Natori and S. Nozoe (Ed.,) *Natural Products Chemistry*, vol.3, Kodansha Ltd., Tokyo (co-published by University Science Books, California) (1983), p.3

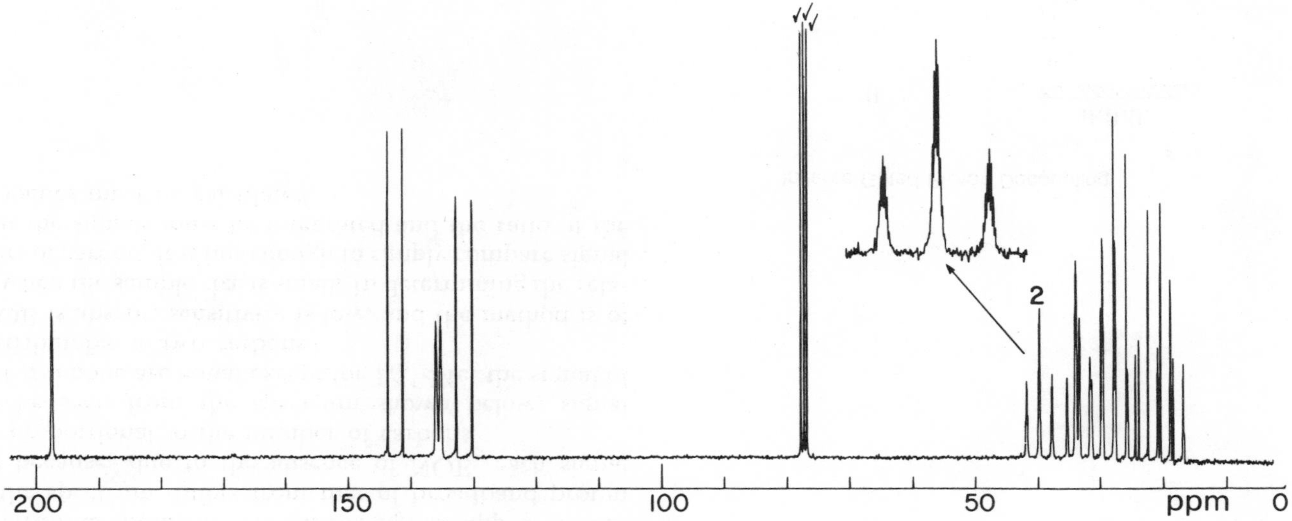

62.5 MHz ^{13}C-NMR ($CDCl_3$)[2)]

20 Inverse Gated Proton Decoupling Spectrum of β-Ionone (No NOE)

When, in reverse of gated proton irradiation, the decoupler is set at ON only during acquisition and at OFF other times, a "no NOE" spectrum is obtained.[1] All carbon signals appear as singlets, but the spectrum differs from that of broadband proton decoupling because, due to the absence of NOE, each signal strength is proportional to the number of carbons.

As may be seen from the spectrum shown below, signal strengths of β-ionone are equal except for 1,1′-Me, the signal of which is attributable to two carbons.

Since NOE is absent, sensitivity is low, and the method is of little value when the sample size is small. In determining the relative numbers of carbon, it is not enough to simply compare signal heights, but the signals must be integrated and the ratio of the integrated values must be calculated.

Inverse Gated Proton Decoupling

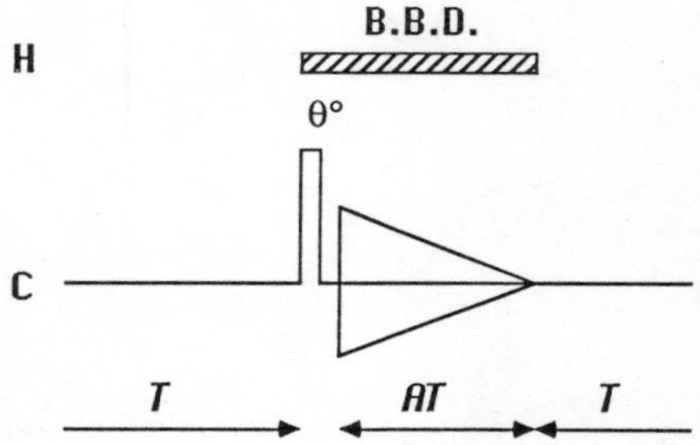

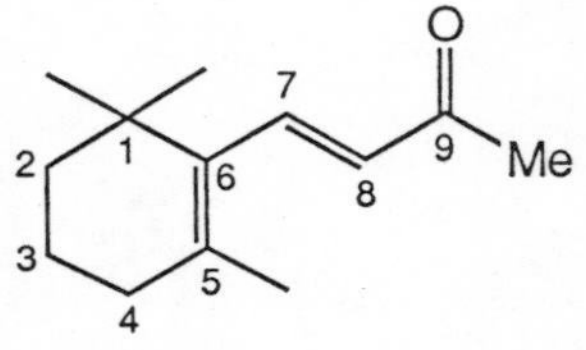

1) R.K. Harris, *Nuclear Magnetic Resonance Spectroscopy — A Physicohemical View*, Pitman Books Ltd., London (1983), p.111-113
2) K. Nakanishi, T. Goto, S. Itô, S. Natori and S. Nozoe (Ed.), *Natural Products Chemistry*, vol.3, Kodansha Ltd., Tokyo (co-published by University Science Books, California) (1983), p.4

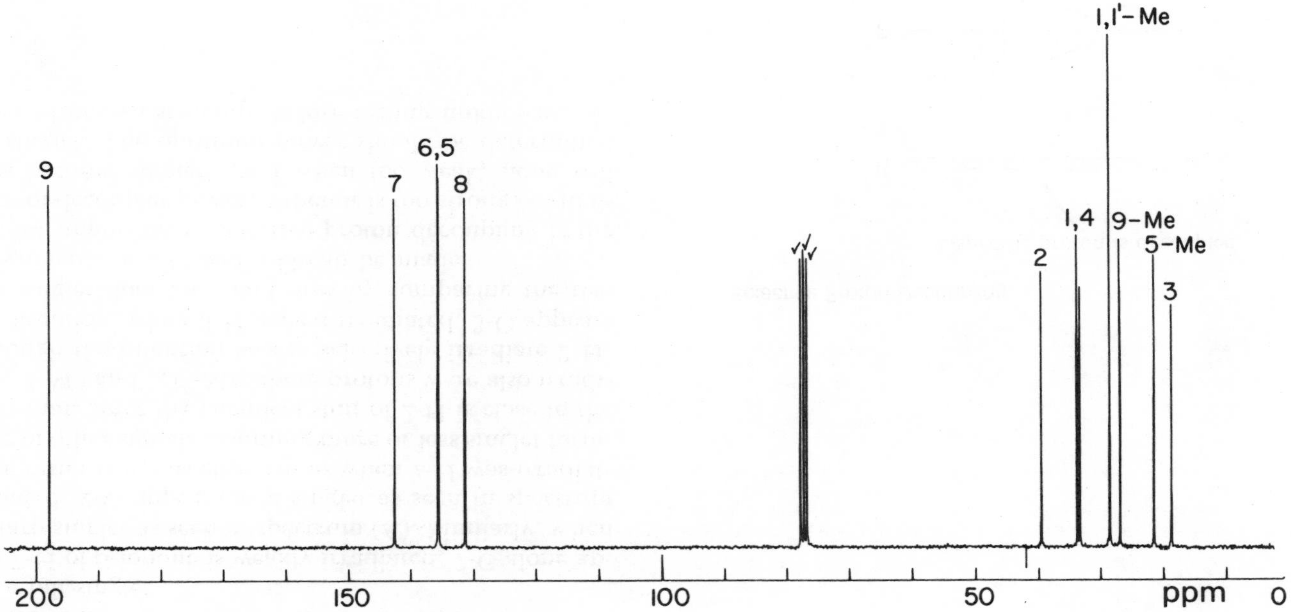

62.5 MHz ^{13}C-NMR ($CDCl_3$)[2)]

21 Selective Proton Decoupling Spectrum of β-Ionone

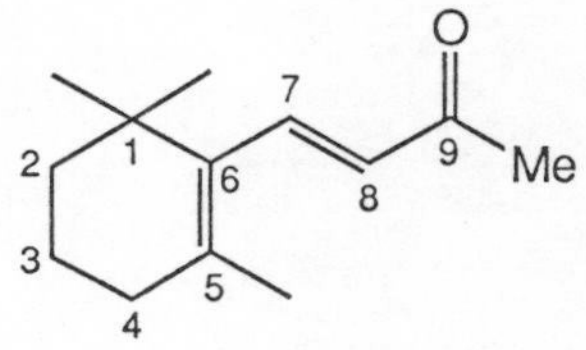

Selective proton decoupling spectrum can be used for carbon signal assignment. Where assignment of protons has been made, selective irradiation of a specific proton at low power results in the appearance of the signal of carbon bonded to that proton in the form of a tall singlet.

When the 7-H of β-ionone is weakly irradiated, 7-C alone appears as a sharp singlet as seen in spectrum (A). Similarly, when 2-H is irradiated, 2-C appears as a singlet as seen in spectrum (B). Here the result is not as clear-cut as when 7-H was irradiated, a number of other signals assuming more or less singlet form. The reason is that, since the chemical shift of 2-H is close to the shifts of 3-H, 5-Me and 1,1′-Me, these protons were also irradiated even though the intention was to selectively irradiate 2-H. Even in this situation, when 3-H is next irradiated, 3-C appears as a sharper singlet than 2-C, and thus by comparing the two spectra, assignments of 2-C and 3-C can be made.

What is most important in selective proton decoupling is the establishment of decoupler power. When it is too strong, virtually all signals become singlets, and when too weak, none will appear as a singlet. The optimum power should be determined using samples of known structure before testing unknowns.

Selective Proton Decoupling

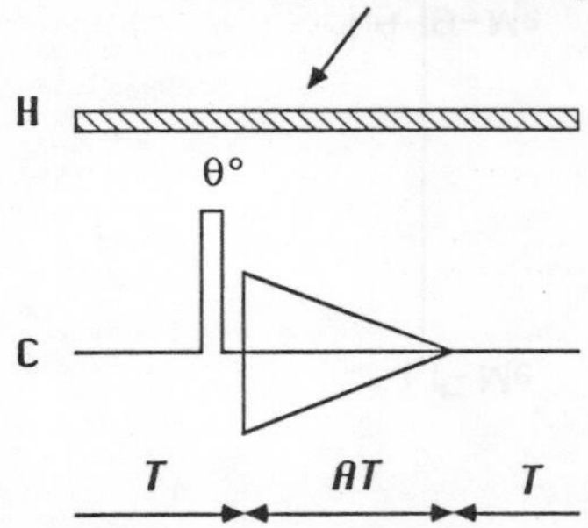

1) K. Nakanishi, T. Goto, S. Itô, S. Natori and S. Nozoe (eds.), *Natural Products Chemistry*, vol.3, Kodansha Ltd., Tokyo (co-published by University Science Books, California) (1983), p.4

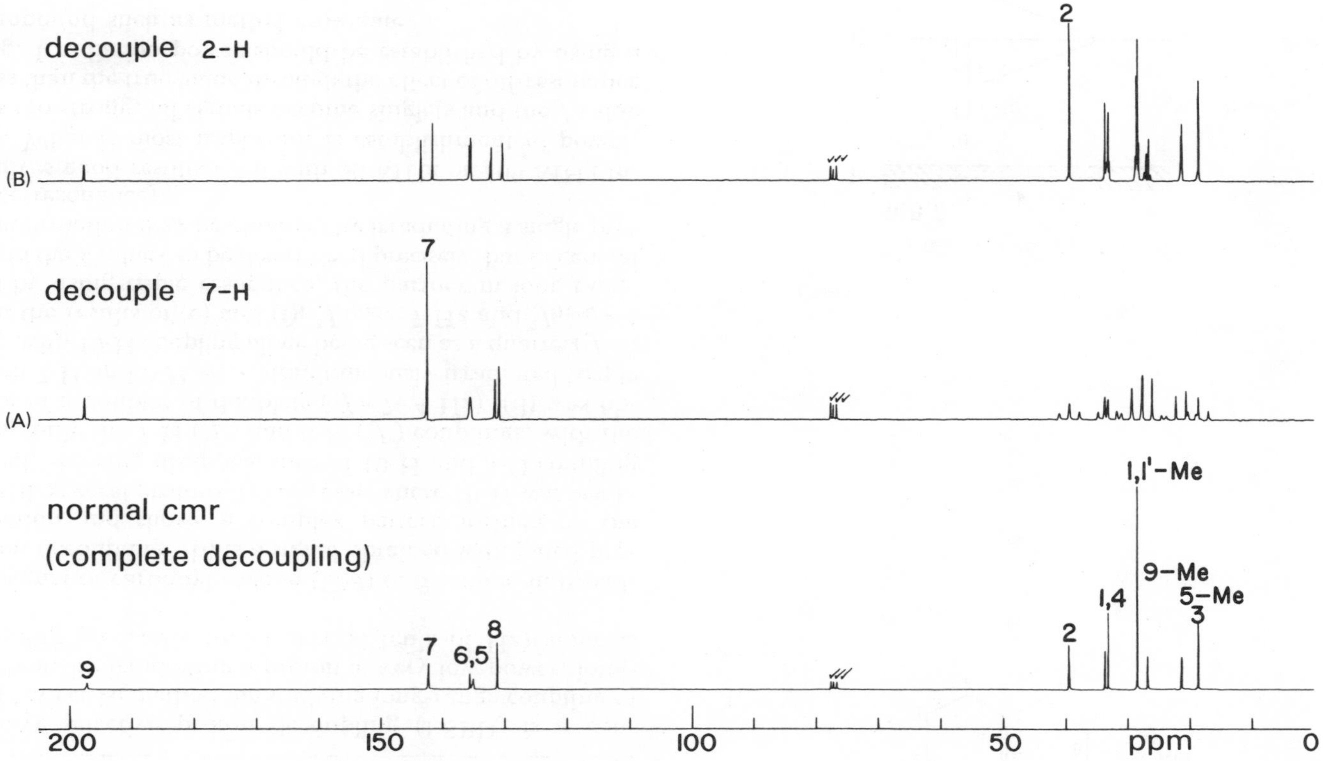

62.5 MHz ^{13}C-NMR ($CDCl_3$)[1]

22 LSPD Spectrum of β-Ionone

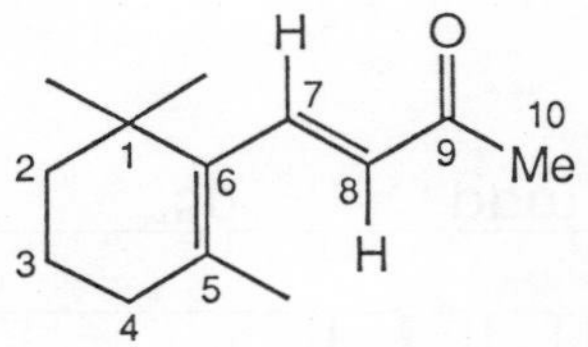

Long-range selective proton decoupling (LSPD) is a convenient and effective method for studying long-range coupling of proton-carbon. By irradiating a proton at very low power, long-range coupling (generally under several ten's of Hz) alone is decoupled.[1,2]

(a) is a signal of carbonyl carbon (C-9) of β-ionone in broad-band proton decoupling. (b) is a signal obtained with gated proton irradiation and shows a complex pattern formed by the coupling with several protons. (c) is a case where 10-H was weakly irradiated, showing disappearance of 10-H and 9-C coupling (^2J) leaving only the 7-H (^3J) and 8-H (^2J) couplings, with the appearance of a doublet of doublets (J = 7, 4 Hz). (d) was obtained when 7-H and 8-H were simultaneously irradiated (triple resonance), with 10-H coupling alone being seen as a quartet (J = 6 Hz). From the results of (e) and (f), $^3J_{7H\text{-}9C}$ = 7 Hz and $^2J_{8H\text{-}9C}$ = 4 Hz. Thus by using triple resonance, the partner in long-range coupling and the J value can be determined precisely, but in general adequate information may be obtained by irradiating a single proton (double resonance).

LSPD gives good results even with 90 MHz or 100 MHz instruments. What is most important is establishment of power. When it is too strong, all signals become singlets and the J value may be less than the true value through the effect of off-resonance decoupling. Irradiation power should be established by using a simple compound such as methyl crotonate.

LSPD

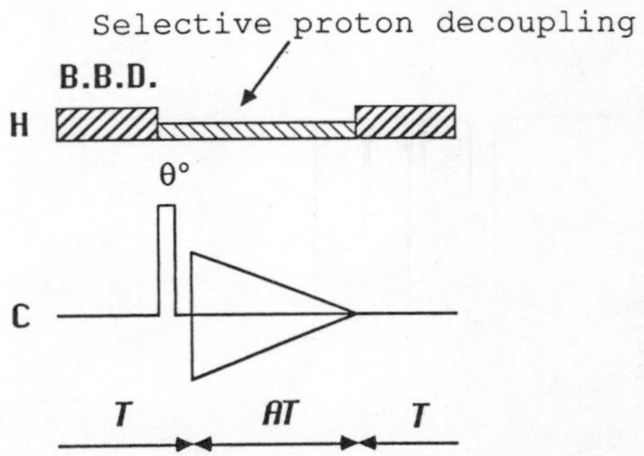

1) S. Takeuchi, J. Uzawa, H. Seto and H. Yonehara, *Tetrahedron Lett.*, **1977**, 2943
2) J. Uzawa and S. Takeuchi, *Orgn. Magn. Reson.*, **11**, 502 (1978)

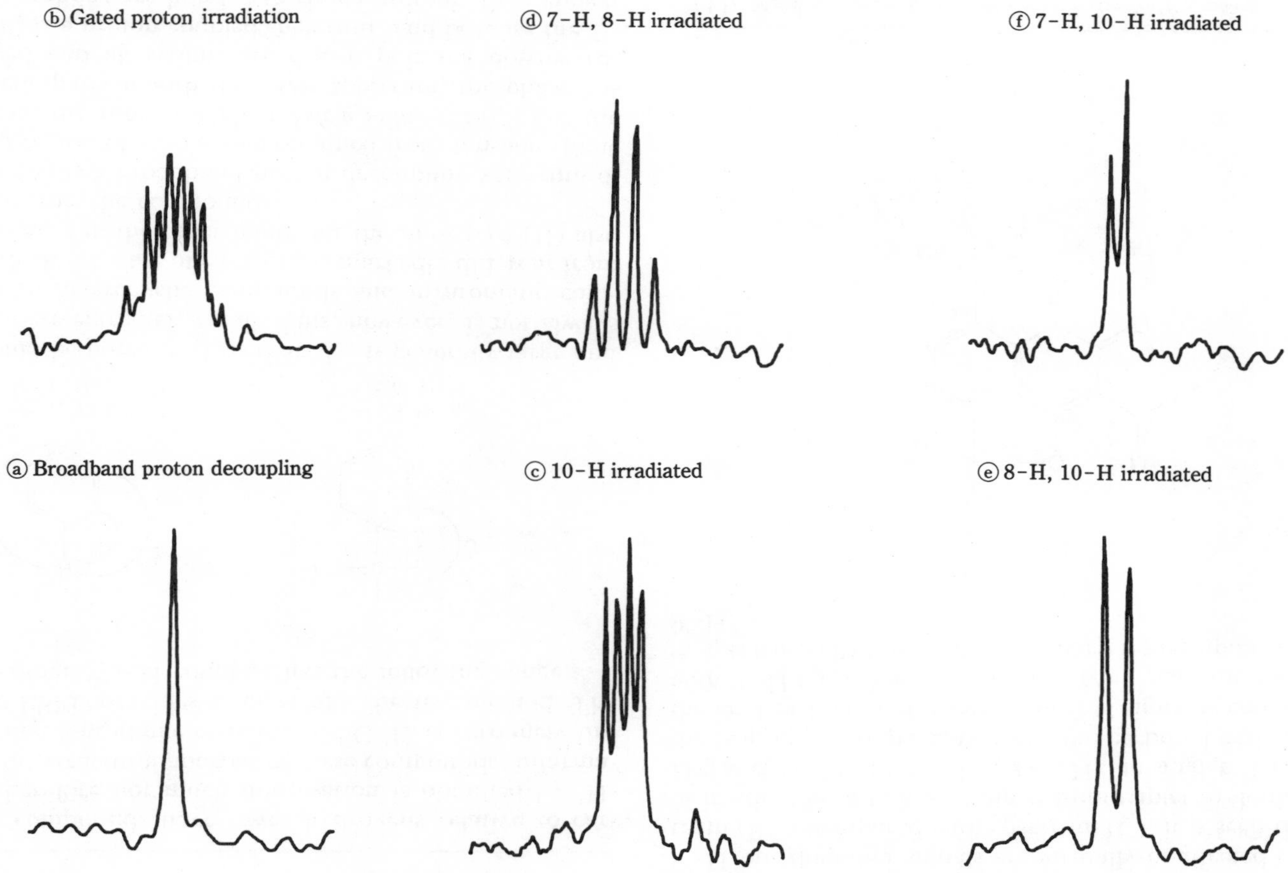

Signals of 9-carbonyl carbon in 25 MHz ^{13}C-NMR ($CDCl_3$) [KM]

23 LSPD Spectrum of Salvilenone

Aromatic compounds have fewer hydrogens relative to carbons, and therefore not much information is obtained by ^{1}H-NMR. For the structural analysis of these compounds, information concerning long-range coupling of ^{13}C-^{1}H is extremely important, and LSPD provides an especially effective method. The size of long-range ^{13}C-^{1}H coupling has the following range.[1)]

(I)

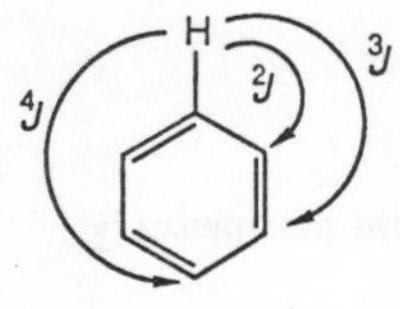

$^2J \simeq 1, ^3J \simeq 5$
$^4J \simeq 1$ (Hz)

(II)

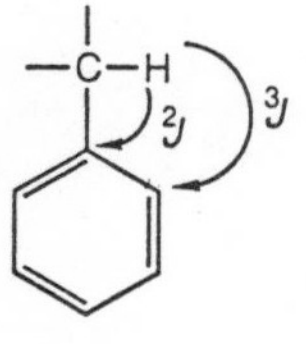

$^2J \simeq 5, ^3J \simeq 5$
$^4J \simeq 0$ (Hz)

What should be noted in (I) is that $^3J_{CH}$ is generally large and $^2J_{CH}$ and $^4J_{CH}$ are extremely small. This, however, is not always the case, as in heterocyclic compounds and in aromatic compounds in which the state of electrons is markedly different from that in benzene. On the other hand, for the most part (II) also applies to heterocyclic compounds.

Spectrum (A) is a broadband proton decoupling spectrum of the low field region of salvilenone obtained from tanshen (tonic prepared from the root of a plant *Salvia miltiorrhiza*).[2)] For the purpose of comparison with the lower spectrum, the phase has been changed and all signals are drawn pointing downward. Spectrum (B) is a proton coupled spectrum, and because the 7-, 6-, 4- and 3-carbons are bonded to single protons, they appear as doublets of large (ca. 160 Hz) $^1J_{CH}$. The 6-C appears as a broad signal through long-range coupling with the 5-Me proton.

Spectrum (C) is an LSPD spectrum obtained by irradiating 9′-H with a very weak power. The strength of the irradiation is just enough for decoupling long-range C-H coupling (7 Hz), so that change is seen in the signal of carbon in long-range coupling with 9′-H but the other signals are virtually unaffected. When spectrum (B) is compared with spectrum (C), it is seen that as result of irradiation, 8-C has changed from triplet to doublet ($J = 5.5$ Hz) and 9a-C from doublet ($J = 6$ Hz) to singlet. This shows that the isopropyl group lies between the carbonyl carbon (8-C) and the enol carbon (9a-C). Since the 9-C signal is coupled not only with 9′-H ($^2J_{CH}$) but also with two Me ($^3J_{CH}$), the signal is broad in spectrum (B) but becomes a sharp septet upon irradiation of 9′-H.

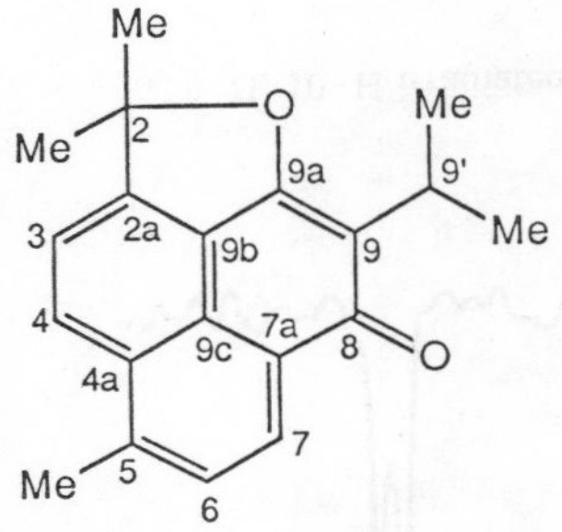

1) J.L. Marshall, *Carbon-Carbon and Carbon-Proton NMR Couplings: Applications to Organic Stereochemistry and Conformational Analysis*, Verlag Chemie International, Florida, (1983), pp.42-51
2) T. Kusumi, T. Ooi, T. Hayashi and H. Kakisawa, *Phytochemistry*, **24**, 2118 (1985)

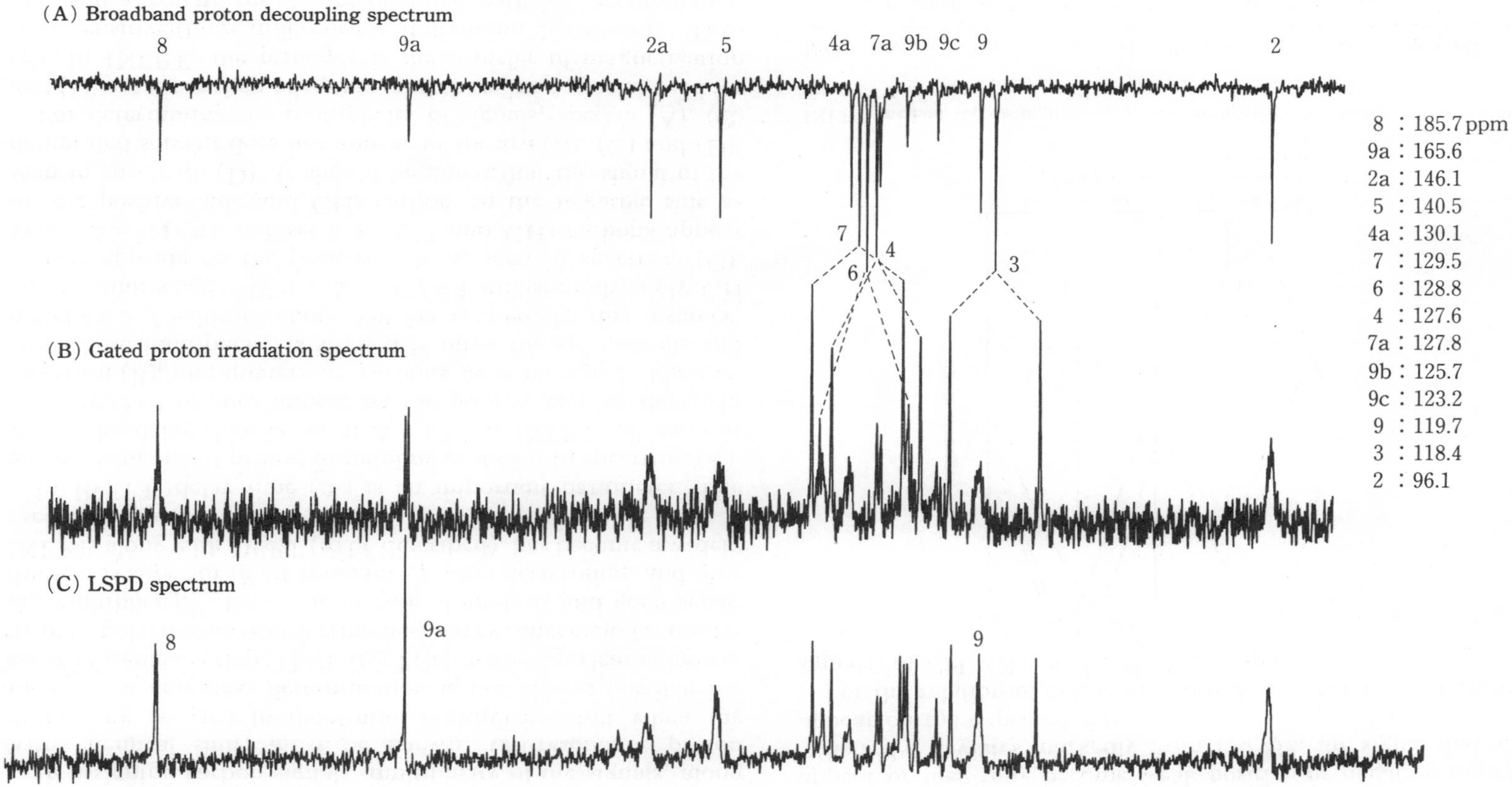

22.5MHz ^{13}C-NMR ($CDCl_3$) [TK]

24 INEPT Spectrum of β-Ionone

In assigning carbon signals, multiplicity of the signals, along with chemical shift, must be known. Off-resonance proton decoupling is used to determine multiplicity, but when the molecule is complex, determination is not always possible because of signal overlap. The INEPT (insensitive nuclear enhancement by polarization transfer) method is very convenient for resolving multiplicity.[1,2] By virtue of ease of analysis and good sensitivity, it is superior to off-resonance proton decoupling, and thus INEPT, along with DEPT (to be described), has become a widely used technique.

In INEPT, delay time (Δ_3) is an important parameter. The normal broadband proton decoupling is shown in spectrum (A). When the delay time is set at $\Delta_3 = 1/4J$ in INEPT, all carbons with attached protons appear on the positive side as shown in spectrum (B), and quaternary carbons show no signal. The carbon whose multiplicity is studied is often the sp^3 carbon, and therefore a J value of about 130 Hz is used. In this instance, $\Delta_3 \simeq 2$ milliseconds. When $\Delta_3 = 1/2J \simeq 4$ milliseconds, only CH carbon appears on the positive side as seen in spectrum (C). When $\Delta = 3/4J \simeq 6$ milliseconds, CH and CH_3 carbons appear on the positive side and CH_2 carbon on the negative side as seen in spectrum (D). It should be noted that the signal of the deuterated solvent does not appear in spectra (B), (C) and (D).

For determining the multiplicity of signals, spectra (A), (C) and (D) are more than adequate, and there is no special need for (B). In INEPT, the principle is the transfer of magnetization from sensitive (large difference in Boltzmann distribution) ^1H to ^{13}C, and adequate results are obtained with less accumulation than for broadband proton decoupling spectrum. Consequently, the time required for obtaining spectra (C) and (D) is far less than that for off-resonance proton decoupling spectroscopy, and the results are often less ambiguous than those obtained by off-resonance proton decoupling experiments.

When signals overlap, interpretation of INEPT requires some care. When CH_2 and CH overlap, the positive and negative signals in INEPT spectrum (D) may cancel each other out and disappear or they may become weak positive or negative signals. When $^1J_{CH}$ deviates markedly from 130 Hz, the signal may not appear on the expected side.

For the relation of delay time to phase (positive and negative signals) of CH, CH_2 and CH_3, see graph.

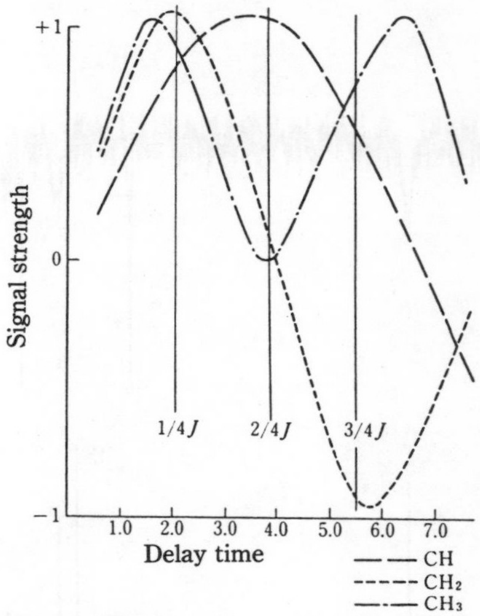

INEPT method for distinguishing atomic groups—Phase change in each atomic group

JEOL Application Note—Application of Pulse Technique Supplement (1983)

1) D.M. Doddrell and D.T. Pegg, *J. Am. Chem. Soc.*, **102**, 6388 (1980)
2) D.P. Burum and R.R. Ernst, *J. Magn. Reson.*, **39**, 163 (1980)
3) K. Nakanishi, T. Goto, S. Itô, S. Natori and S. Nozoe (eds.), *Natural Products Chemistry*, vol.3, Kodansha Ltd., Tokyo (co-published by University Science Books, California) (1983), p.6

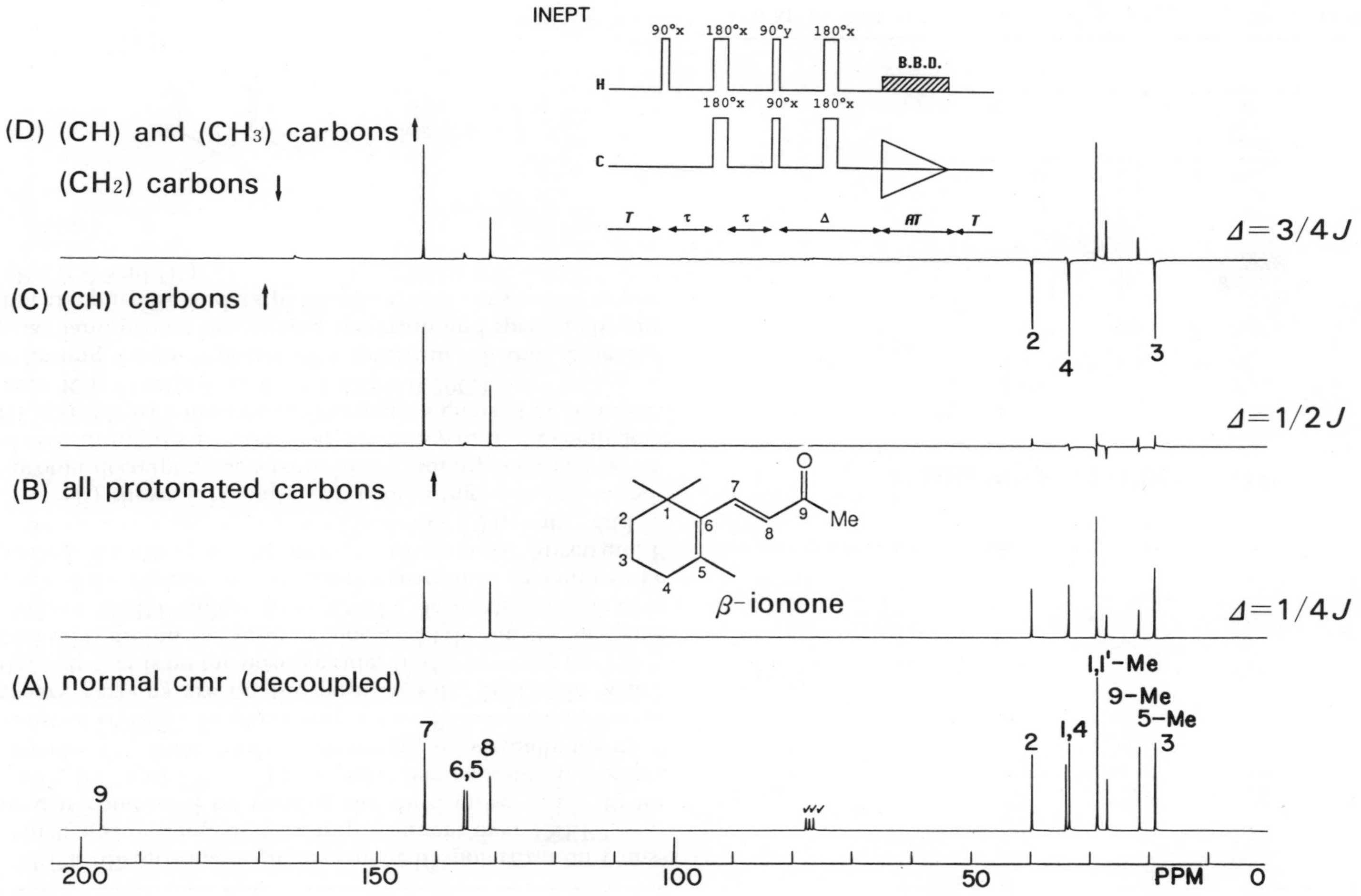

62.5MHz ^{13}C-NMR ($CDCl_3$)[3]

25 DEPT Spectrum of β-Ionone

In DEPT (distortionless enhancement by polarlization transfer), which was developed as an improvement over INEPT, experiment is conducted by varying the third pulse width (θ) for ^1H at 45°, 90° and 135°.[1] These pulse widths apply to spectra (A), (B) and (C), respectively. The results are virtually identical with those of INEPT. In (A) in which $\theta = 45°$, all carbons except quaternary carbons are on the positive side. In (B) in which $\theta = 90°$, only CH is on the positive side. In (C) in which $\theta = 135°$, CH and CH_3 are on the positive side and CH_2 on the negative side. Where DEPT differs from INEPT is in the quantitative character of the signals. By editing the spectrum, carbon can be categorized according to the number of protons. Spectrum P (= B) shows CH only, Q (= A − C) shows CH_2 only, and R (= A + C − B) shows CH_3 only. By comparing with the broadband proton decoupling spectrum below, multiplicity can be assigned. In editing the spectra, it is necessary to use a coefficient for (B) and (C) to eliminate unnecessary signals. For instance, spectrum R is obtained as $\{A - 1.45B + 0.79C\}$.

For finding the multiplicity of a spectrum, all that is needed are broadband proton decoupling spectrum and spectra (B) and (C). For the purpose of data publication, it is convenient to prepare (P), (Q) and (R).

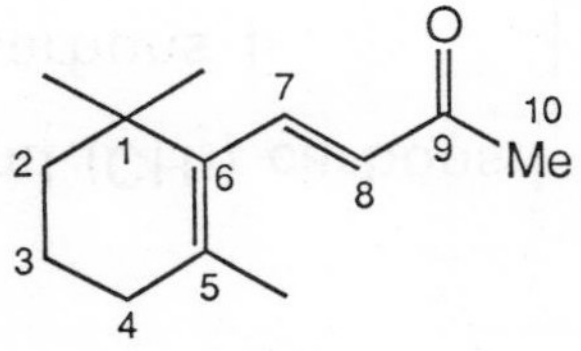

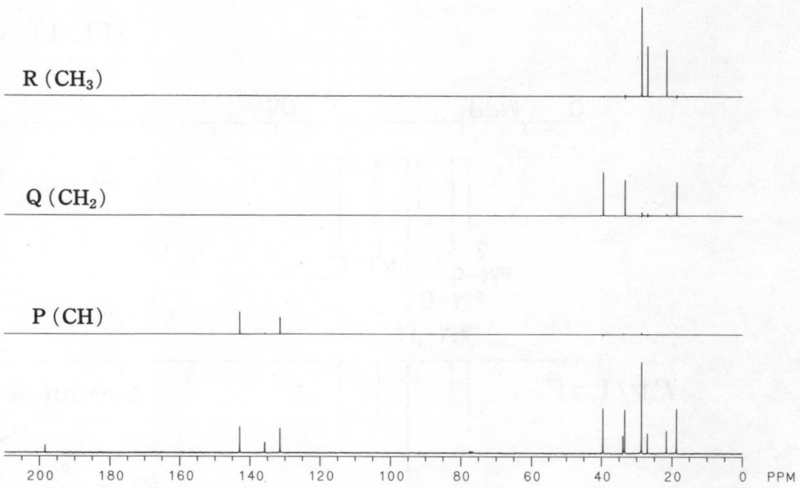

75 MHz ^{13}C-NMR ($CDCl_3$) [KM]

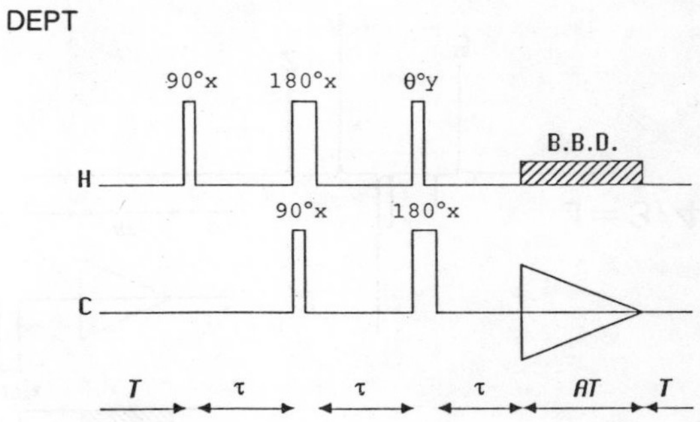

1) D.M. Doddrell, D.T. Pegg and M.R. Bendall, *J. Magn. Reson.*, **48**, 323 (1982)

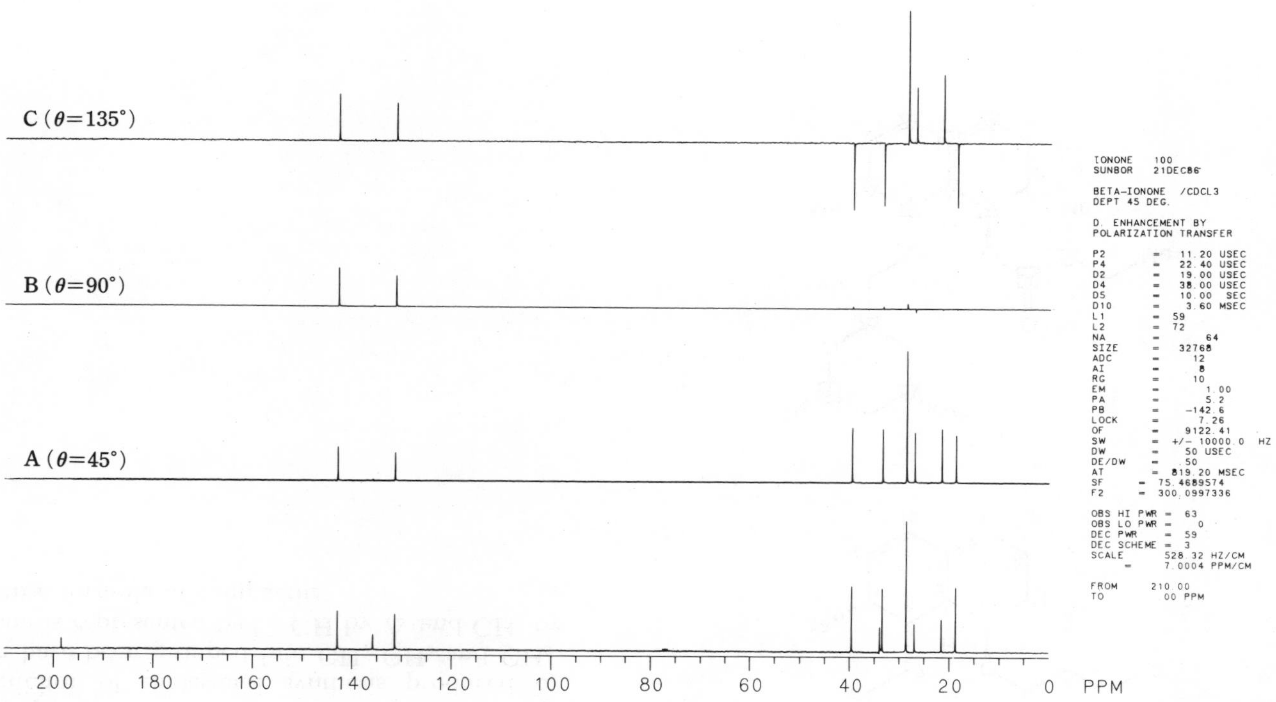

75 MHz ^{13}C-NMR ($CDCl_3$) [KM]

26 DEPT Spectrum of Compactin

The spectra represent ^{13}C signals of compactin[1] (ML-236B),[2,3] an inhibitor of cholesterol synthesis produced by *Actinomyces,* which have been grouped into CH, CH_2 and CH_3. Quaternary carbon is represented by □, CH by △ and CH_3 by Me in the structural formula of compactin.

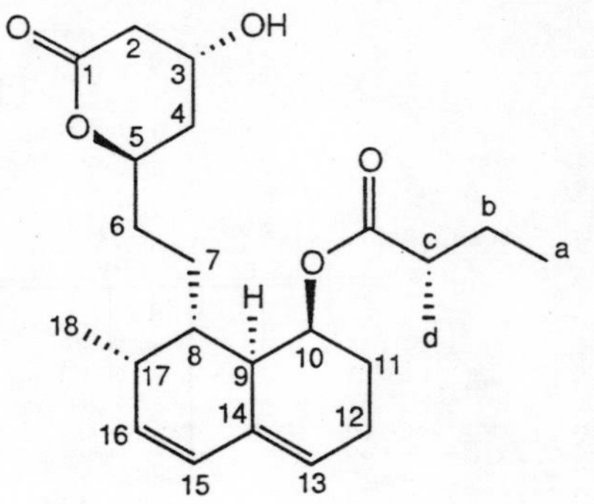

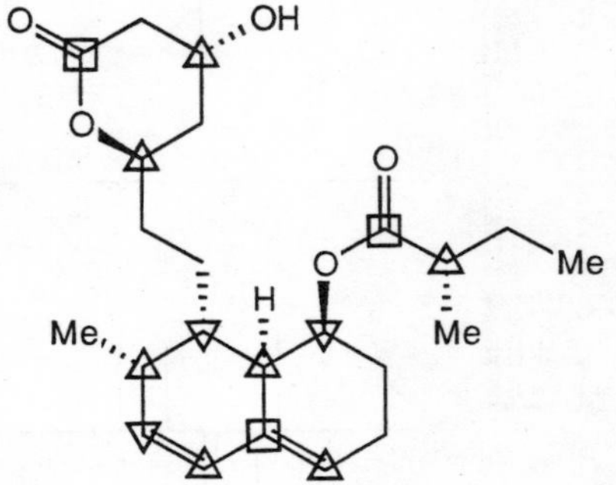

1) A.G. Brown, T.C. Smale, T.J, King, R. Hasenkamp and R.H. Thompson, *J. Chem. Soc., Perkin Trans. 1.*, **1976**, 1165
2) A. Endo, M. Kuroda and Y. Tsujita, *J. Antibiot.*, **29**, 1346 (1976)
3) M. Hirama and M. Uei, *J. Am. Chem. Soc.*, **104**, 4251 (1982); *Idem, Tetrahedron Lett.*, **23**, 5307 (1982)

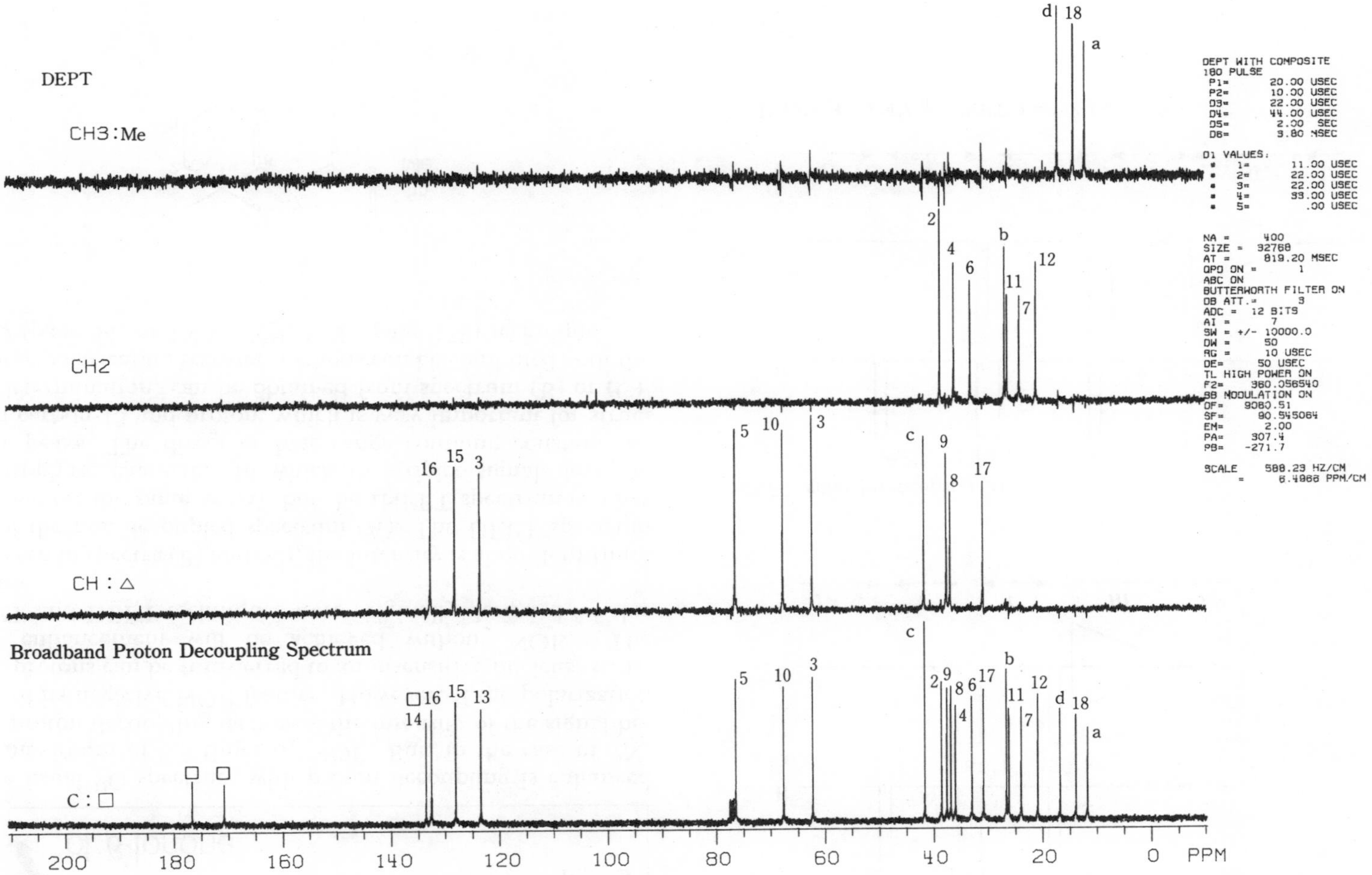

90MHz ^{13}C-NMR ($CDCl_3$) [TI]

27 INEPT and DEPT Without Decoupling of β-Ionone

The usual ^{13}C spectrum with proton decoupling is enhanced by a maximum of 2.9 times by NOE. But, in the case of ^{15}N, usual proton decoupling decreases the intensity of the signal because of its negative NOE feature. However, if the polarization of the protons can be transferred to an insensitive nucleus, sensitivity enhancement will be achieved without NOE.[1] The INEPT and DEPT techniques were originally designed for this purpose.

As seen in spectra (B) and (C), the intensity is about four times that of the non-decoupled spectrum (A). The DEPT spectrum (B) is almost the same as (A), but the INEPT spectrum (C) has an anti-phase character, in which the triplet signals lose the center peaks. The direct or long-range coupling constant between carbon-13 and proton, which is very important for structure determination, can be obtained from spectrum (B) or (C).

These polarization transfer methods can be combined with the LSPD (page 44) or INADEQUATE (page 178) technique.

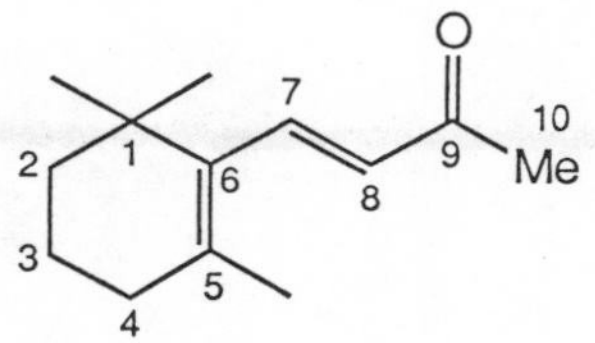

INEPT (C)

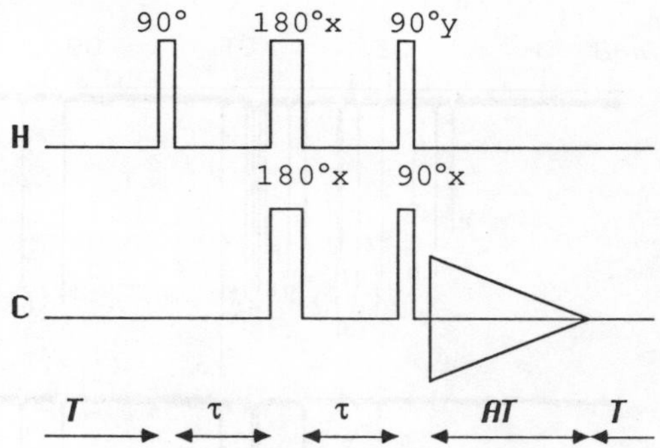

DEPT without decoupling (B)

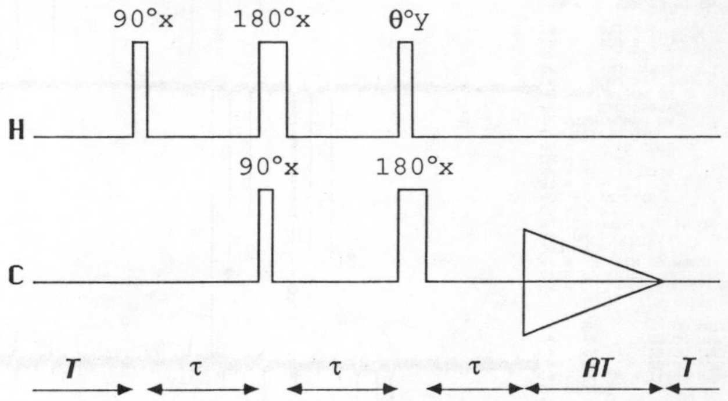

INEPT $\tau = 1/4\,^1J_{CH}$: DEPT $\tau = 1/2\,^1J_{CH}$

1) G.A. Morris and R. Freeman, *J. Am. Chem. Soc.*, **101**, 761 (1979)

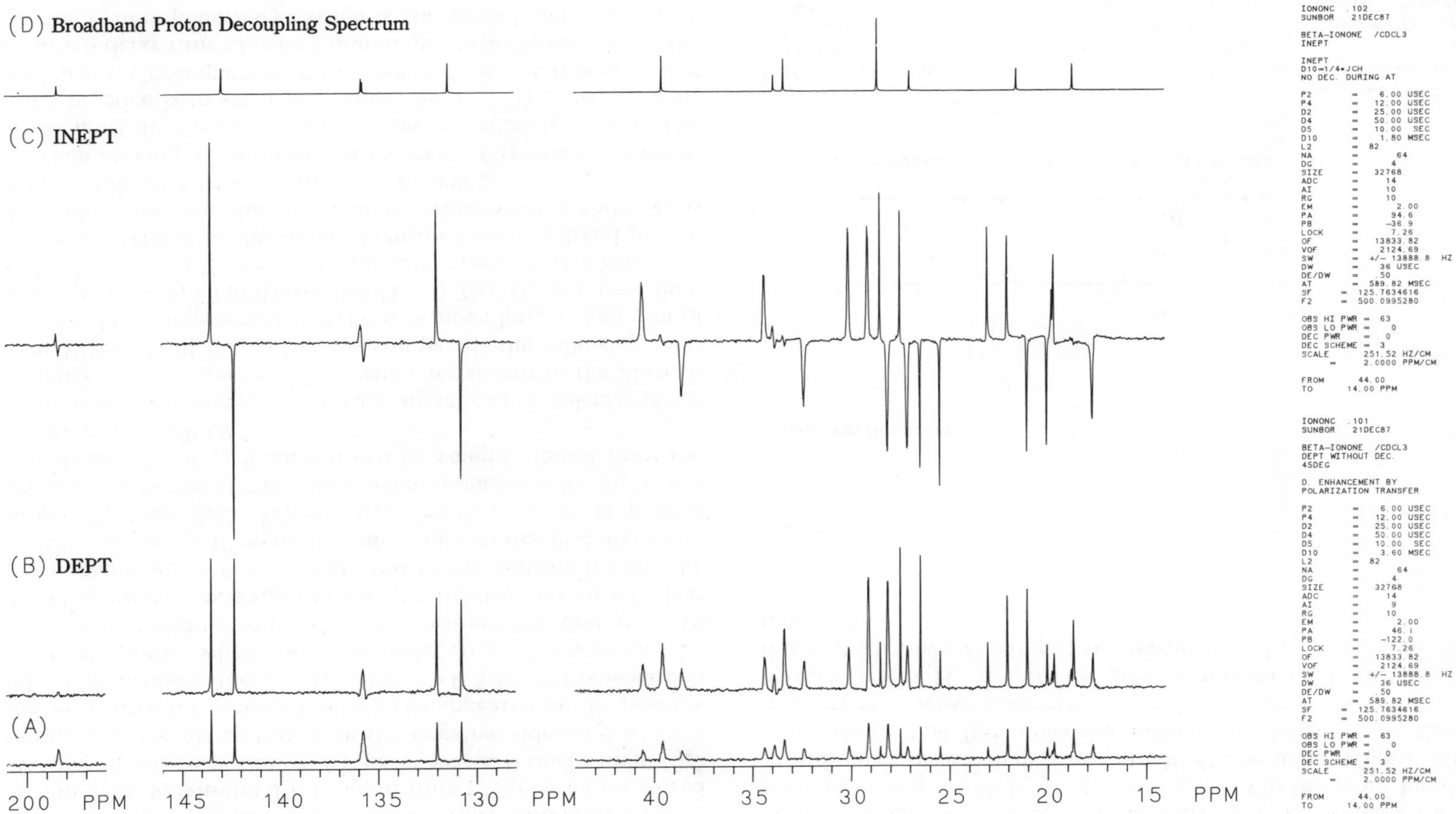

125 MHz ^{13}C-NMR ($CDCl_3$) [TI]

28 T_1 (Longitudinal Relaxation / Spin-lattice Relaxation) of β-Ionone

Using the pulse sequence shown in the illustration (see Part III section 1.7), longitudinal relaxation time (T_1) can be measured by varying the delay time τ.[1] When τ was extremely small (0.5 second), all signals appeared on the negative side, and when it was very large (20 seconds), all signals appeared on the positive side. For instance, the 7-C signal at $\tau = 0.5$ second is negative, but it disappears when $\tau = 2.5$ seconds. At $\tau = 3.5$ seconds, the 7-C signal is slightly restored to the positive side, and at $\tau = 10$ seconds, restoration is approximately compelete. As for 9-C, it is still negative at $\tau = 6.5$ seconds, and barely returns toward the positive side at $\tau = 10$ seconds. This demonstrates that 9-C has a longer T_1 than 7-C. When each signal intensity is plotted against τ, τ_0 (value of τ at which signal intensity is 0) can be obtained, and T_1 of each carbon can be roughly found from the equation $T_1 = \tau_0/0.69$.

In newer instruments, T_1 of each signal can be calculated automatically by computer. The results are shown in the illustration. When T_1 of the carbons is compared, the following facts emerge: 1) T_1 of quaternary carbons is much longer than that of other carbons. 2) $2T_1(CH_2) \simeq T_1(CH)$. 3) $T_1(CH_3)$ is longer than $T_1(CH)$. (T_1 of 1,1′-Me is exceptionally very short.) When T_1 values are related to the signal intensities in broadband proton decoupling spectra, the signal of a carbon with a short T_1 is strong while that with a long T_1 is weaker.

When seeking T_1 by inversion recovery, T (recovery interval) of the illustrated pulse sequence must be sufficiently long. The usual practice is to set it at 5 times the $T_1(5T_1)$ expected to be the longest. Consequently, in the case of β-ionone, it is necessary to have a delay time of over 1 minute per experiment. When the precise T_1 of a quaternary carbon is not needed, delay time may be set at 10-30 seconds. T_1 values obtained in this way have little absolute significance, but may be used to compare T_1 within the same molecule.

In general, T_1 of ^{13}C is markedly affected by 1) the number of attached ^1H and 2) the velocity of molecular movement. ^1H brings about longitudinal relaxation most effectively, therefore T_1 of a carbon near multiple ^1H is short. This is the reason for the relationship $T_1(CH_2) < T_1(CH) \ll T_1(C)$. On the other hand, in molecules of less than a few hundred molecular weight, one may assume that fast molecular movement→long T_1 and slow molecular movement→short T_1. The fact that $T_1(CH_3) > T_1(CH_2)$ despite the greater number of ^1H in CH_3 is due to CH_3 always being at the terminal and in rapid free rotation.

Inversion Recovery

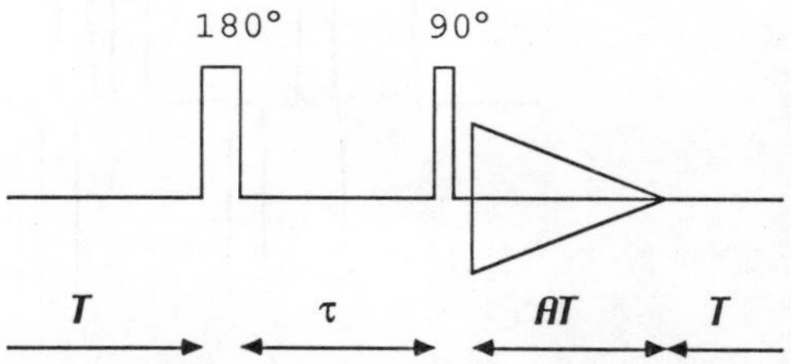

τ : Recovery Time.   This parameter is varied.

1) H. Günther (translated by R.W. Gleason), *NMR Spectroscopy—An Introduction*, John Wiley & Sons, Ltd., New York (1980), pp.218-220, 225-227
2) K. Nakanishi, T. Goto, S. Itô, S. Natori and S. Nozoe (eds.), *Natural Products Chemistry*, vol.3, Kodansha Ltd., Tokyo (co-published by University Science Books, California) (1983), p.7

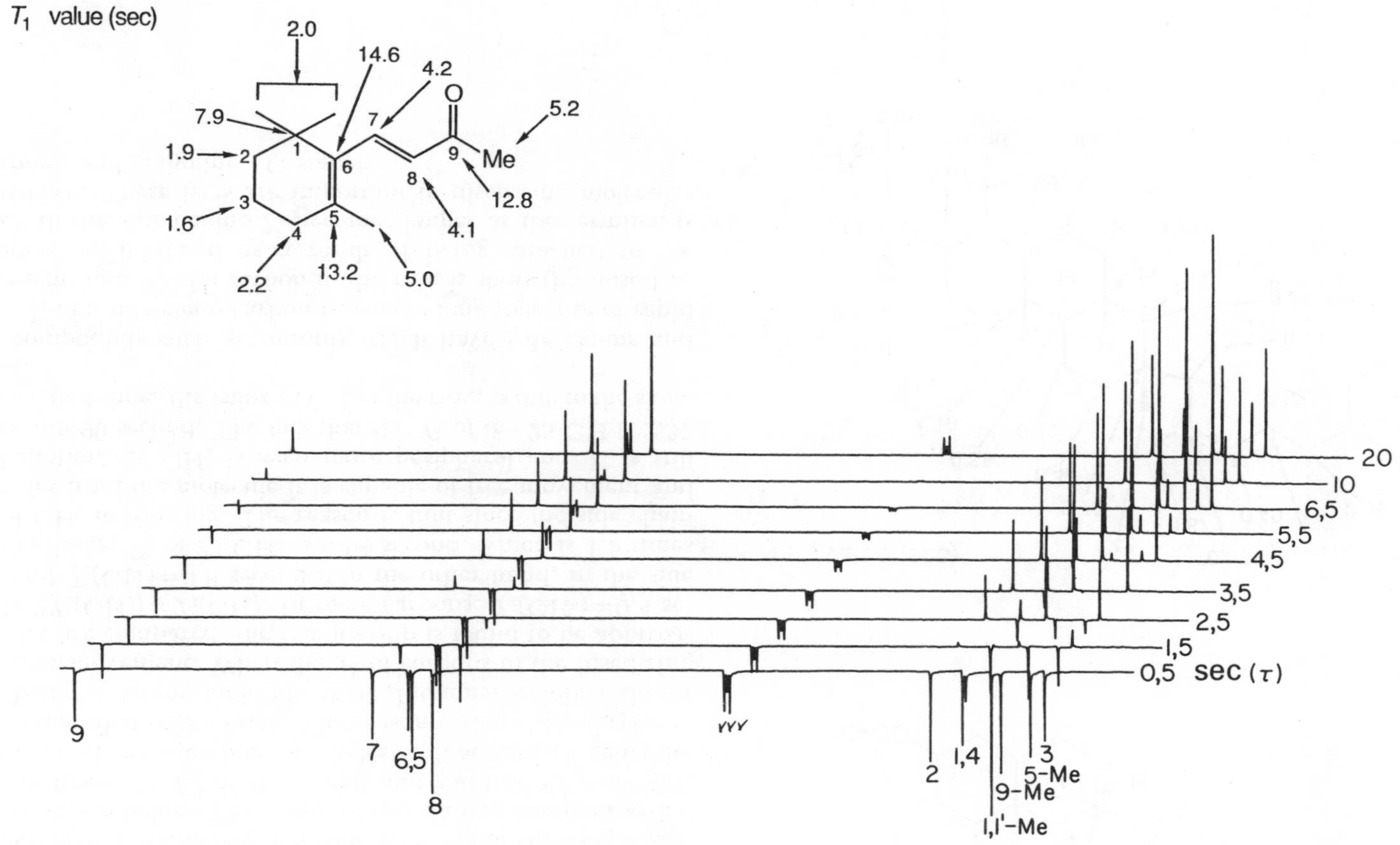

62.5 MHz ^{13}C-NMR ($CDCl_3$)[2)]

29 T_1 of Cholesteryl Acetate

The spectra obtained by inversion recovery of cholesteryl acetate are shown below. The group of spectra was obtained with a recovery interval (T) of 30 seconds and τ at 0.05-12.8 second.

The T_1 of each carbon of cholesteryl acetate is generally shorter than that of β-ionone. The reason is that cholesteryl acetate, being a larger molecule than β-ionone, exhibits slower molecular movement. When the T_1 of carbons in the fused ring structure are compared, the relationship is found to be approximately $2T_1(CH_2) \simeq T_1(CH)$. In these carbons, $T_1(CH_2) \simeq 0.4$ second and $T_1(CH) \simeq 0.8$ second. On the other hand, in the side chain carbons, T_1 of 23-CH_2 is 0.64 second, which is 1.5 times that of CH_2 in the ring. The reason is that since the side chain protrudes from the molecule it is capable of free movement and rapid motion. 24-CH_2 is even more peripheral and T_1 is still longer at 0.99 second. The fact that the T_1 of the 25-CH is 2.37 seconds, or 3 times the value of CH in the ring, is due to the same reason.

In compounds such as steroids which have side chains and rings, T_1 of a side chain carbon is usually long (because of rapid movement) and T_1 of a carbon in the ring is short (because free movement is inhibited as a result of being attached to the ring).[1] In the side chain, T_1 becomes longer as the terminal is approached. These facts are important in discussing molecular movement and assigning ^{13}C signals.

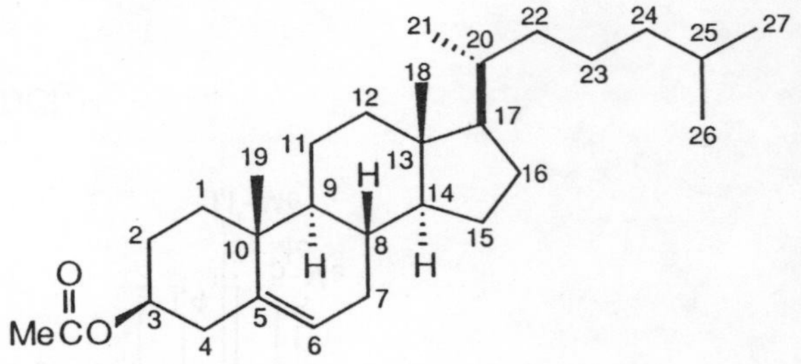

T_1 value (sec)

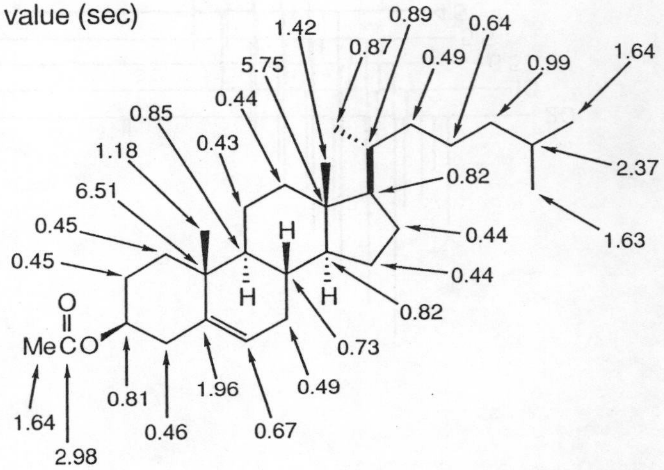

1) A. Allerhand, D. Doddrell and R. Komoroski, *J. Chem. Phys.*, **55**, 189 (1971)

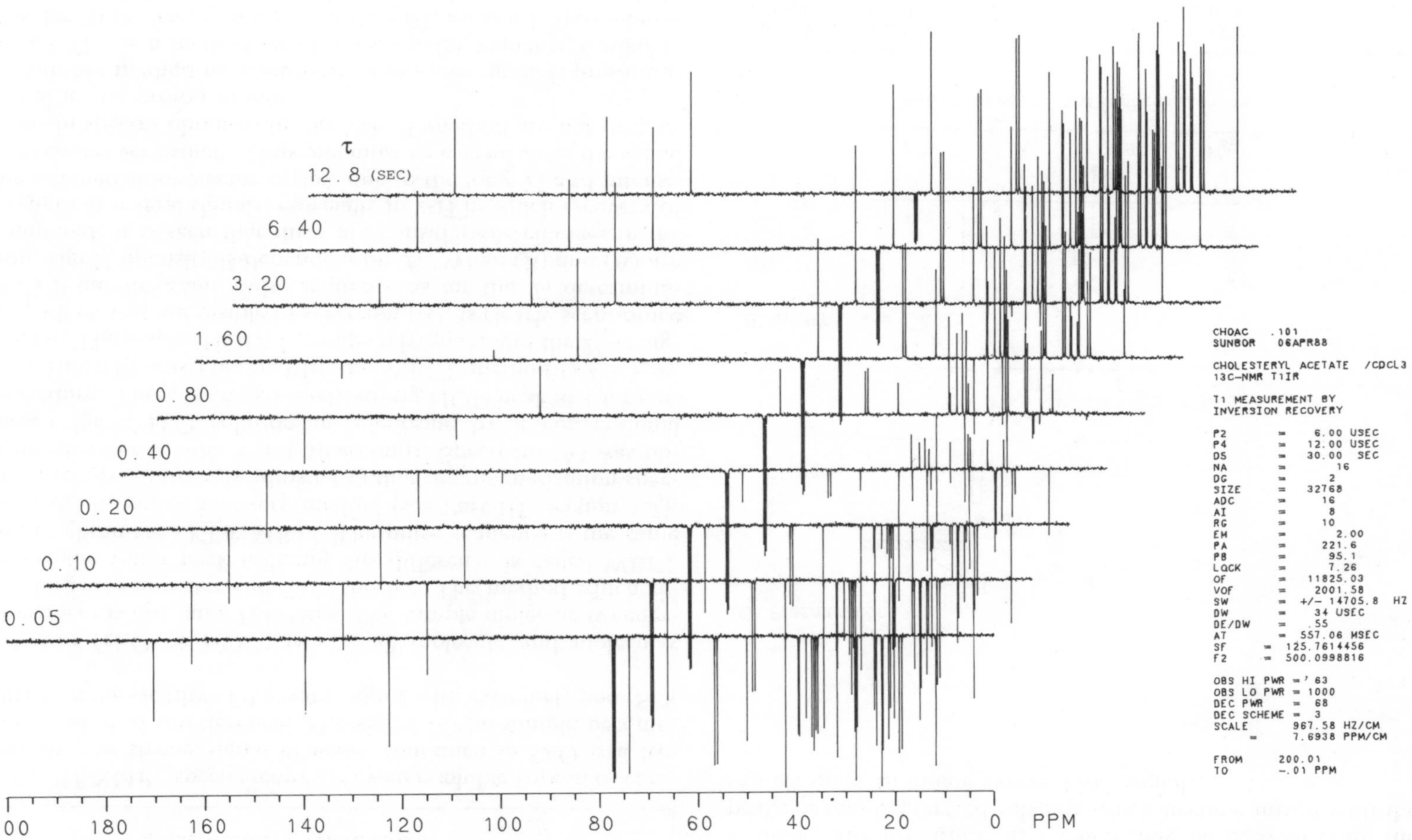

125 MHz ^{13}C-NMR (CDCl$_3$)

30 Removal of Huge Signals (WEFT and Presaturation Methods)

In ^{1}H-NMR spectroscopy of water-soluble organic compounds, the strong signal of water contained in D_2O is a frequent cause of interference. The signal of the sample becomes buried in the vicinity of the water signal with extremely poor S/N ratio.

Water (H_2O or HOD) is a small molecule and molecular movement is fast, and T_1 is long. The sample molecule is generally larger than water and T_1 is shorter. The method which removes the water peak utilizing this difference is called WEFT (water eliminated FT-NMR).[1] The pulse sequence is the same as in the inversion recovery method (see Part III, section 1.7), and τ (recovery time) is adjusted so that the magnetization spectrum of water is zero. τ is 1-10 seconds. Spectrum (A) was obtained for a D_2O solution of adenosine by a conventional procedure. There is an extremely strong HOD peak at 4.6 ppm. Spectrum (B) was obtained by the WEFT method ($\tau = 3.5$ seconds). The water (HOD) is completely gone, and the 2′-H signal, which was not visible in spectrum (A), is clearly seen. Since WEFT has the same pulse sequence as for the T_1 determination, signal intensity is dependent on T_1. When (B) and (A) are compared, it is seen that there are considerable changes in the strength of several signals, especially in 2-H in which recovery of the magnetization vector is slow due to the long T_1 and intensity becomes very small. Thus one must be careful since the signal areas in spectra obtained by the WEFT method are not proportional to the proton number.

Another method for eliminating the water signal is presaturation.[2] This is a method which uses a pulse sequence similar to that for NOE determination (see Part III, section 1.5) to remove water signal by saturation. Spectrum (C) was obtained by presaturation. Signal intensities are identical with those of spectrum (A) except for distortion at 4.6 ppm caused by irradiation.

WEFT may be used to remove not only water peaks but also peaks of solvents such as chloroform, methanol and DMSO. Similarly the presaturation method may be used to erase the peaks of non-deuterated solvents which become mixed with the sample or to eliminate excess TMS signals.

(C) Presaturation

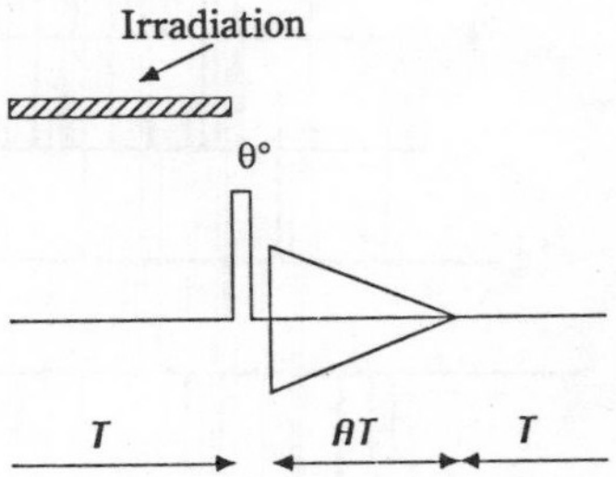

(B) WEFT

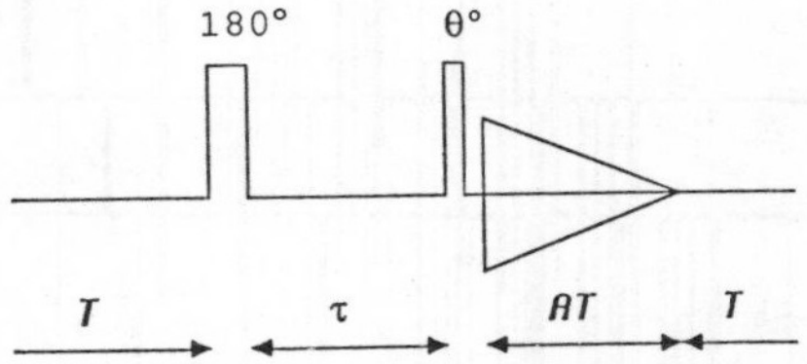

τ: Recovery Time. This parameter is varied.

1) S.L. Patt and B.D. Sykes, *J. Chem. Phys.*, **56**, 3182 (1972)
2) (Application for 2DFT-NMR): W.R. Croasmun and R. K. Carlson (Ed.), *Two-Dimensional NMR Spectroscopy—Applications for Chemists and Biochemists*, VCH Publishers, Inc., New York (1987), pp.162-163

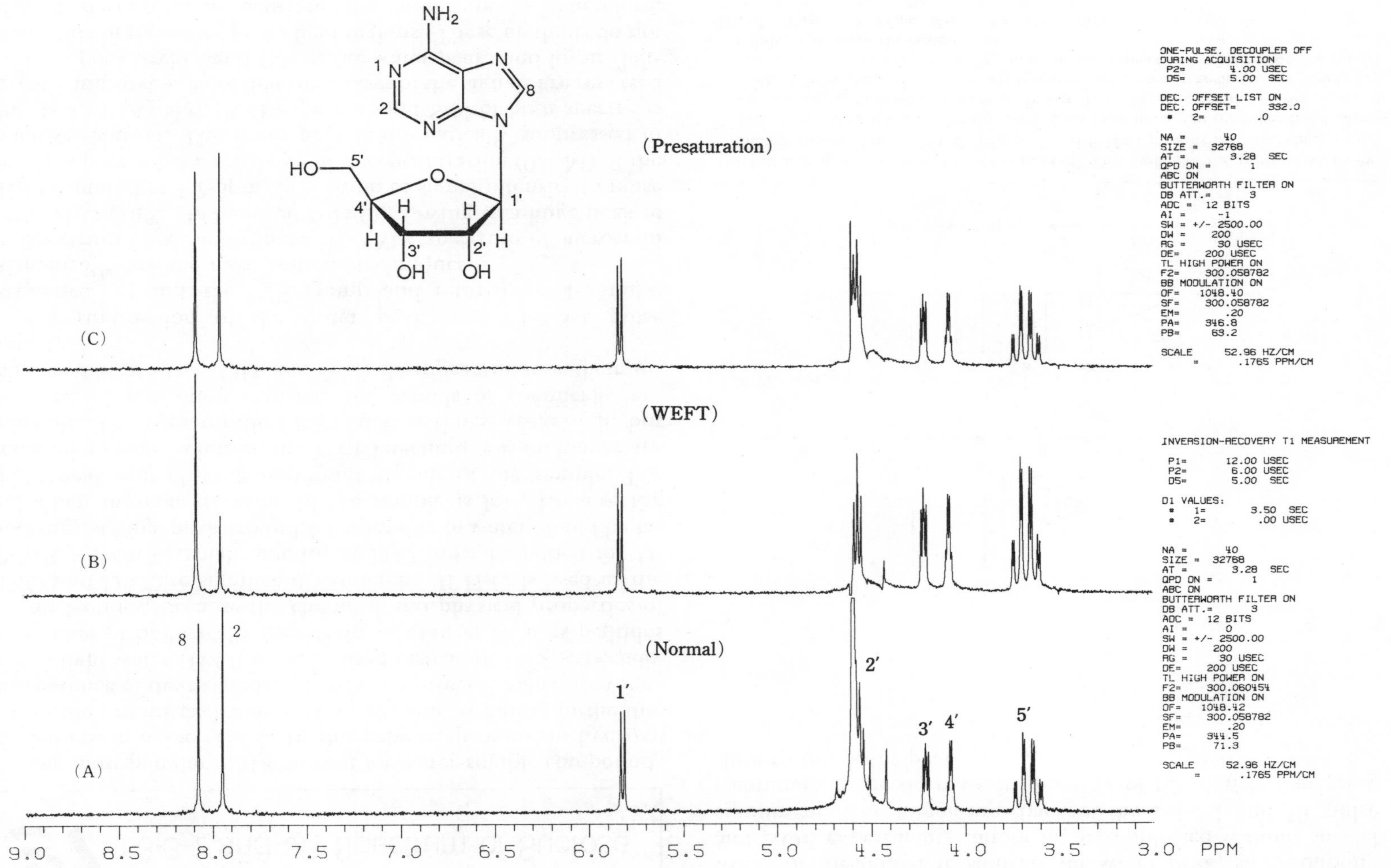

300MHz ^{1}H-NMR (D_2O) [HN]

31 1-3-3-1 and JR Spectrum of Sucrose

The most popular NMR solvent for water-soluble compounds is deuterium oxide (D_2O). In this solvent, protons on hydroxy or amino groups exchange with deuterium, resulting in the disappearance of their signals. There are, however, many occasions when light water (H_2O) must be used instead of D_2O, especially in studies of biologically important substances such as peptides or nucleotides, because the chemical and physical properties of H_2O and D_2O are significantly different. If H_2O is used as the NMR solvent (a small amount of D_2O must be added for D-locking), a huge peak around 4.8 ppm due to water would be fatal when the concentration of the sample is low, because the H_2O peak will obscure important signals of the sample. For such an aqueous solution, the WEFT technique is no longer applicable. The presaturation method sometimes works well, but because of saturation transfer, the signals of chemically exchangeable protons (OH and NH) disappear or lose their intensity.[1)]

For suppression of the water peak, the "1-3-3-1 pulse sequence"[2, 3)] and the "JR (jump and return), or 1-1 pulse sequence"[4)] are the most suitable techniques.

Spectrum C is the ordinary ^{1}H-NMR spectrum of sucrose in 90% H_2O (10% D_2O and 90% H_2O). Owing to a huge peak of H_2O centered at 4.7 ppm, S/N (ratio of signal intensity to noise level) is poor for the relatively high concentration (0.1 M) of the sucrose solution. The water peak is dramatically suppressed in the 1-3-3-1 (A) and JR (B) spectra, and S/N of both spectra is greatly improved. Note that the phases of the signals are reversed in the upper (right-hand side of the water peak) and lower (left-hand side of the water peak) field regions. These methods do not use a decoupler to saturate the water peak. Therefore, decoupling and NOE experiments can be done within the whole pulse sequence. The presaturation method, on the other hand, works the decoupler to saturate the water peak, so decoupling and NOE experiments cannot be accomplished without special equipment. The major drawback of the 1-3-3-1 and JR pulse techniques is the decrease in intensity of the signals appearing close to the water peak.

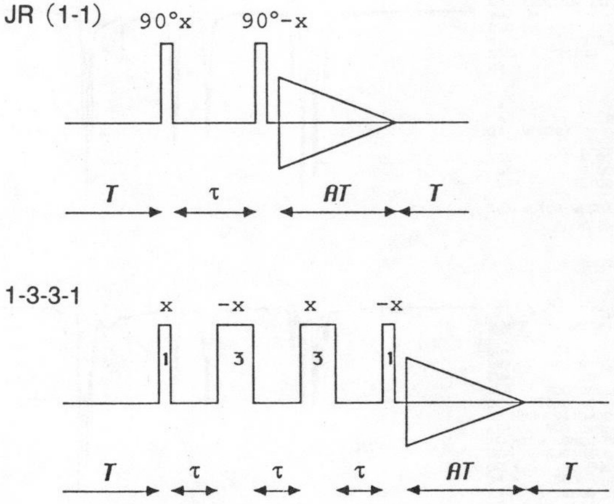

1) This description requires further comment. The signals of the protons which exchange rapidly (on an NMR time scale) with H_2O protons are 'absorbed' in the H_2O signal because of the large excess of water protons. On the other hand, slowly exchanging protons will show their signals at (or close to) their own chemical shifts. Under these circumstances, however, saturation transfer to the slowly exchanging OH or NH can occur when the water peak is irradiated for several seconds to saturate the water signal.

2) D.L. Turner, *J. Magn. Reson.*, **54**, 146 (1983)

3) P.J. Hore, *J. Magn. Reson.*, **55**, 283 (1983)

4) P. Plateau and M. Guéron, *J. Am. Chem. Soc.*, **104**, 7310 (1982)

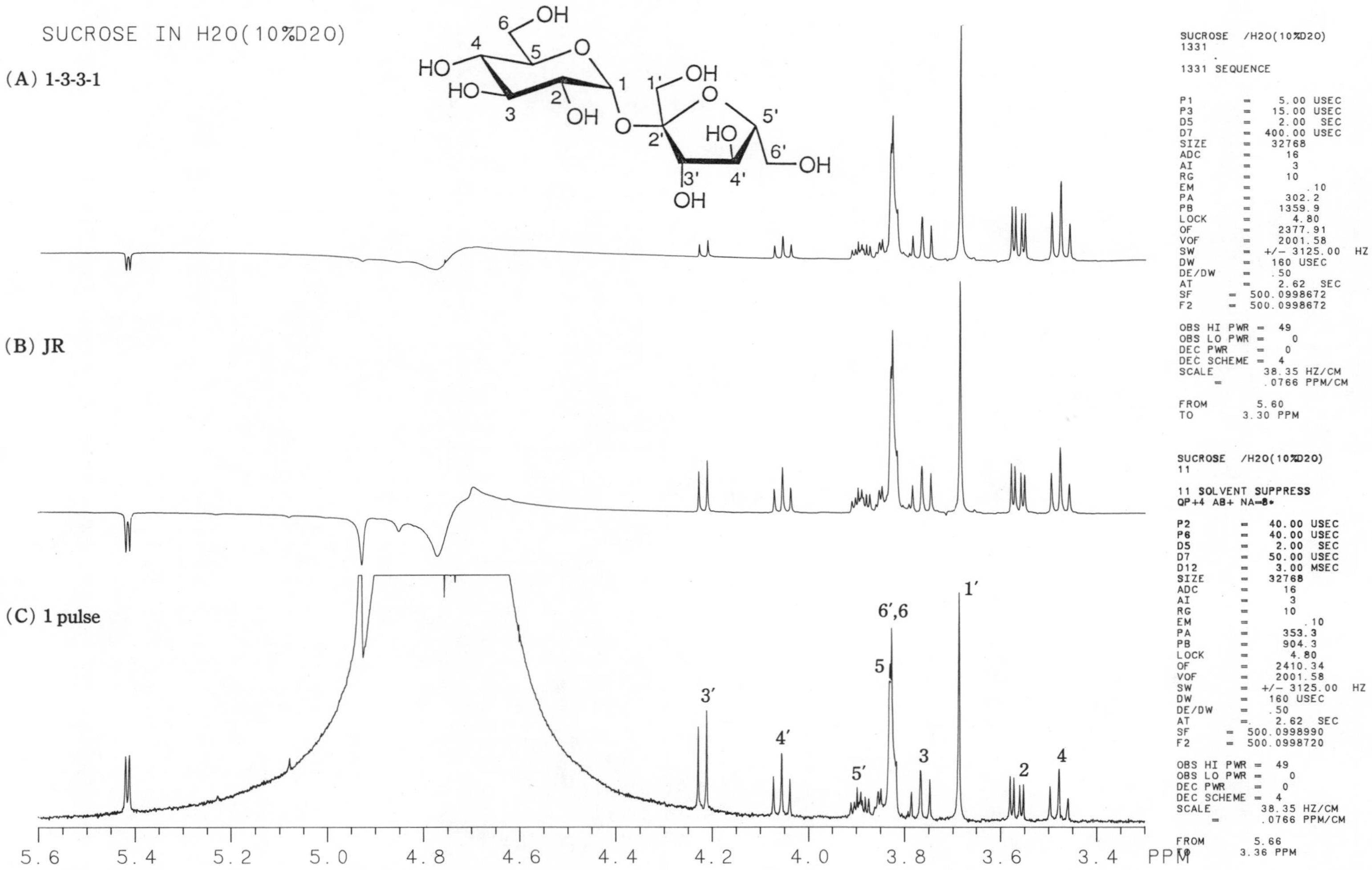

τ=D7

500 MHz ^{1}H-NMR (H_2O) [TI]

Part II

Two-dimensional FT-NMR

32 J-Resolved Spectrum of α-Santonin

J-resolved spectra[1, 2] of several protons of α-santonin were prepared by three different methods. (A) was obtained by the stacked plot method. Because the height of the peak is directly discernible and the spectrum is visually attractive, the majority of spectra at the time of the introduction of 2-dimensional FT-NMR were prepared by the stacked plot technique. Today the contour plot (B) is made in virtually all instances. (A) and (B) provide about the same amount of information, only the manner of presentation being different. (C) shows cross sections of 9β-H and 13-H peaks. These may be regarded as being viewed at the point indicated by ★ in spectrum (A). The use of *J*-resolved spectra is discussed in the following pages.

(B)

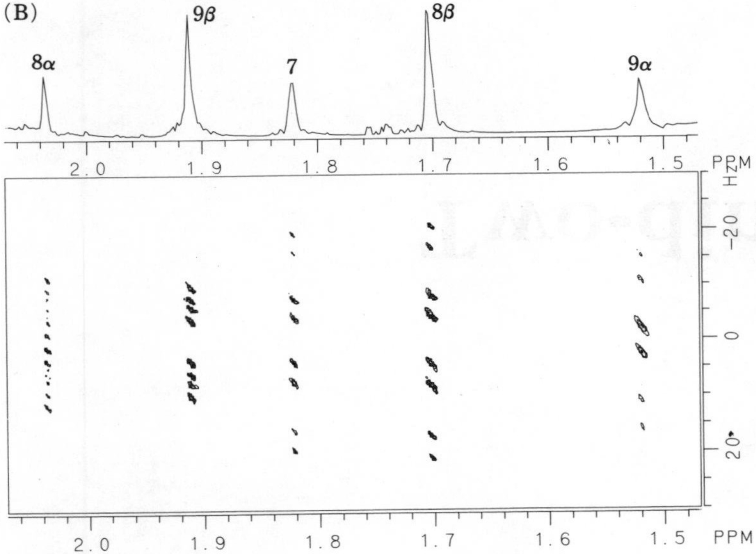

(C)

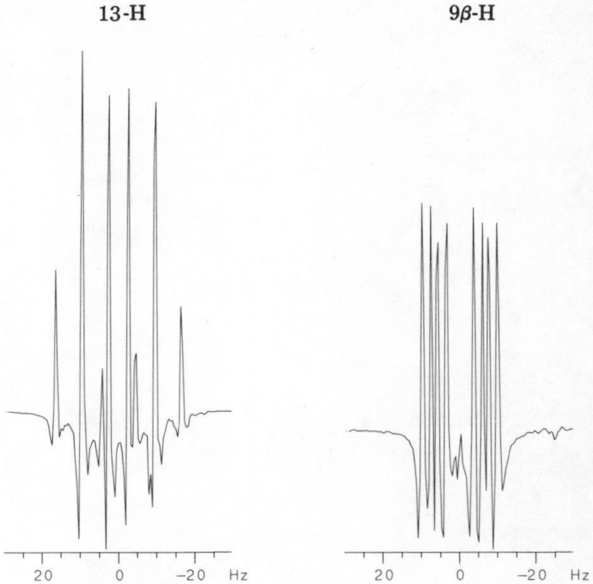

J-Resolved Spectroscopy

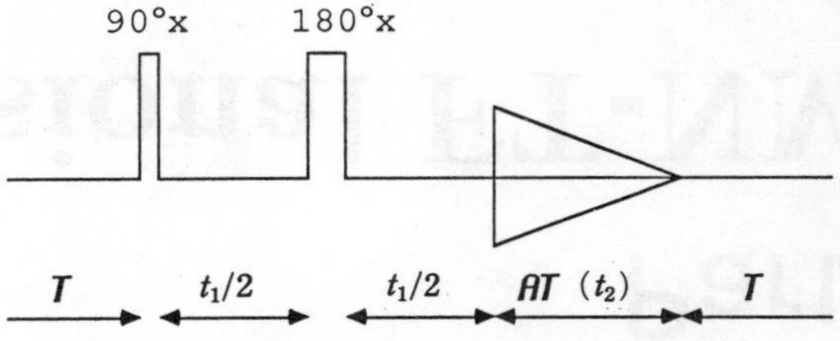

1) Ad Bax, *Two-Dimensional Nuclear Magnetic Resonance in Liquids*, Delft University Press, Delft, Holland (1982), pp.110-119
2) R.R. Ernst, G. Bodenhausen and A. Wokaun, *Principles of Nuclear Magnetic Resonance in One and Two Dimensions*, Oxford University Press, Oxford (1987), pp.360-366

(A)

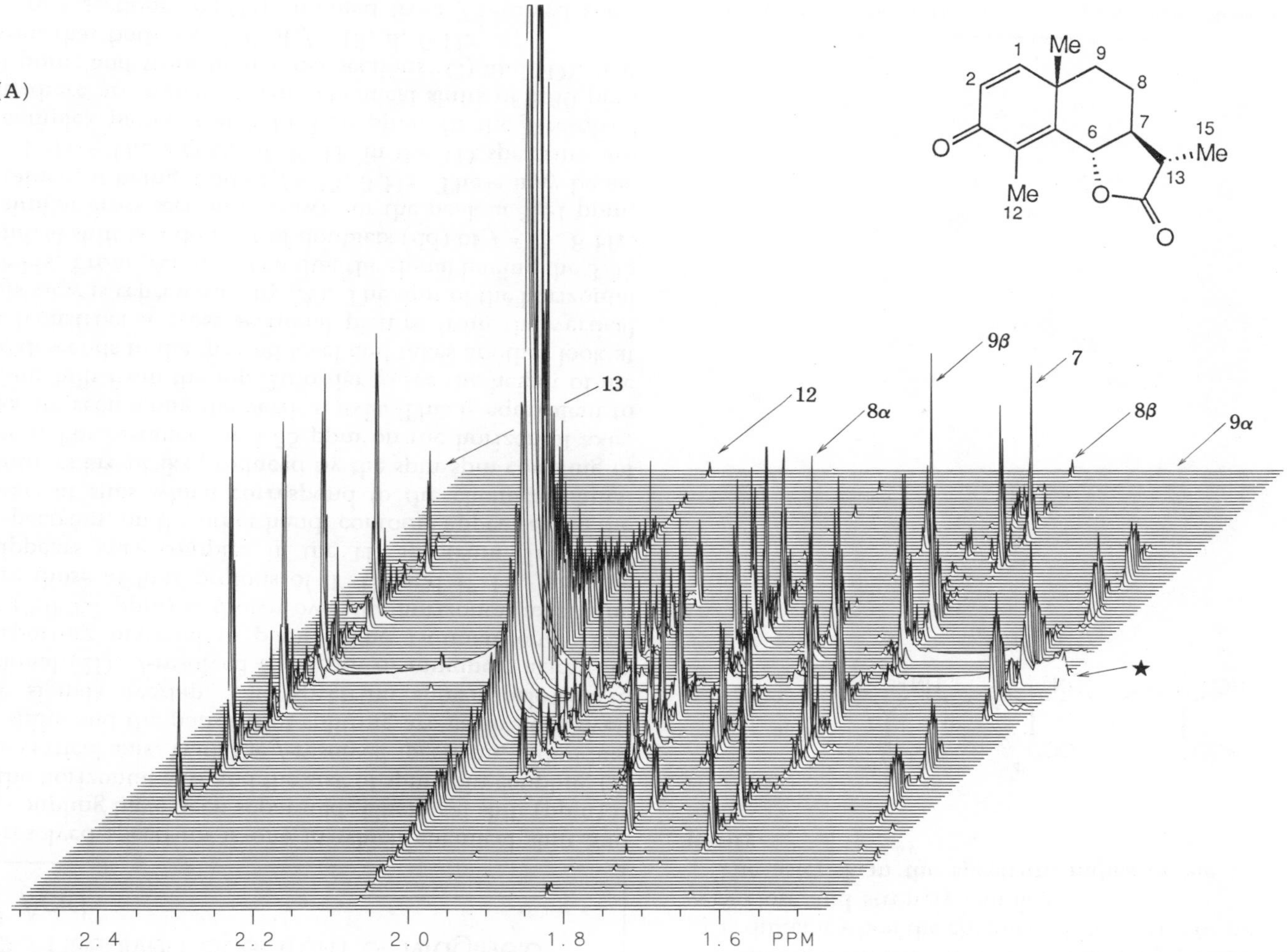

500 MHz ^{1}H-NMR ($CDCl_3$) [TI]

33 J-Resolved Spectrum of Mugineic Acid

The *J*-resolved spectrum is one in which chemical shift and spin-spin coupling have been separated. Chemical shift (ppm) is taken as the horizontal axis and the size of spin-spin coupling (*J*; Hz) as the vertical axis. With the *J*-resolved spectrum, individual chemical shifts and the patterns of splitting are easily seen even when the signals overlap. The spectrum shown below is a 2-dimensional (2D) *J*-resolved spectrum of mugineic acid,[1] an iron-transporting material in plants. The 1-dimensional (1D) spectrum (3.0-3.7 ppm) is shown over the horizontal axis. The signals are those of four protons of 1′-H_2 and 1″-H_2, and the pattern appears very complex in the 1D spectrum. In the *J*-resolved spectrum, on the other hand, contours appear along the vertical axis at sites which correspond to the chemical shifts. These contours are peaks produced by the spin-spin coupling of each proton. For instance, at 3.55 ppm on the horizontal axis, four peaks are seen along the vertical axis. This is equivalent to viewing four hills from the top. In order to see the height of the peak, one descends to the ground level and takes another look at the peak (construct a cross sectional picture from the vertical axis). This view is represented by (A). The unit of the horizontal axis is the Hz. From (A), it is seen that the signal having the 3.55 ppm chemical shift is a doublet of doublets (dd) of $J = 13$, 8 Hz. When a similar cross section is drawn for the peak at 3.41 ppm, (B) is obtained, it being a dd of $J = 13$, 3 Hz. These may be assigned to 1′-H_2. The signals of 1″-H_2 in the 1D spectrum appear as complex patterns at 3.14-3.36 ppm. In the *J*-resolved spectrum, there are signals having chemical shifts of 3.30 ppm and 3.20 ppm, and from their cross sections (C) and (D), it is readily seen that both are ddd of $J = 13$, 8, 6 Hz.

When cross sections (A)-(D) obtained from *J*-resolved spectrum are compared with the 1D spectrum, it is seen that the former are narrower and of better resolution.

It should be noted that interpretation by *J*-resolved spectroscopy is difficult when the chemical shifts of two or more protons are very close and strongly coupled.

The asterisk in the spectrum indicates the signal of an impurity.

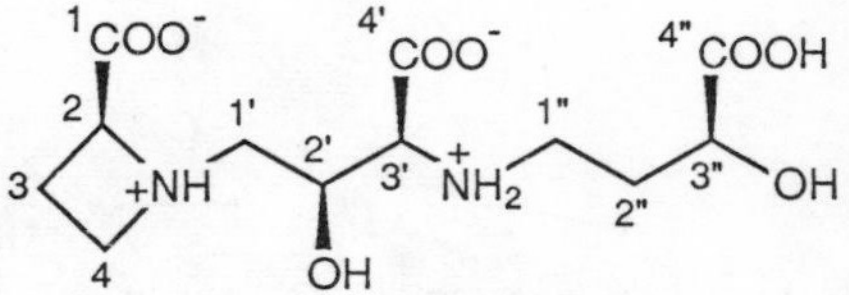

1) T. Iwashita, Y. Mino, H. Naoki, Y. Sugiura and K. Nomoto, *Biochemistry*, **22**, 4842 (1983)

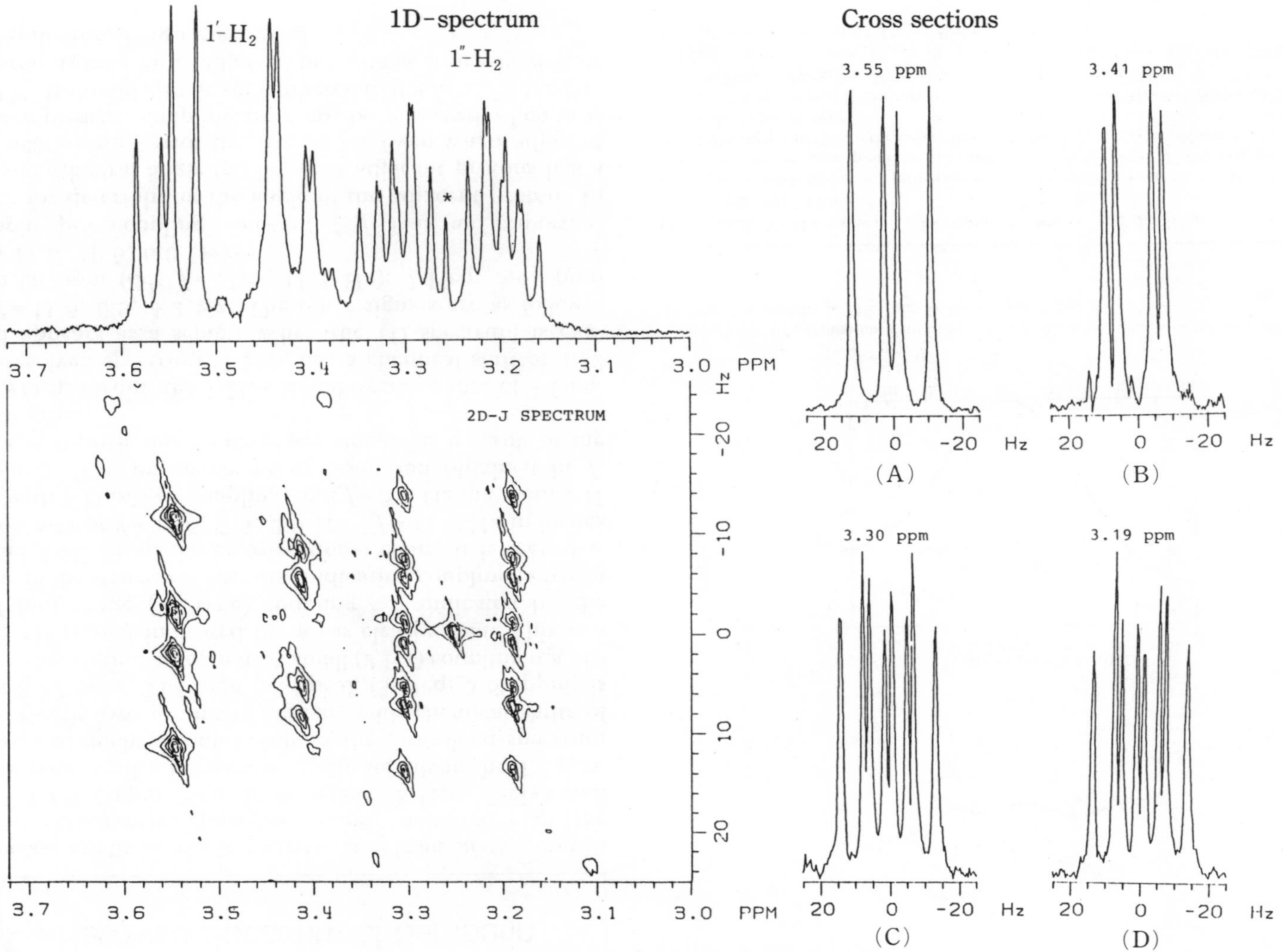

360MHz ^{1}H-NMR (D_2O) [TI$^-$]

34 *J*-Resolved Spectrum of Semburin

The spectra are those of semburin isolated from *Swertia japonica* Makino (Gentianaceae; Japanese name, semburi)[1], in the region of 3.2-4.2 ppm. Signals of 3-Hax, 3-Heq, 7-Hax and 7-Heq are seen. In the 1D spectrum, the signals in the 3.7 ppm region are extremely complex, but in the *J*-resolved spectrum there are clearly two groups of signals, with chemical shifts of 3.69 and 3.72 ppm. The high field peak (3-Heq, 3.69 ppm) is shown in cross section at top left. A small (2 Hz) coupling not observed in 1D (signal indicated by ★) is clearly seen. This is a result of long-range (W type) coupling [2, 3] indicated by the heavy line in the structural formula, indicating coupling between 3-Heq and 5-H. From the cross sectional chart, it is seen that 3-Heq is a ddd of $J = 11.6$, 5.5, 2.0 Hz. $J = 11.6$ Hz indicates coupling with 3-Hax (*gem*-coupling) and $J = 5.5$ Hz that with 4-H (*vic*-coupling). The improvement in resolution obtained in *J*-resolved spectrum is due to narrower signals as a result of the refocusing effect.

In the 1D spectrum, the 7-Hax signal overlaps that of 3-Heq. In the *J*-resolved spectrum, 7-Hax has a chemical shift of 3.72 ppm and appears as a septet. When the 1D spectrum is examined, $J = 11.6$, 6.9, 4.2 Hz. The other signals are as follows. 3-Hax: 4.08 ppm (dd, $J = 11.6$, 11.8 Hz); 7-Heq: 3.41 ppm (ddd, $J = 11.6$, 10.0, 6.0 Hz).

The spin-spin coupling constant (J) is an important parameter for determining the angle of the adjacent proton. In general, the dihedral angle (ϕ) between adjacent protons has a Karplus relationship[4] with the size of J.[5] Even when adjacent protons are present, coupling may not be observed when ϕ is close to 90°. It should also be remembered that when $J = 1$-2 Hz, the two protons may be 1) adjacent but with ϕ at nearly 90°, or 2) may result from long-range coupling.

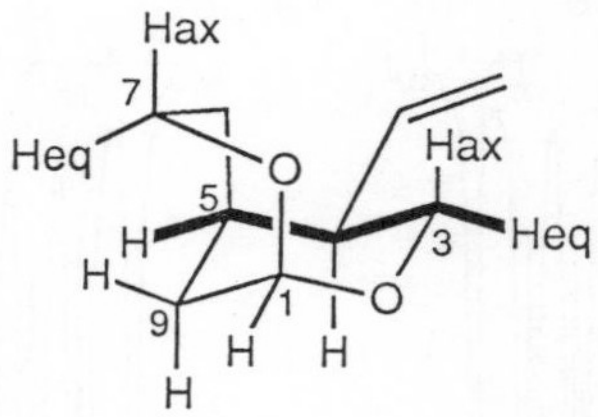

[Karplus relationship]

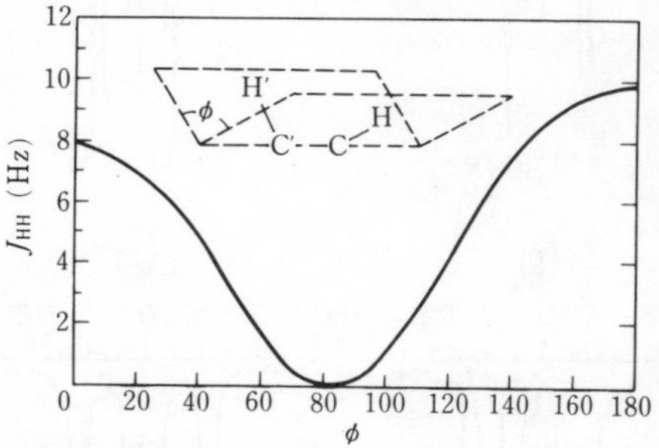

(Reproduced by permission from Silverstein, R. M. *et al.*, *Spectrometric Identification of Organic Compounds*, p. 210, John Wiley & Sons (1981))

---

1) T. Sakai, Y. Nakagawa, T. Iwashita, H. Naoki, and T. Sakan, *Bull. Chem. Soc. Jpn.*, **56**, 3477 (1983)
2) In the cyclohexane type conformation, the equatorial protons in the 1,3 relationship are in the same plane as indicated by the heavy line. These protons undergo 1-2 Hz coupling. Since the heavy line is W-shaped, this type of long-range coupling is called the W type.
3) L.M. Jackman and S. Sternhell, *Applications of Nuclear Magnetic Resonance Spectroscopy in Organic Chemistry*, Pergamon Press Ltd., Oxford (1969), pp.334-341
4) M. Karplus, *J. Chem. Phys.*, **30**, 11 (1959); *J. Am. Chem. Soc.*, **85**, 2870 (1963)
5) A.A. Bothner-By, *Adv. Magn. Reson.*, **1**, 195 (1965)

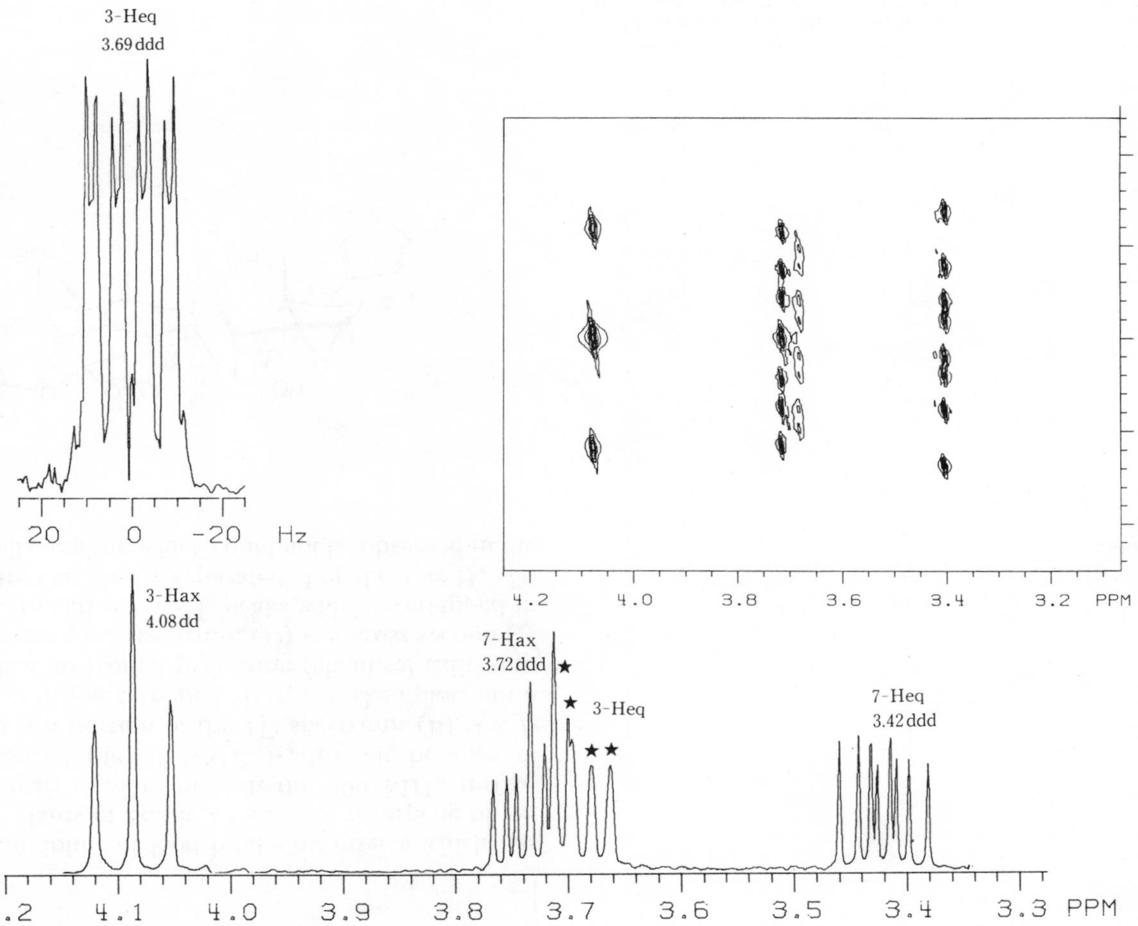

360MHz ^{1}H-NMR (C_6D_6) [HN]

35 *J*-Resolved Spectrum of Trichilin-A

Trichilin-A, an inhibitor of food intake by insects which has been isolated from plants of South Africa, is a triterpene of extremely complex structure.[1] Even with the 300 MHz instrument, interpretation of the ^{1}H-NMR is difficult because of signal overlap. (A) is a portion of the 1D spectrum. (B) is a *J*-resolved spectrum of the same region. It is a stacked plot, not a contour plot. (C) is a horizontal projection (chemical shift) and represents a homodecoupled spectrum. (D) is a cross section (*J* axis of each peak). In (D) the 3'-H_a peaks which overlapped in the 1D spectrum are completely separated. For the 16α-H, (D) clearly shows a small coupling which could not be observed in the 1D spectrum.[2]

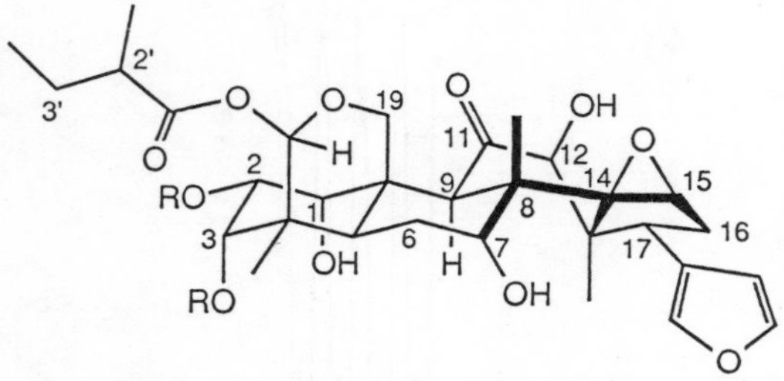

1) M. Nakatani, J.C. James and K. Nakanishi, *J. Am. Chem. Soc.*, **103**, 1228 (1981)
2) K. Nakanishi, T. Goto, S. Itô, S. Natori and S. Nozoe (eds.), *Natural Product Chemistry*, vol.3, Kodansha Ltd., Tokyo (co-published by University Science Books, California) (1983), pp.160-161

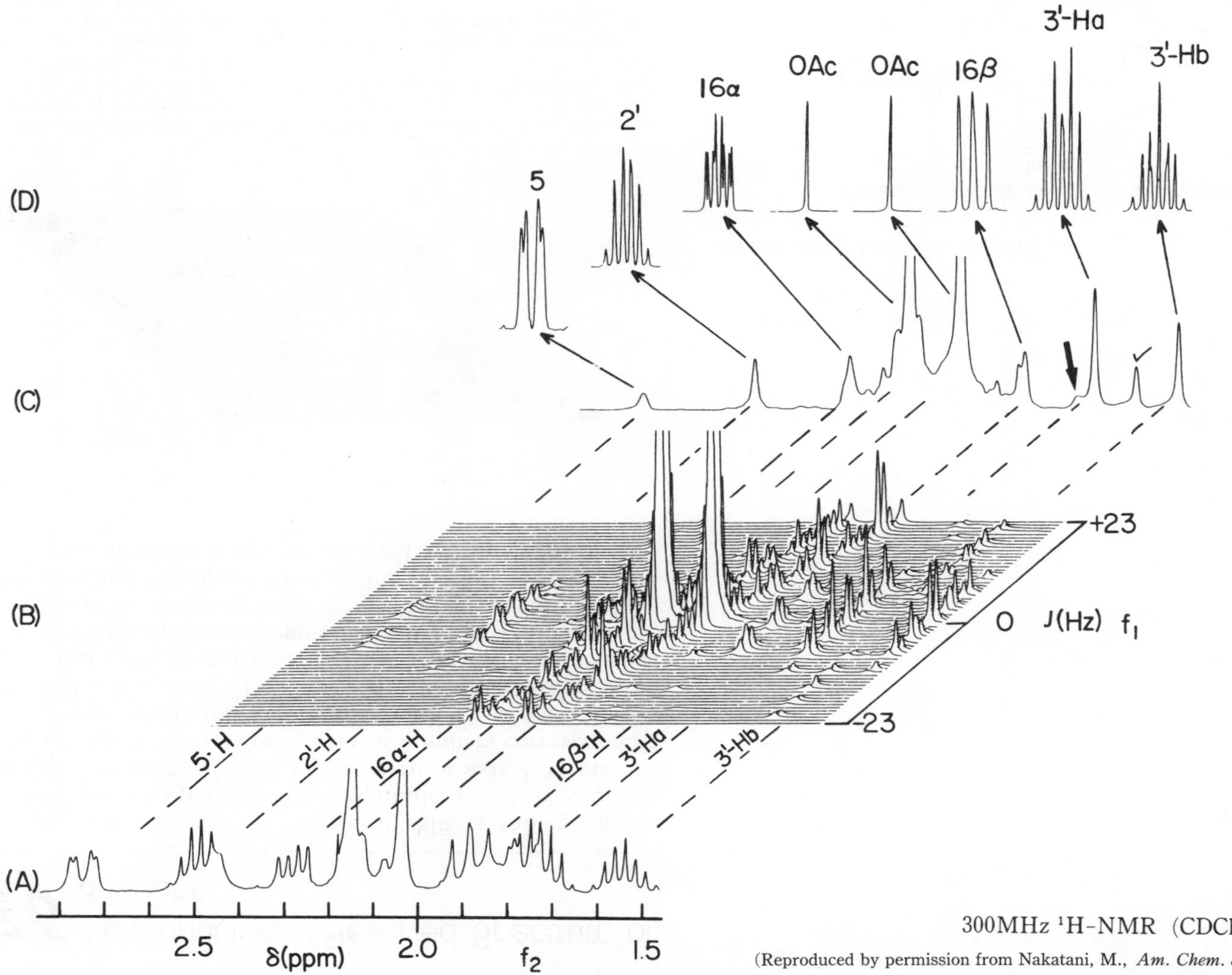

300MHz ^{1}H-NMR ($CDCl_3$)

(Reproduced by permission from Nakatani, M., *Am. Chem. Soc.*, **103**, 1228(1981))

36 Heteronuclear *J*-Resolved Spectrum of β-Ionone

In spectrum (A), the ^{13}C chemical shift is shown on the horizontal axis in ppm and the magnitude of ^{13}C-^{1}H coupling (C-H direct coupling ^{1}J; Hz) on the vertical axis.[1] As may be seen in spectrum (A), contour peaks in numbers corresponding to the number of protons are seen for each ^{13}C signal (CH_3: 4 peaks; CH_2: 3 peaks; CH: 2 peaks). Where it differs from homonuclear (^{1}H) *J*-resolved spectrum is on the large vertical (*J* axis) scale, the maximum width being 500 Hz. This is because $^{1}J_{CH}$ is about ten times as large as the ^{1}H-^{1}H J value. (B) is a stacked plot and (C) shows cross sections at each carbon. From (C), $^{1}J_{CH}$ for all carbons can be obtained accurately.[2, 3]

(B)

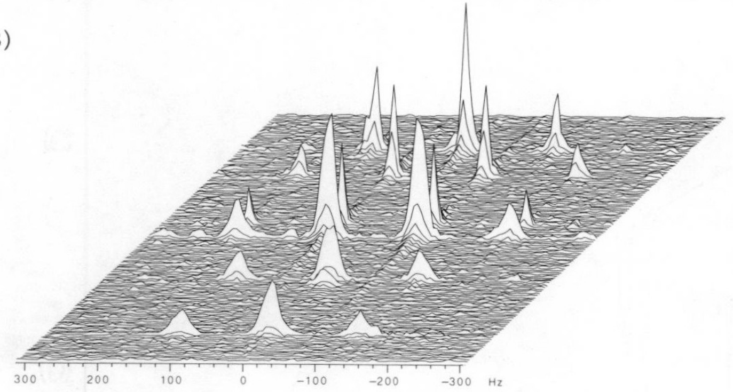

(C)

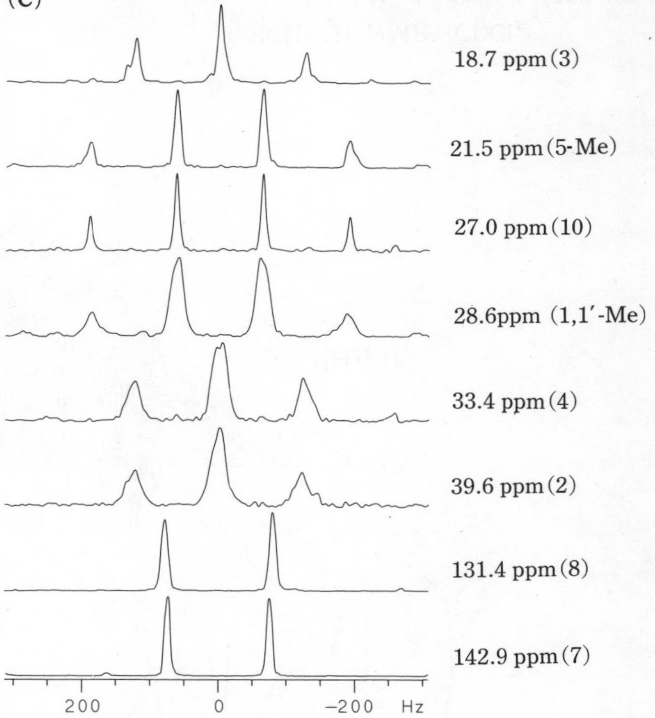

Heteronuclear *J*-Resolved Spectroscopy

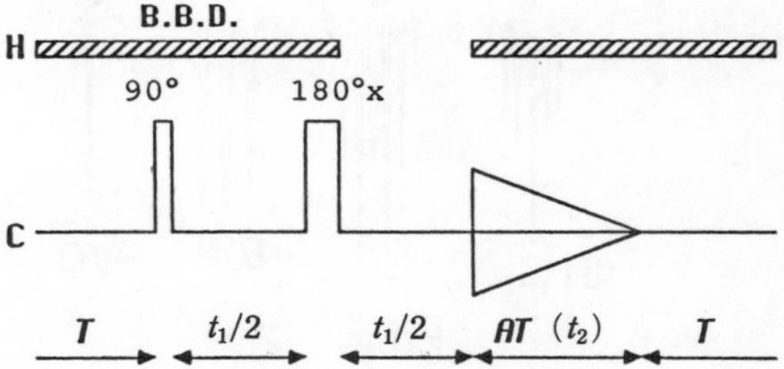

1) Ad Bax, *Two-Dimensional Nuclear Magnetic Resonance in Liquids*, Delft University Press, Delft, Holland (1982), pp.99-110
2) G.C. Levy (Ed.) *Topics in Carbon-13 NMR Spectroscopy*, vol. 4, John Wiley & Sons, New York (1984), pp.220-227
3) R.R. Ernst, G. Bodenhansen and A. Wokaun, *Principles of Nuclear Magnetic Resonance in One and Two Dimensions*, Oxford University Press, Oxford (1987), pp.358-360, pp.366-374

(A)

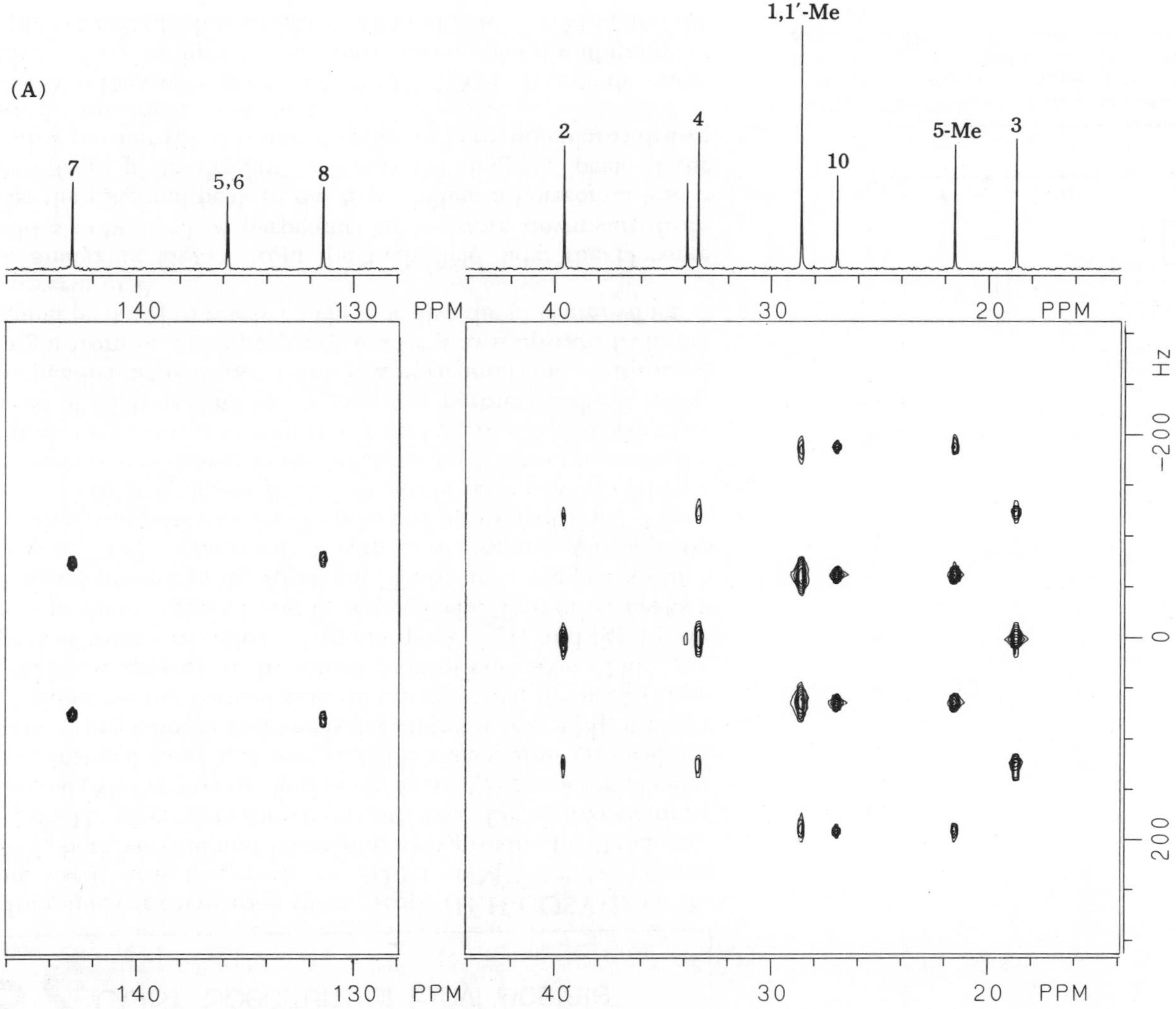

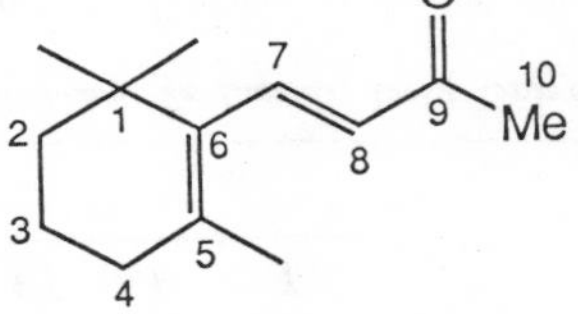

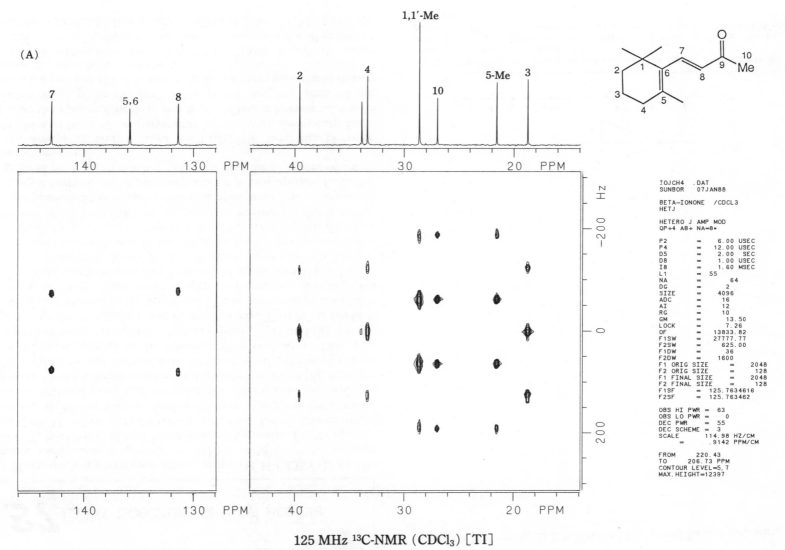

125 MHz ^{13}C-NMR ($CDCl_3$) [TI]

37 COSY Spectrum of Ethyl Acetate

Homonuclear correlation spectroscopy (H,H-COSY) is a technique used most frequently in 2D-FT NMR.[1,2] In normal COSY, both vertical and horizontal axes provide for ^1H chemical shift. 1D spectra are shown on both axes. Let us first examine the signal (Me) at 3 on the horizontal axis. When a perpendicular line is dropped from 3, it intersects the diagonal where peak [3] is seen. When a horizontal line is drawn to the left of [3], it meets the 3 signal on the vertical axis. In other words, the peak [3] on the diagonal appears at the intersection of both axes. This type of peak is therefore called "diagonal peak." [1] and [2] are diagonal peaks of signals 1 and 2, respectively. Two other peaks p and p′ are present in the spectrum. These are cross peaks which appear in COSY as a result of spin-spin coupling. Cross peaks must always appear as a pair in matched positions (p, p′). When a perpendicular is drawn from 2, it meets cross peak p. When a horizontal line is drawn to the left from p (A route) it meets the 1 signal. From this it is seen that 1 and 2 are coupled. Another method of analysis (finding the coupling partner) is the B route. A perpendicular is drawn from 2, a horizontal line is drawn to the right from p, and a perpendicular is drawn upward from the diagonal peak [1] to reach 1. There is also the C route, which is the reverse of B.

In analyzing COSY from the high field side, the D route should also be used. A perpendicular is drawn downward from 1 past the diagonal peak to reach p′. When a horizontal line is drawn from p′ to the left, it meets the diagonal peak of the coupling partner [2] of 1, and therefore a perpendicular is drawn upward from there to reach 2.

COSY is the easiest method in 2D-FT NMR. It provides satisfactory spectra in three to four hours when several milligrams of sample is available. It is a basic method of 2D-FT NMR, and the various COSY methods are discussed in detail in the following sections.

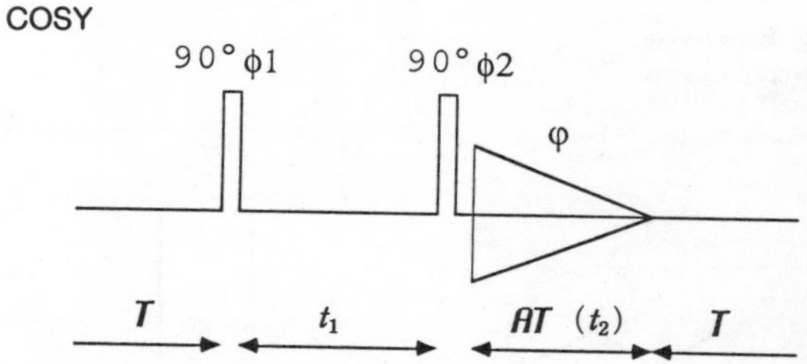

1) Ad Bax, *Two-Dimensional Nuclear Magnetic Resonance in Liquids*, Delft University Press, Delft, Holland (1982), pp.69-93
2) R.R. Ernst, G. Bodenhansen and A. Wokaun, *Principles of Nuclear Magnetic Resonance in One and Two Dimensions*, Oxford University Press, Oxford (1987), pp.400-448

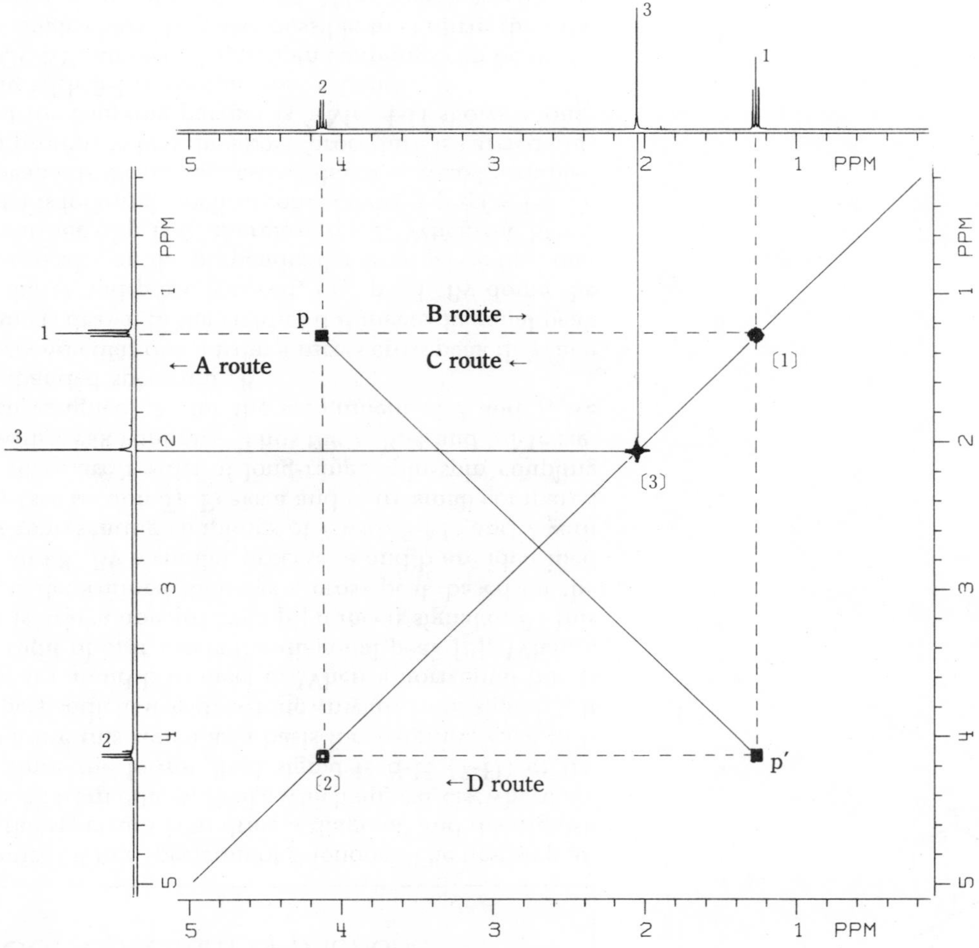

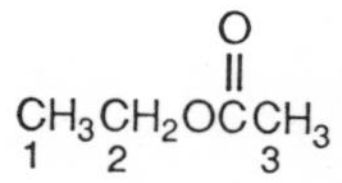


```
COSY(P-TYPE)
NA=4 OR 16,QP=+0,AB=-
I8=DW
 P2=        7.20 USEC
 D5=        1.00  SEC
 D8=      560.00 USEC

I8=     560.00 USEC

NA =        16
SIZE =    1024
AT =      286.72 MSEC
OPD ON =         1
ABC ON
BUTTERWORTH FILTER ON
DB ATT.=         3
ADC =  12 BITS
AI =        9
SW = +/- 892.857
DW =      560
RG =        10 USEC
DE=     560 USEC
TL HIGH POWER ON
F2=    360.056514
BB MODULATION OFF
OF=    934.11
SF=    360.054917

SCALE       89.28 HZ/CM
     =        .2479 PPM/CM
CONTOUR LEVEL=16
(YS=13878)
```

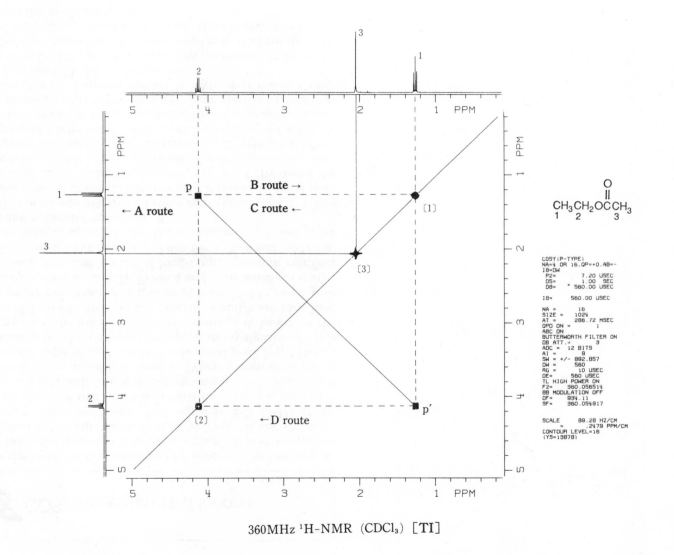

360MHz ^{1}H-NMR ($CDCl_3$) [TI]

38 COSY Spectrum of β-Ionone

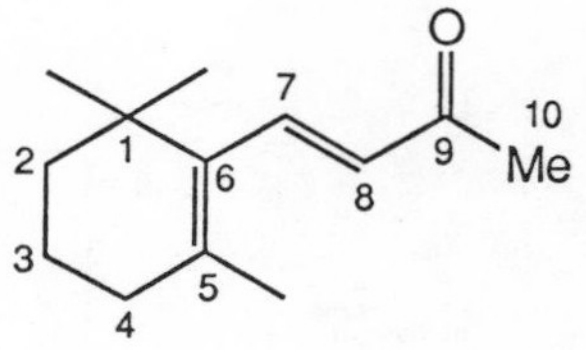

(A) is the entire COSY spectrum of β-ionone. The first step after obtaining the spectrum is to draw a diagonal and distinguish the diagonal peak from others. Peaks which appear elsewhere are cross peaks. Since the lowest field signal is β-H (7-H) of the enone, we shall use this signal as a basis for assigning each proton. When a perpendicular is drawn downward from signal 7, it passes cross peaks a and b to meet c. When a horizontal line is drawn to the right of c, it meets the diagonal peak [8]. When a perpendicular is drawn upward from [8] it meets signal 8. By this procedure, it is determined that c is a cross peak based on the coupling of 7 and 8. By a similar process, a and b are identified as cross peaks representing couplings of 7 with 5-Me and 7 with 4, respectively (see section 5). Peaks a and b are small compared with c, since these are results of long-range spin-spin coupling (J = 1-2 Hz) with weak contours. Thus the 7, 8, 4 and 5-Me signals have been assigned. As for the assignment of 2 and 3, we refer to the expanded spectrum (B).

In (B), a perpendicular down from 4 meets cross peak d. When a horizontal line is drawn to the right of d it meets diagonal peak p, indicating that 4 and p are coupled, *i.e.*, p = 3. By doing the same with cross peak f on the perpendicular from p, we find that the coupling partner of p is q, therefore q = 2. When the heavy line starting at 4 is followed, we find connectivity 4-p-q (= 4-3-2).

The cross peak e is weak, suggesting that 4 is weakly coupled with a certain proton. When the above procedure is carried out, it is found that the coupling partner is 5-Me. 4-H shows a long-range coupling with 5-Me proton over 4 bonds.

Thus with COSY, almost all spin-spin couplings can be determined from a single chart. It is also possible to confirm the existence of long-range coupling of about 1Hz, together with the identity of the coupling partner.

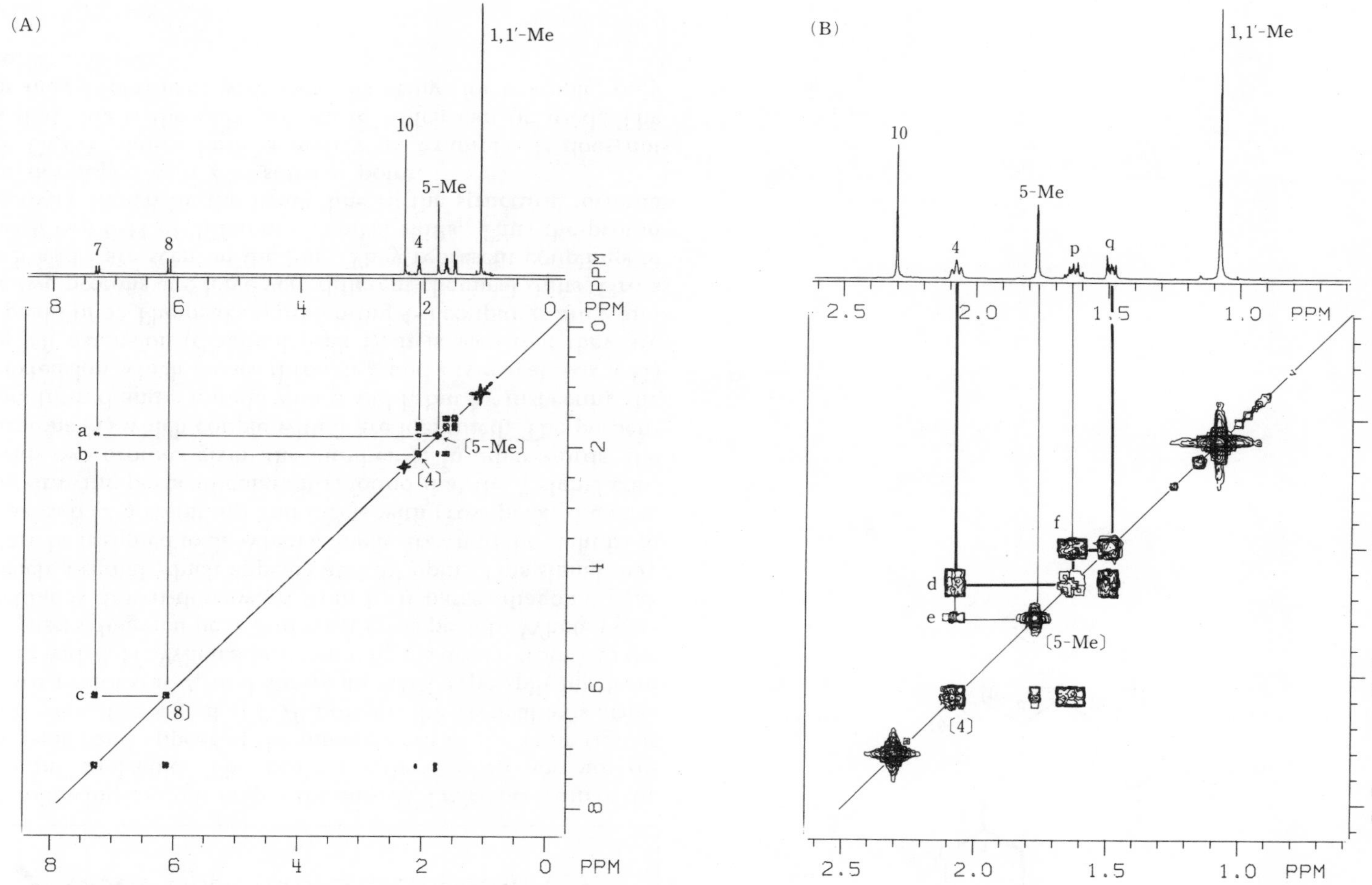

360MHz ¹H-NMR (CDCl₃) [TI]

39 COSY Spectrum of Compactin (I)

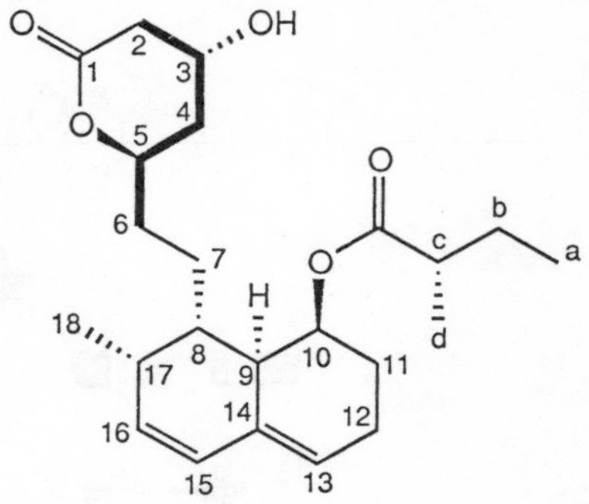

The following contour map is the entire COSY spectrum of the compactin[1] molecule. The peaks on the broken line are diagonal peaks and appear at the intersections of the same signals on both axes. The signal at 2.70 ppm on the vertical axis arises from two protons at 2, and shows an ABX type splitting from coupling with 3-H. When a horizontal line is drawn from 2 to the left, it passes diagonal peak a to meet cross peak b. When a perpendicular is drawn downward from b, it passes diagonal peak c to reach a signal which appears at 4.39 ppm. This signal may therefore be assigned to 3. When a line is drawn to the right from c, it passes b (2-3 coupling) and meets with cross peaks d and e, and by drawing perpendiculars it is found that the 3 signal couples with two protons given the number 4. In other words, the two protons (4) which couple with 3 are identified. The perpendiculars from d and e mingle with g and i, but by inspecting the right extension which passes through g and i (vertical axis 5-H) or the left extension (diagonal peak f), it is seen that they are cross peaks of 5. The peaks representing 4-5 couplings are g and i. The two protons at 4 have very different chemical shifts. Cross peaks h and j are seen on the line. They represent couplings of 5-H with two 6-H of different chemical shifts. Thus the proton connectivity shown by the heavy line in the structural formula can be developed with 2 as starting point.

The COSY shown here is merely an example. It does not mean that this is the only procedure which can be used. The reader may experiment with methods using, for example, only the horizontal axis.

1) See section 26

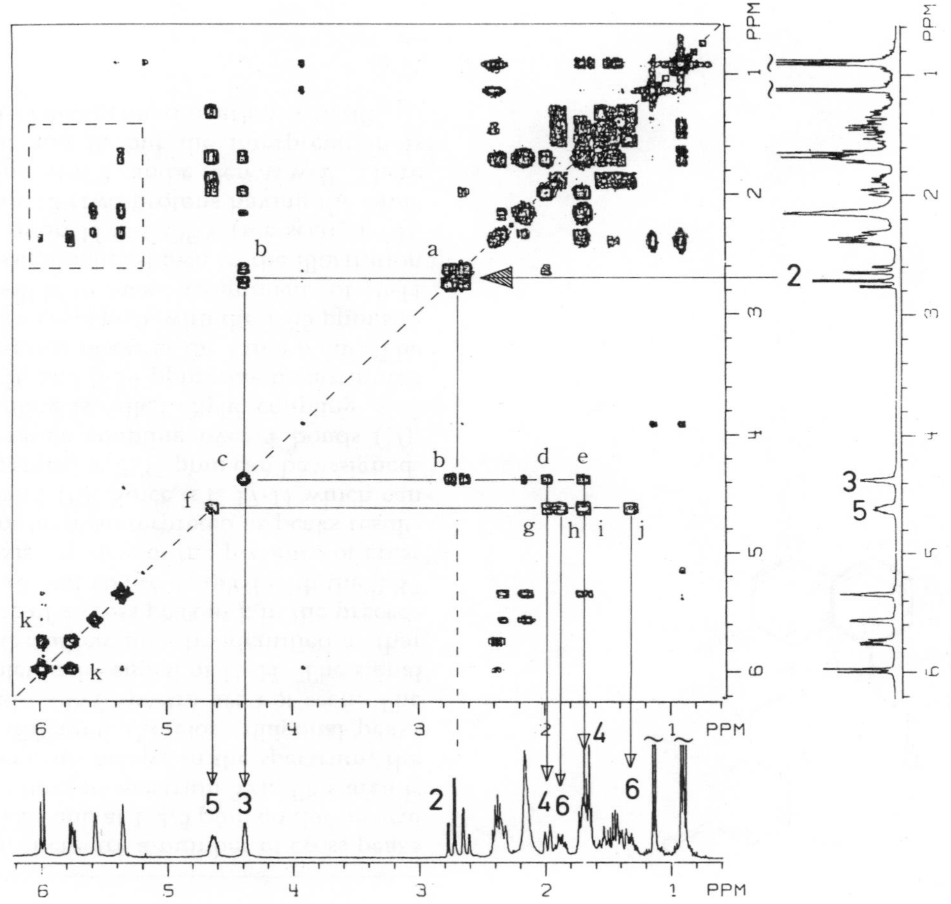

360MHz ^{1}H-NMR (CDCl$_3$) [TI]

40 COSY Spectrum of Compactin (II)

In the preceding spectrum, there are a number of cross peaks at 5-6 ppm on the horizontal axis and at 1-2.5 ppm on the vertical axis (area enclosed by broken lines in spectrum 39). This area is included in the expanded spectrum below. In the spectrum, the ranges of the two axes are different, therefore diagonal peaks which are shown in the usual COSY spectra are not seen. The lowest field (5.98 ppm) doublet is the signal of 15-H. The signal at 5.75 ppm to the right of the above may be identified as that of 16-H based on the presence of a cross peak at k in the preceding spectrum. It is seen that 15 and 16 are coupled with the 2.37 ppm signal on the vertical axis, in view of the presence of cross peaks l and m (these must not be misinterpreted as peaks resulting from the coupling of 15 and 16). Since it is 17-H which can couple with 15 and 16-H, the signal at 2.37 ppm can be assigned. 15-H and 17-H show long-range coupling over 4 bonds (^4J). This type of long-range coupling is called allylic coupling.

The low field signals at 5.56 and 5.34 ppm may be attributed to 10-H or 13-H. Both show cross peaks at the same point. The signal at 5.34 ppm also shows a cross peak with the 1.65 ppm signal. At this point it is impossible to make assignments of 10-H and 13-H. In the end, the assignments shown in the illustration could be made on the basis of an H,C-COSY (see section 70). It is seen that 13 couples with 12 (two protons having the same chemical shift); allylic coupling with 9 can be seen as well. There is also coupling between 10 and 9, but the interpretation is difficult since one side of 11-H happens to overlap with 12.

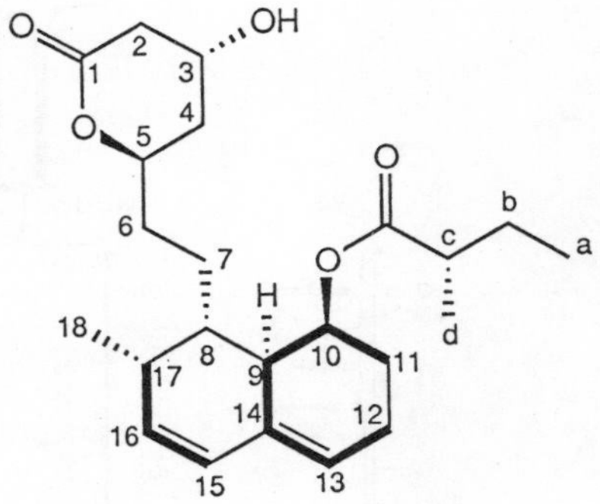

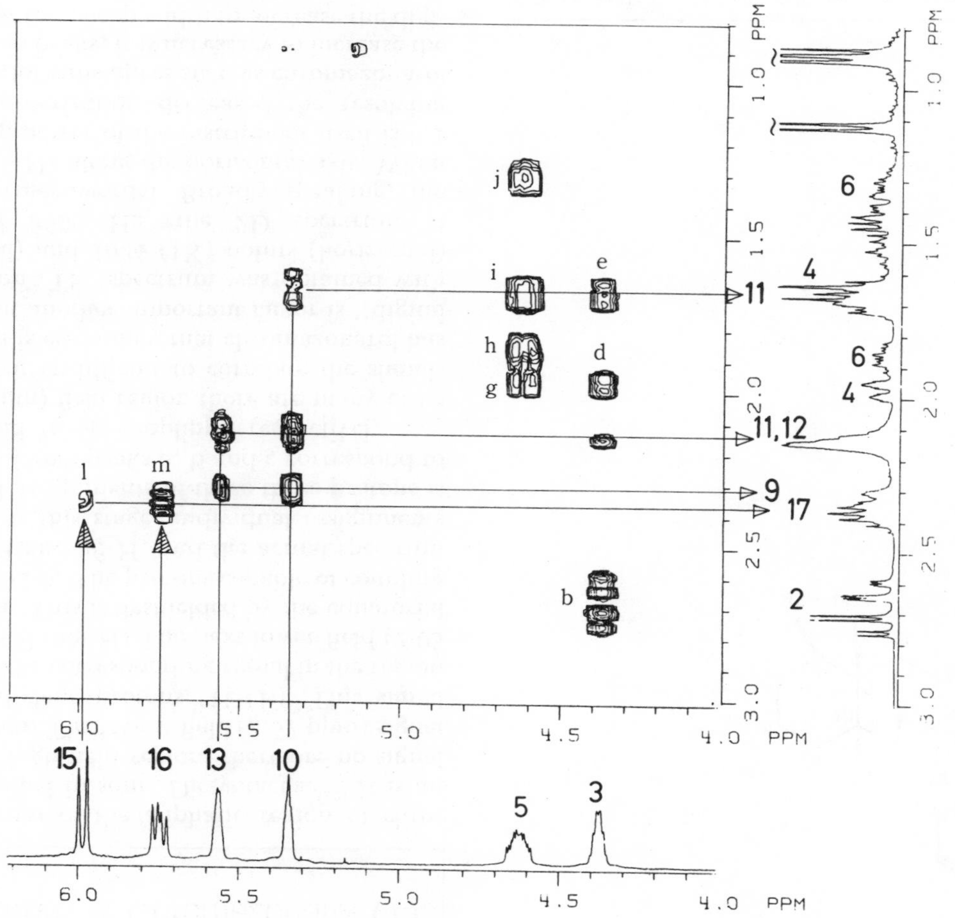

360MHz ^{1}H-NMR ($CDCl_3$) [TI]

41 COSY Spectrum of Chromazonarol (I)

This is a COSY spectrum of the aliphatic region of chromazonarol, which is contained in some Dictyotacea.[1, 2] It is an expanded spectrum of the high field region, therefore no signal of an aromatic proton is seen. The lowest field (2.57 ppm) signal is that of the benzyl methylene protons (11-H_2). This signal shows a single cross peak. The corresponding signal in the region of 1.62 ppm may be assigned to 9-H. The next lower field (2.03 ppm) signal is that of 7α-H. This is deshielded by the equatorial oxygen functional group at C-8. The protons capable of coupling with 7α-H are 7β-H, 6α-H and 6β-H, and the actual spectrum shows three cross peaks. At this stage, individual assignments cannot be made. The final assignments of these three protons is shown on the vertical axis. Cross peaks a, b and c correspond to 7α-6β couplings, 7α-7β and 7α-6α couplings, respectively.

In the higher (0.8-1.8 ppm) field region there are many cross peaks which overlap, and it is difficult to correlate the signals with the peaks. The reason is essentially that chromazonarol has many aliphatic protons, but another important factor is "digital resolution" of the spectrum. The spectrum was obtained with 512 (0.5K) points (vertical) and 1024 (1K) points (horizontal) with a sweep width of 3600 Hz, the 2D spectrum at $0.5K \times 1K = 0.5M$ words (megawords). Broadly speaking, the digital resolution is about 6 Hz along the horizontal axis. When we recall that the resolving power of the instrument itself is 0.1 Hz,this means that computerization decreased the resolving power to about 1/60. Thus for substances such as chromazonarol which give complicated cross peaks, it is necessary to increase the number of points or narrow the sweep width to increase the digital resolution.

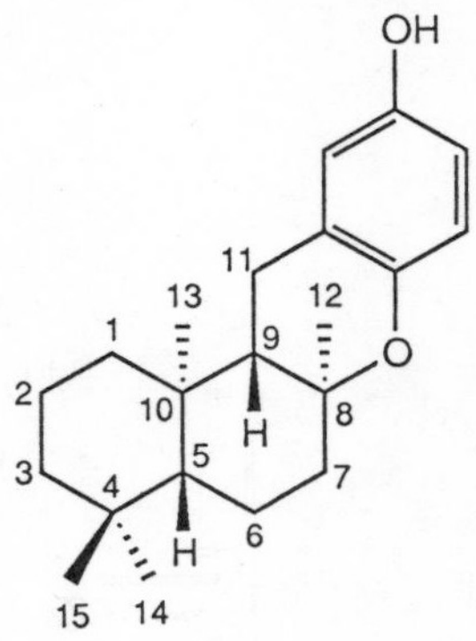

1) P. Djura, D.B. Stierle, B. Sullivan, J. Faulkner, E. Arnold and J. Clardy, *J. Org. Chem.*, **45**, 1435 (1980)

2) M.-N. Dave, T. Kusumi, M. Ishitsuka, T. Iwashita and H. Kakisawa, *Heterocycles*, **22**, 2301 (1984)

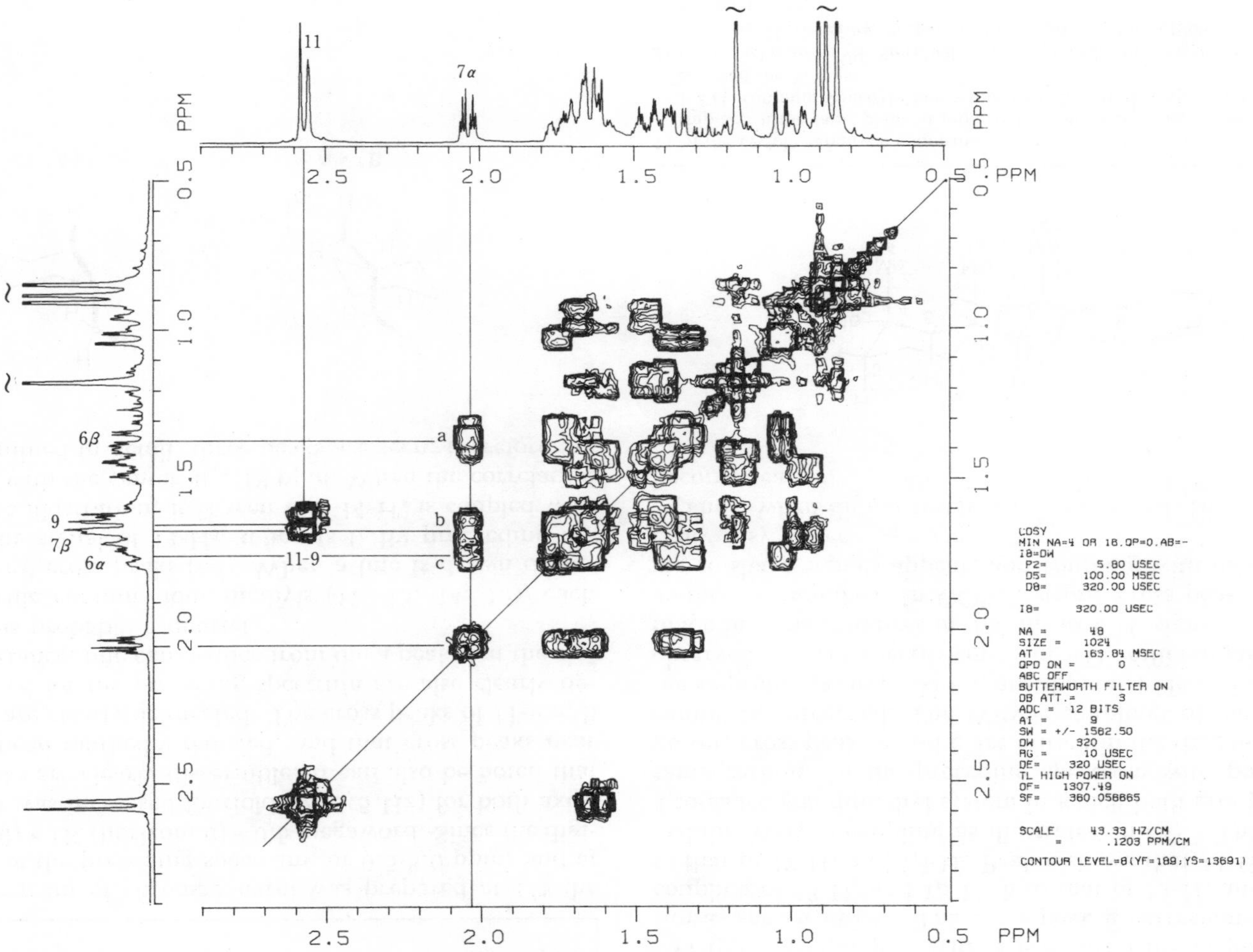

360MHz ^1H-NMR ($CDCl_3$) [TI]

42 COSY Spectrum of Chromazonarol (II)

COSY spectrum of chromazonarol was prepared at 1/4 the sweep width of the preceding spectrum, or 0.5-3.0 ppm, and at 0.5K (vertical) × 1K (horizontal) = 0.5 megaword. Since the digital resolution was increased fourfold (*ca.* 1.5 Hz) for both axes, the cross peaks are clearly discernible. It can also be noted that overlap has been markedly reduced, and that cross peaks near the diagonal are clearly delineated. The cross peaks of 11-9 a, b and c discussed for the preceding spectrum are also clearly defined. For instance, one can deduce from the a peak that the signal of 6α-H is probably a quartet.

The molecule contains four methyls (12, 13, 14, 15), each showing several cross peaks (e-i). When a line is drawn downward from the signal of 14-H_3, it meets f. By proceeding according to the illustration, it is seen that 14-H_3 is coupled with proton 3β-H with the center at 1.18 ppm. When the correlation peak f is examined in detail, three peaks are seen; therefore the coupling partner of 14-H_3 may be presumed to be a triplet. Actually 14-H_3 shows a long-range coupling with 3β-H. In general, protons of an axial Me on a cyclohexane ring enter into "W-type" coupling[1, 2] with a proton on the neighboring axial site as shown in (A). The cross peak g corresponds to W-type coupling of 13-H_3 and 1β-H, h to that of 13-H_3 and 9-H, and i to that of 12-H_3 and 7β-H. Peak e shows that 14-H_3 and 15-H_3 exihibit W-type coupling as illustrated in (B).[3] This establishes a so-called *gem*-dimethyl system in which both groups are on the same carbon. In the preceding spectrum with poor resolving power, cross peaks e and g are buried in the diagonal peaks and cannot be discerned. The W-type couplings of the *ax*-CH_3 and the *ax*-proton (as in peaks e-i) are about 0.1 Hz and are not easily observed by 1D spectroscopy (in 1D, differentiation may be made in some instances in that the *ax*-CH_3 signals are not as tall as those of *eq*-CH_3). In COSY, strong cross peaks corresponding to sharp signals appear, and coupling with methyls is especially easy to see.

Thus, when digital resolution is increased, signal assignment becomes easier.

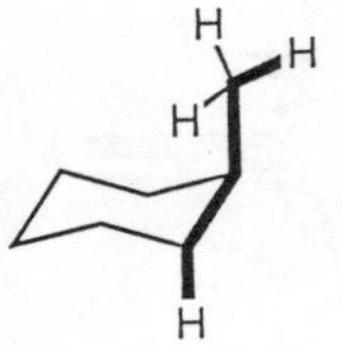

(A)

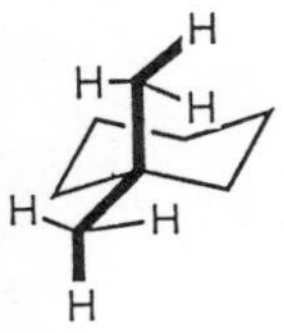

(B)

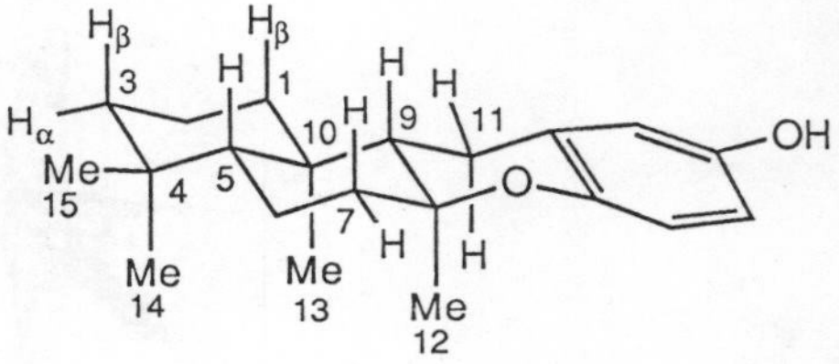

1) In the cyclohexane type conformation, the equatorial protons in the 1,3 relationship are in the same plane as indicated by the heavy line. These protons undergo 1-2 Hz coupling. Since the heavy line is W-shaped, this type of long-range coupling is called the W type.

2) L.M. Jackman and S. Sternhell, *Applications of Nuclear Magnetic Resonance Spectroscopy in Organic Chemistry*, Pergamon Press Ltd., Oxford (1969), pp.334-341

3) See section 46.

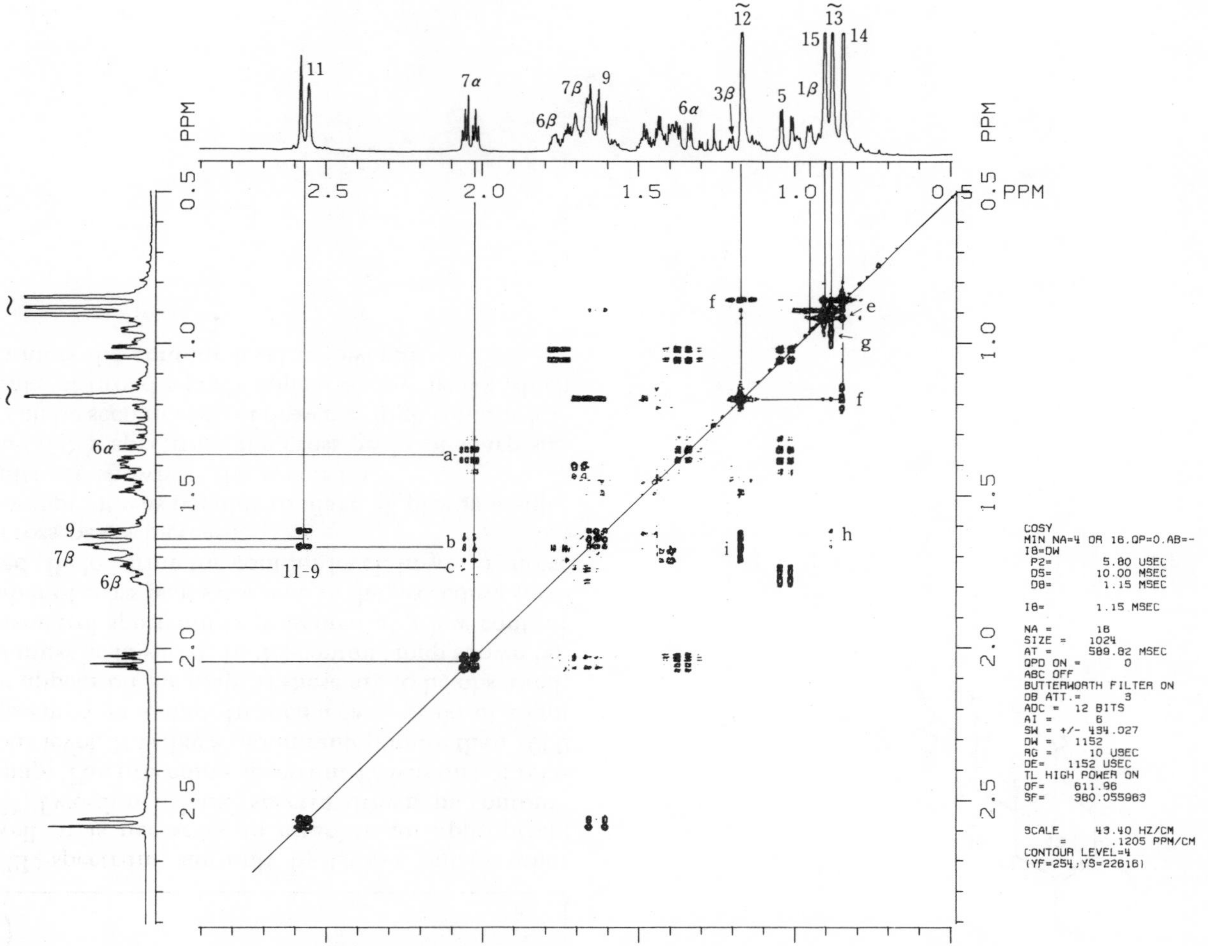

360MHz ^{1}H-NMR ($CDCl_3$) [TI]

43 COSY Spectrum of Chromazonarol (III)

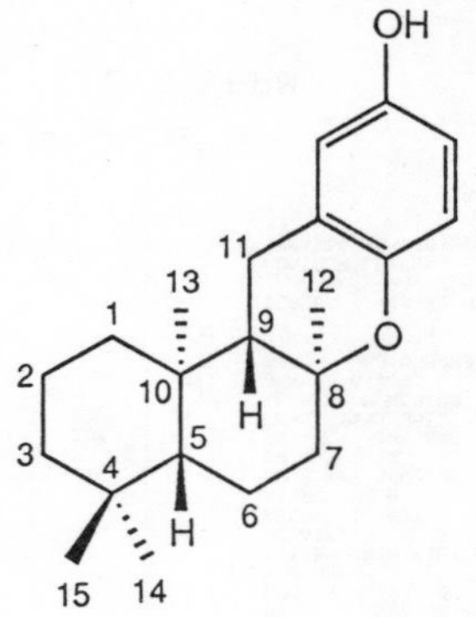

In preparing 2D-spectrum, not only by COSY but by other techniques as well, it is necessary to establish an appropriate "contour level." Two-dimensional spectra drawn in contours are like a relief map. The preceding spectrum 42 was one of relatively high contour level. It is like a mountaintop more than 1000 m in height represented on a map. In such a case, peaks of about 500 m would not appear on the map. If these are to be observed, the contour level must be lowered. In the contour map shown below, the chromazonarol spectrum is presented at a low contour level, and a number of cross peaks not seen in the preceding spectrum are observed. By lowering the contour level, however, noise and overlap of cross peaks increase.

By this spectroscopy, it was possible to make all proton assignments. The results are shown in the spectrum.

Generally in a COSY spectrum, the cross peaks of sharp signals are tall and can be seen in spectra drawn at high contour levels, but cross peaks of broad signals appear as low peaks which cannot be seen unless the contour level is lowered.

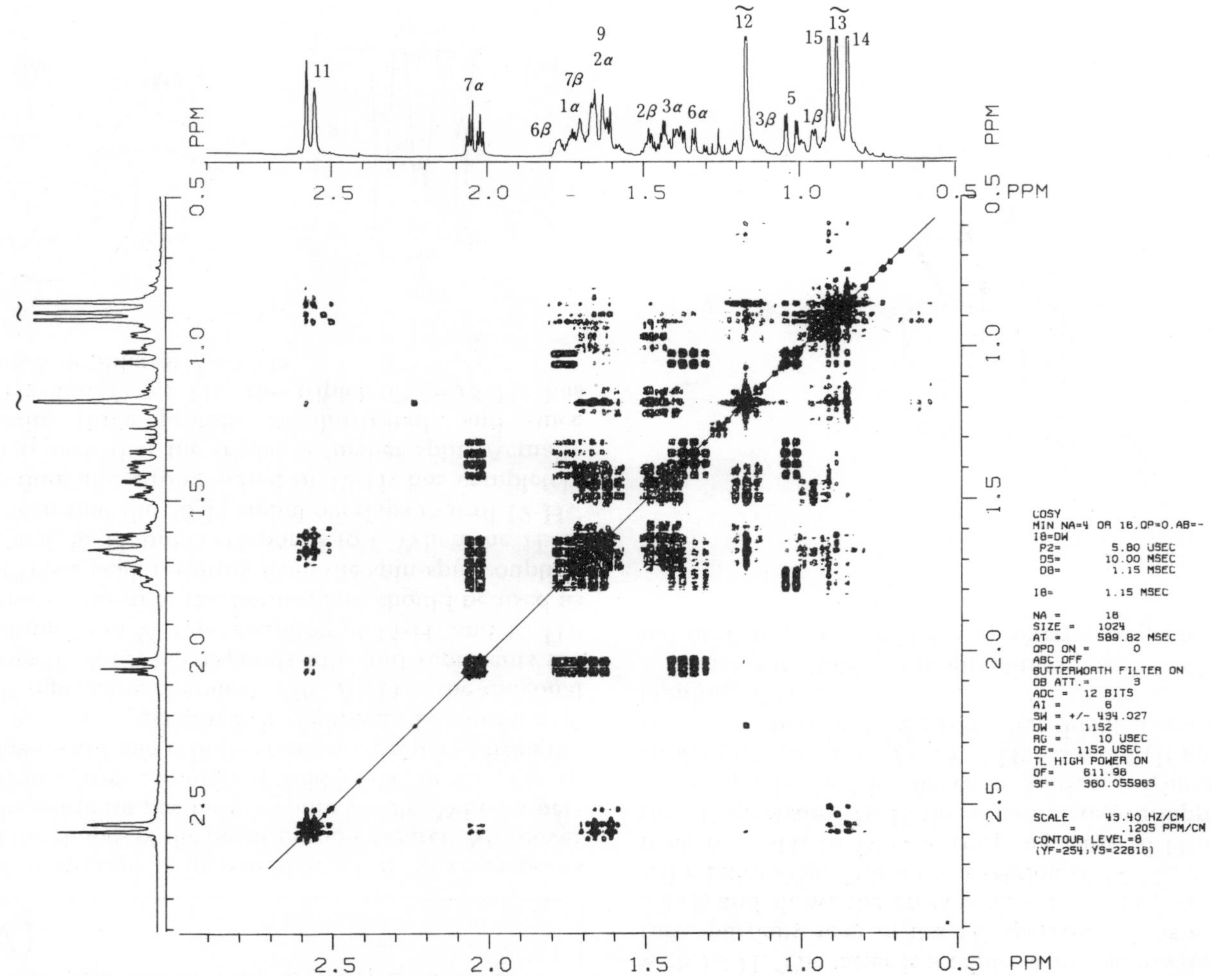

360MHz ^{1}H-NMR ($CDCl_3$) [TI]

44 COSY Spectrum of Chromazonarol (IV)

When digital resolution is increased, overlap of cross peaks diminishes and the shapes of the peaks become clearer. Moreover the signals of the coupling partners become visible. When a perpendicular is drawn from the highest field 14-H_3 in the preceding spectrum downward and cross sections are prepared from the cross peaks on the line, spectrum A is obtained. The numerical value at the left represents chemical shift. A-(1) is the diagonal peak of 14-H_3 itself. A-(2) corresponds to e and represents the cross peak resulting from W-type coupling of 14-H_3 and 15-H_3. The 1D spectrum is shown at the bottom and should be used as a reference. A-(3) is a peak resulting from the spin-spin coupling between 14-H_3 and 3β-H and corresponds to f. When the 1D is inspected, it is seen that the 3β-H signal overlaps that of 12-H_3. In the cross section, the strong signal of 12-H_3 has completely disappeared. It is seen that the triplet is further split. Actually 3-H couples with three protons as illustrated, and since $J_{gem} = J_{trans} = 13$ Hz and $J_{cis} = 4$ Hz, the triplet of $J = 13$ Hz has split further into a doublet of $J = 4$ Hz.

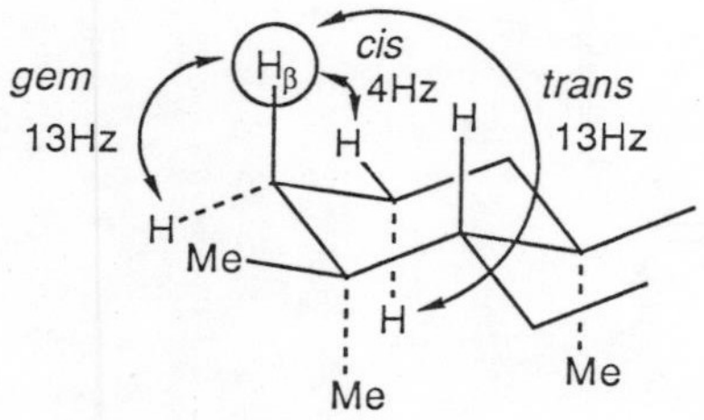

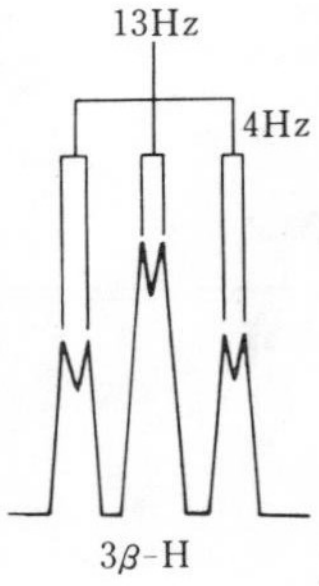

B is a cross section of the 13-H_3 cross peak. B-(1) is a diagonal peak and B-(2) is a cross peak produced through coupling with 1β-H. The latter is a triplet which shows doublet fine structure, partially seen in the 1D spectrum. C is a cross section of 15-H_3 and shows the cross peak with 14-H_3, which is in a geminal relationship. D is a cross section of 12-H_3, showing the cross peak of 12-H_3 in W-type coupling with 7β-H at 1.67 ppm. In the 1D spectrum, 7β-H shows complete overlapping with several other signals, but in the cross section a completely separated triplet (actually a td, $J = 13$, 3 Hz) is seen. (It appears as though there is a cross peak at 14-H_3, but this is a result of 3β-H overlapping 12-H_3.)

Thus with COSY, it is sometimes possible to take out the signal of a single proton from an overlapping group.

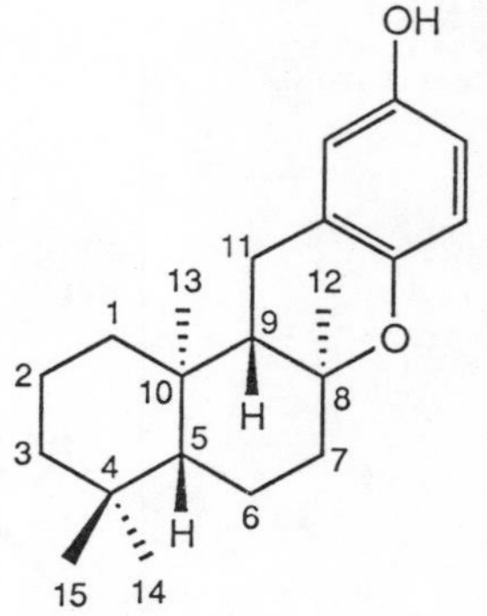

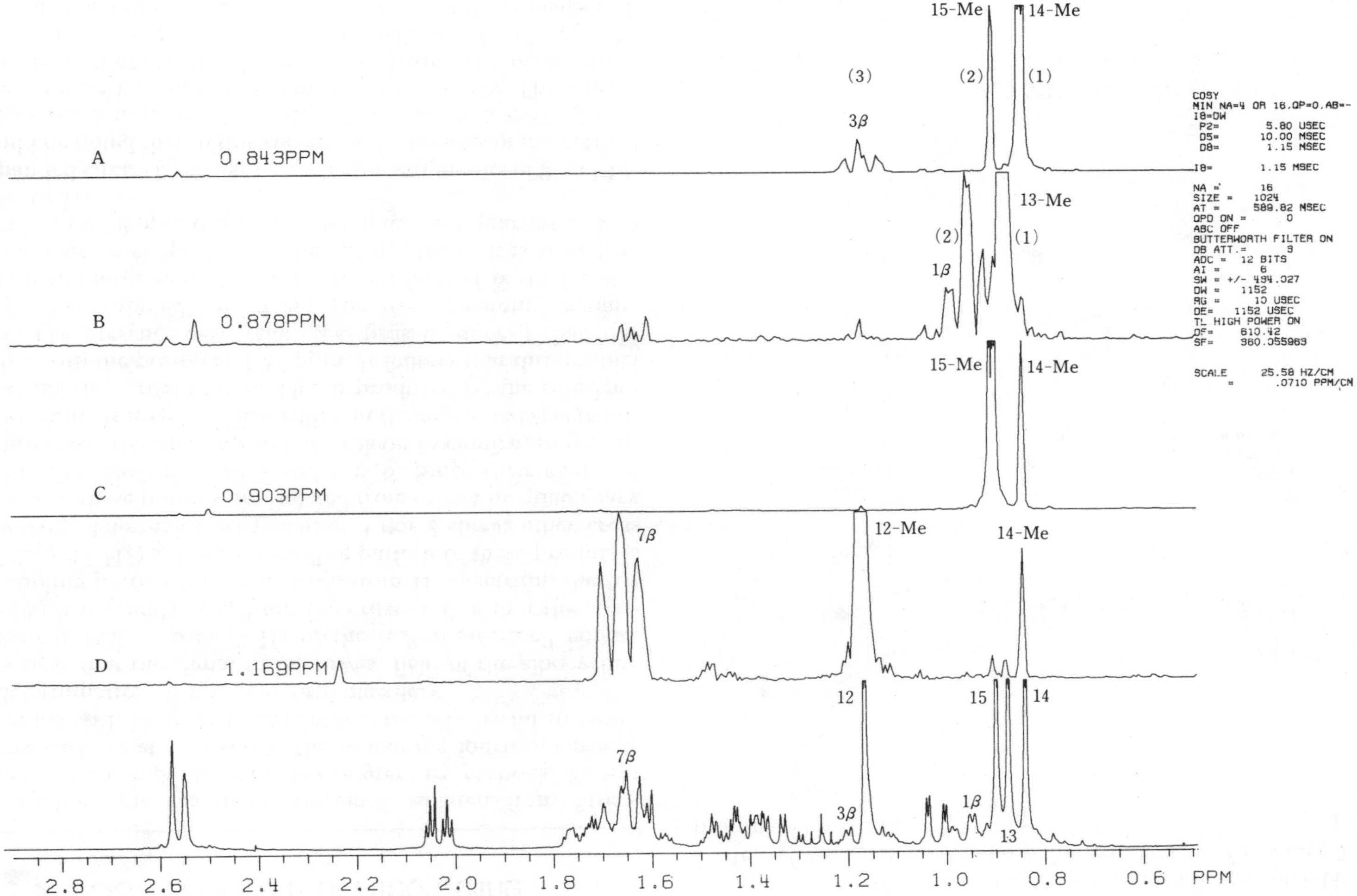

360MHz ^{1}H-NMR ($CDCl_3$) [TI]

45 COSY Spectrum of Albocycline

Albocycline, the macrolide antibiotic isolated from Streptomyces,[1] is a compound with few quaternary carbons. Except for three carbons at 1, 4 and 8, the remaining fourteen carbons are bonded with hydrogen. COSY is extremely useful in structure determination of this type of molecule.

It is clear that the signal in the lowest field of the albocycline spectrum is that of β-H (3-H) of the α,β-unsaturated enone. From (A) it is readily seen from the cross peak a that the spin-spin coupling partner for 3 is 2. Even from 1D spectrum, the AB type (J_{AB} = 16 Hz) spin-spin coupling pattern of these protons is clearly seen. The reason why neither 3 nor 2 shows other cross peaks is that these protons are isolated from others by quaternary carbons. The cross peak of 5 and 6 is b. Since their chemical shifts are close, the existence of b is difficult to confirm in the entire spectrum. It is clearly discernible in the expanded spectrum. 6 shows another cross peak c. This is produced by the spin-spin coupling with the proton at 4.12 ppm. It follows that this doublet is 7-H. The presence of a weak cross peak d suggests that 5-H is long-range coupled with 7-H. The two remaining signals, based on the molecular structure, should be 9-H or 13-H, and from the pattern of splitting in the 1D spectrum, it is clear that the triplet at 5.28 ppm is 9-H and the doublet of quartets at 4.55 ppm is 13-H.

Expanded chart (B) shows even clearer assigments of 9 and 13. It should be noted that in this spectrum the regions of the vertical and horizontal axes are different. The cross peak e of 13 corresponds to the 1.22 ppm signal on the vertical axis. This means that the methyl group of this doublet is 13-Me. The peak f is the result of spin-spin coupling between 13-H and 12-H, and the 1.50 ppm signal may be assigned to 12-H. The cross peaks g, h and i are assigned to 9, g being the cross peak with a methyl group at 1.67 ppm, from which the signal of 8-Me may be assigned, while h and i are cross peaks of the allylic methylene protons 10-H and 10′-H adjacent to 9-H, and j is that of 7-H and 8-Me. These cross peaks show the association of isolated 5-6-7 protons with 8-Me. [TK]

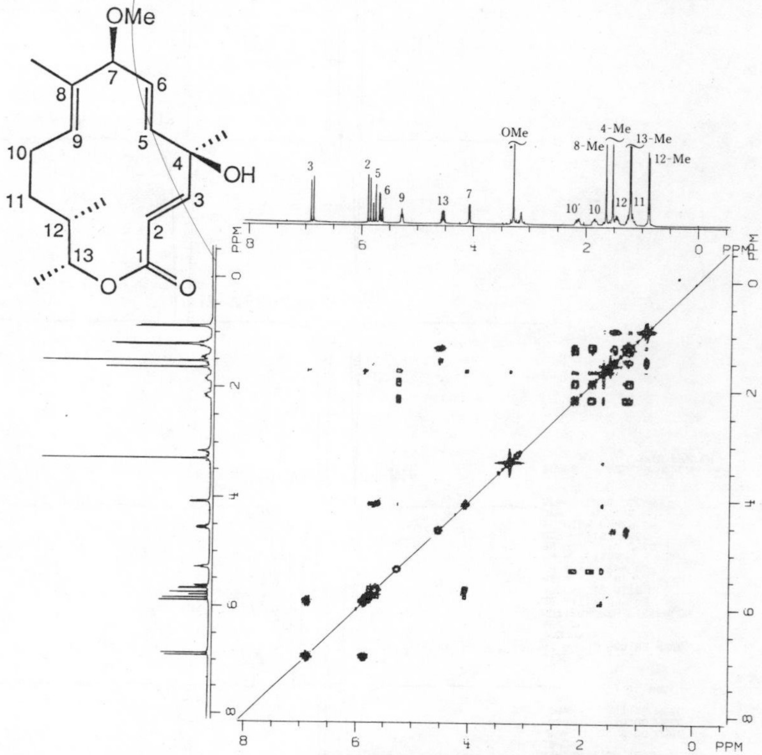

360 MHz ^{1}H-NMR ($CDCl_3$)

(Reproduced by permission from Harada, K. *et al.*, *J. Antibiotics*, **37**, 1192 (1984))

1) K. Harada, F. Nishida, H. Takagi, M. Suzuki, and T. Iwashita, *J. Antibiotics*, **37**, 1187 (1984)

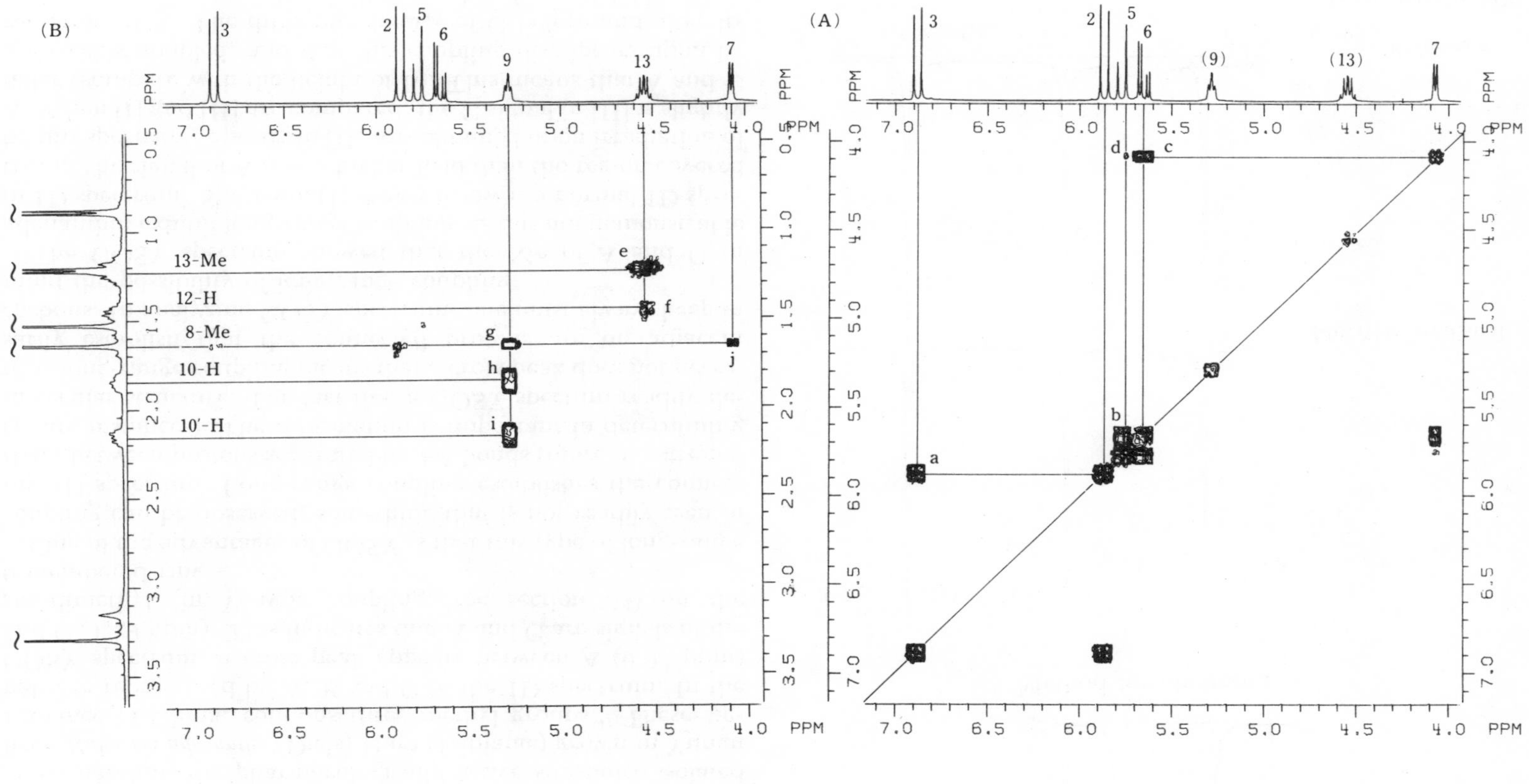

360MHz ¹H-NMR (CDCl$_3$) [TI]

46 COSY Spectrum of Adenanthin

Adenanthin, the pharmacologically active substance isolated from *Rabdosia adenantha* (Diels) Hara (Labiatae) grown in Yunan Province of China, contains three methyl groups.[1)] These signals are represented by A, B and C in the 1D spectrum. In the COSY spectrum, a cross peak appears between A (0.93 ppm) and C (1.29 ppm). This indicates that A and C are signals of the *gem*-dimethyl (in W-type coupling; see section 34) on the 6-membered ring.

One of the advantages of COSY is that this type of long-range coupling can be observed, something that is not readily seen in the 1D spectrum. Long-range coupling establishes the connectivity between protons separated by 4-5 bonds (more in extremely rare instances). The information is important in determining molecular structure. The fact that a COSY spectum readily detects long-range coupling means that a cross peak does not necessarily establish that the connected protons are on adjacent carbons. In analyzing COSY spectrum, one must always keep in mind the possibility of long-range coupling.

The COSY spectrum showed that the Me of A and C in adenanthin exhibit long-range coupling. Is this not demonstrable in 1D spectrum? Spectrum [I] shown below is a normal 1D spectrum. The signal of A is at a higher field than the region covered by this spectrum. Spectrum [II] was obtained upon irradiation of A. When [I] and [II] are compared, the C signal in [II] is slightly taller (compare with the height of B). This means that A and C are weakly coupled, and that this coupling disappears upon irradiation of A. The difference in $\Delta_{h/2}$ of C before and after irradiation was only 0.3 Hz, indicating that $J_{A\text{-}C}$ is 0.1 Hz. Thus in this instance, COSY proved superior to 1D spectrum in detecting long-range coupling.

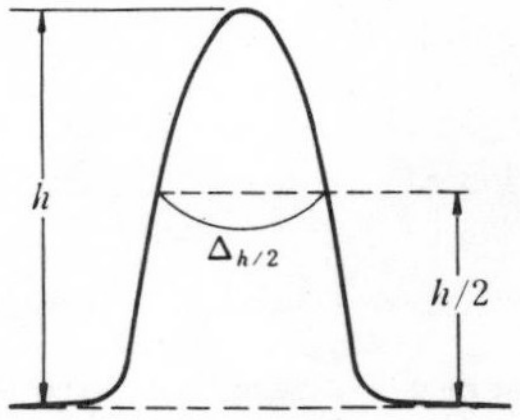

Method for obtaining $\Delta_{h/2}$

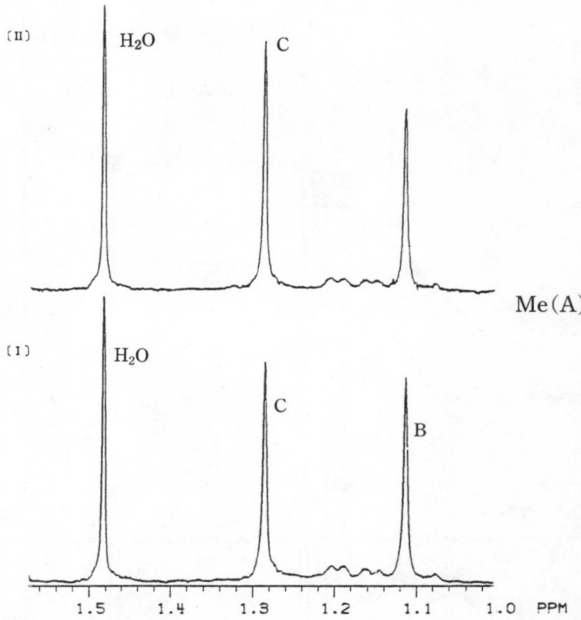

Me(A) is irradiated.

Reference

360MHz ^{1}H-NMR($CDCl_3$)

1) Y.-l. Xu, H.-d. Sun, D.-z. Wang, T. Iwashita, H. Komura, M. Kozuka, K. Naya and I. Kubo, *Tetrahedron Lett.*, **28**, 499 (1987)

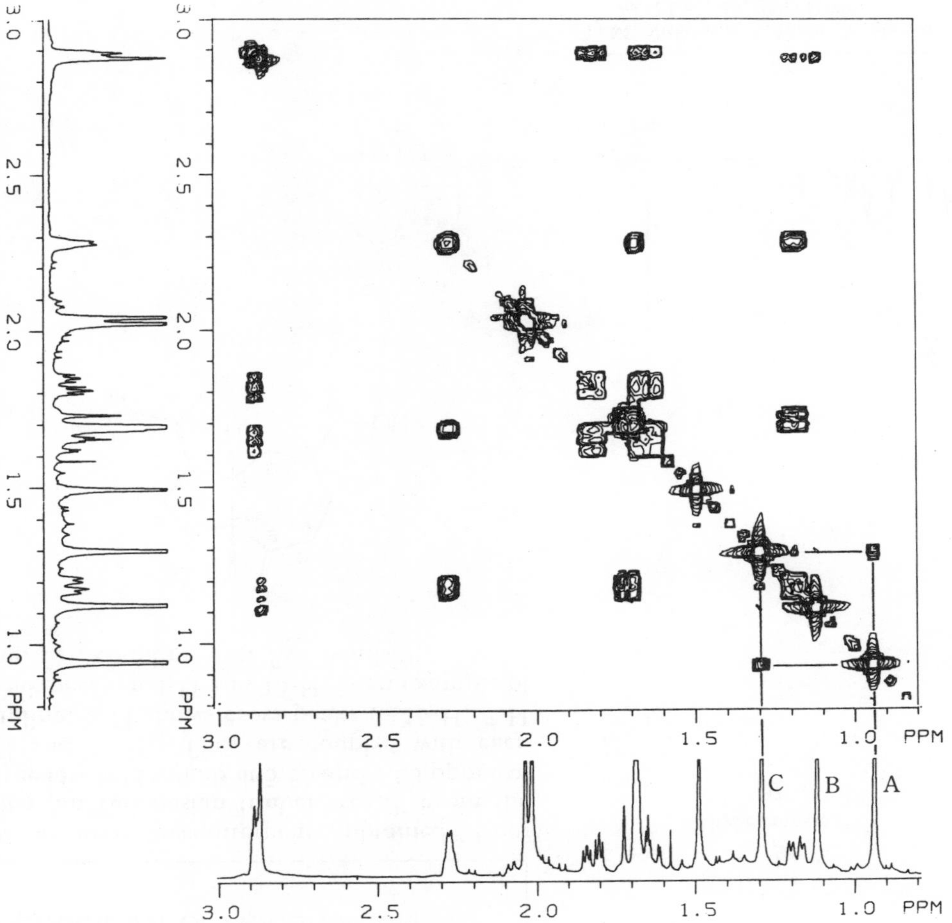

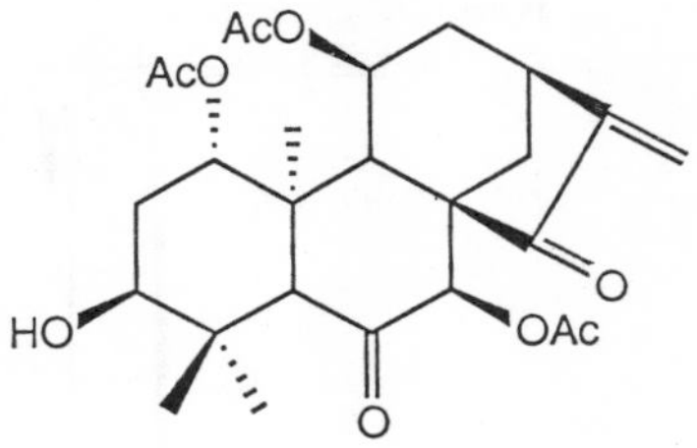


```
HOMONUCLEAR J-CORRELATION
NA=4/16 DP+0 AB-
 P2-        5.90 USEC
 D5-        1.00  SEC
 D6-         .00 USEC
 D8-      360.00 USEC

I8-      360.00 USEC

NA -        64
SIZE -    1024
AT -      184.32 MSEC
DPD ON -        0
ABC OFF
BUTTERWORTH FILTER ON
DB ATT.-        3
ADC -  12 BITS
AI -      10
SW - +/- 1388.88
DW -    360
RG -      10 USEC
DE-    360 USEC
TL HIGH POWER ON
OF-    1347.71
SF-    360.058285

SCALE       39.87 HZ/CM
     -        .1107 PPM/CM
```

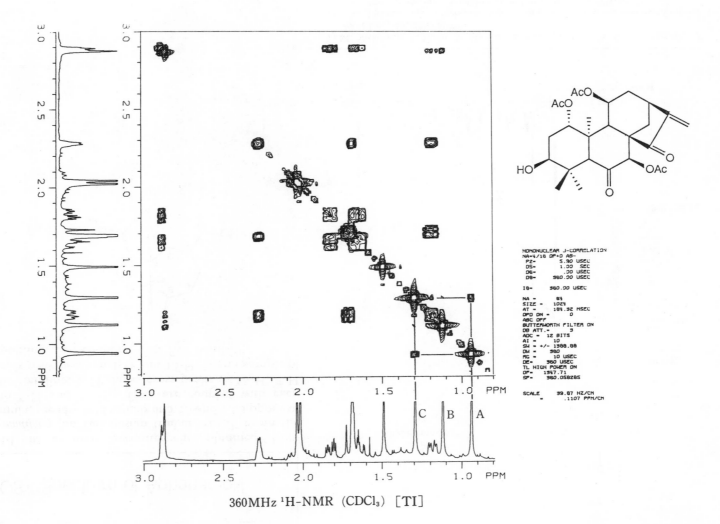

360MHz ^{1}H-NMR (CDCl$_3$) [TI]

47 COSY Spectrum of Aphanamol-I

Aphanamol-I is a toxic sesquiterpene obtained from *Aphanamixis grandifolia* (an Indonesian timber tree).[1] From the structural formula, the low field signals at 5.55 and 4.03 ppm can be assigned to 5-H and 15-H_2. They are coupled with each other by allylic-coupling. 5-H shows cross peaks to 15-H, 7-H and 4-H. A cross peak between 4-H and 15-H_2 is an example of homoallylic coupling, *i.e.*, coupling over five bonds.[2]

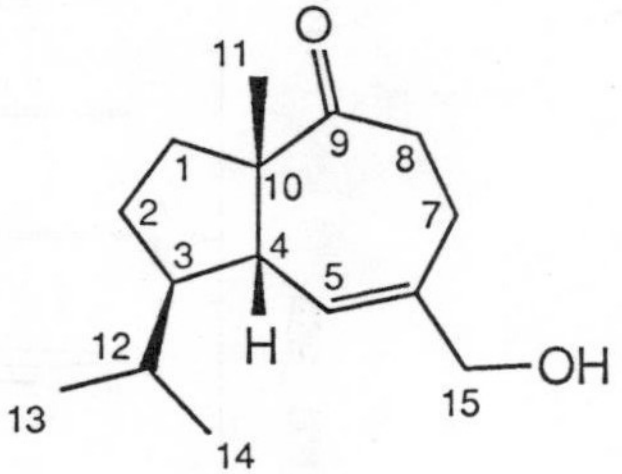

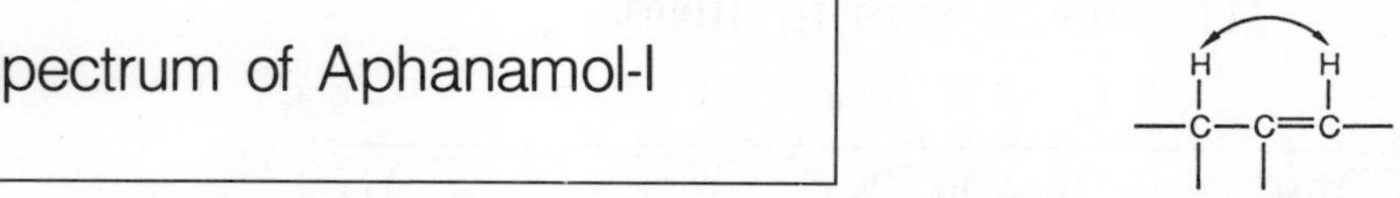

Allylic coupling[3]
(^4J)

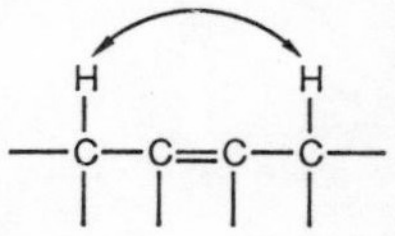

Homoallylic coupling
(^5J)

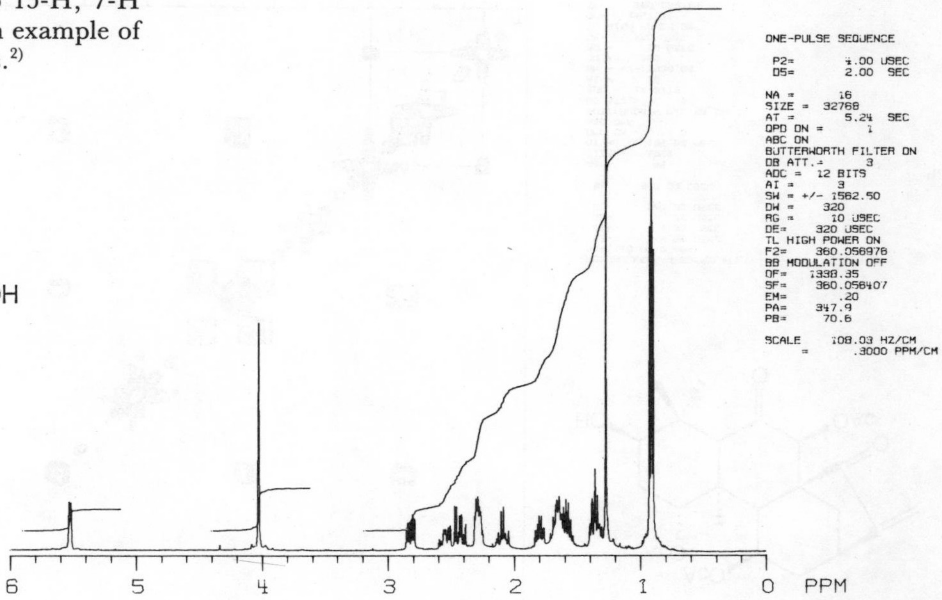

1) M. Nishizawa, A. Inoue, Y. Hayashi, S. Sastrapradja, S. Kosela, T. Iwashita, *J. Org. Chem.*, **49**, 3660 (1984)

2) L.M. Jackman and S. Sternhell, *Applications of Nuclear Magnetic Resonance Spectroscopy in Organic Chemistry*, Pergamon Press Ltd., Oxford (1969), pp.316-328

3) M. Barfield, R.J. Spear, S. Sternhell, *Chem. Rev.*, **76**, 593 (1976)

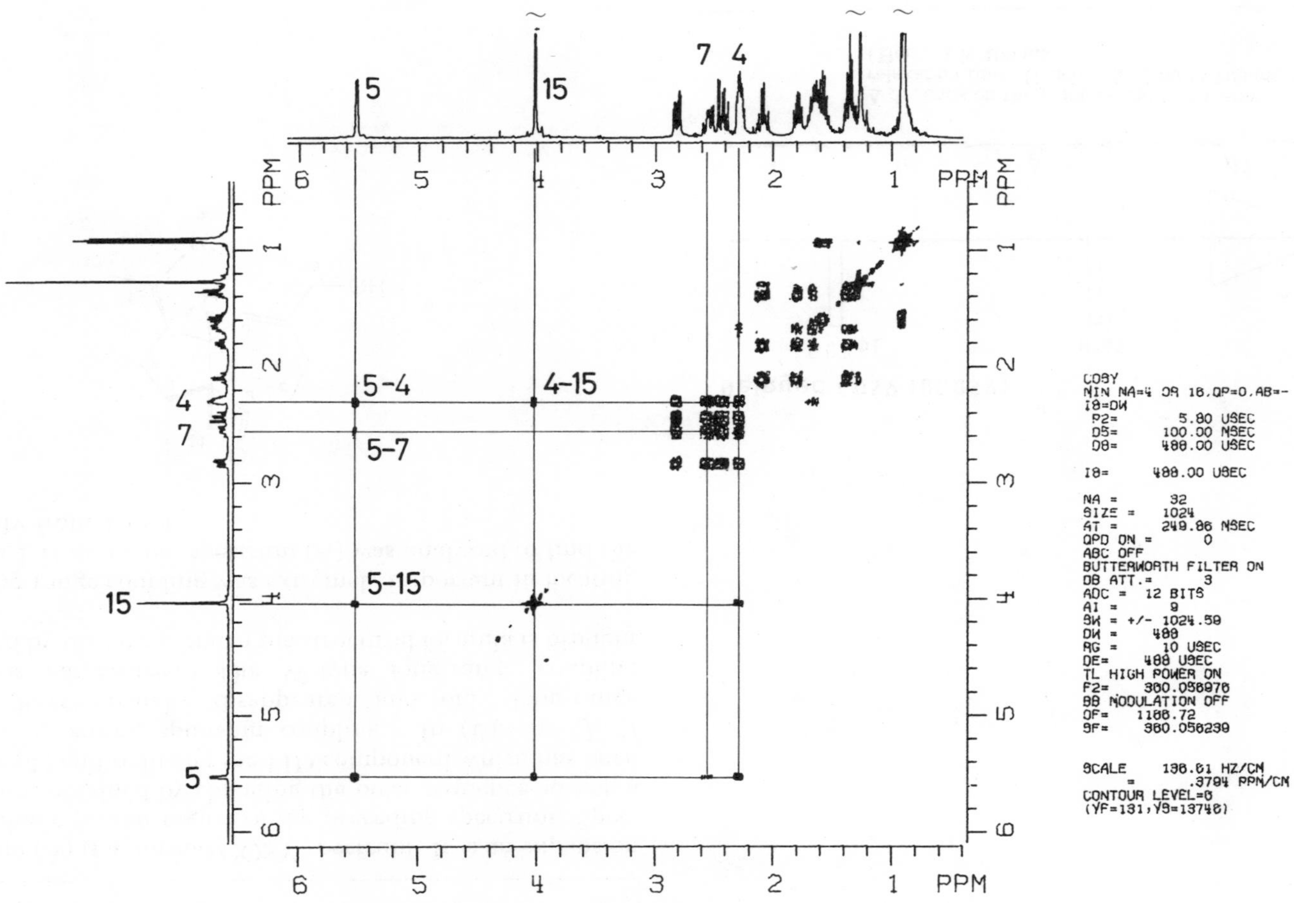

360MHz ^{1}H-NMR ($CDCl_3$) [TI]

48 DCOSY Spectrum Emphasizing Long-range Coupling (Aphanamol-I)

Spectrum (A) is a normal COSY spectrum. It is an expansion of the aliphatic proton region of the preceding spectrum. Spectrum (B) was obtained by changing the pulse sequence adding a delay time (Δ) and reducing the FID component which has been modulated by strong spin-spin coupling.[1] In (B), the ^2J, ^3J couplings have virtually disappeared and only long-range coupling is emphasized. The W-type long-range coupling represented by the heavy line in the structural formula is brought out.

This long-range coupling was extremely important in locating 1-H. With 1-H as a clue, spectrum (A) was analyzed to find the connectivity from 1 to 4.

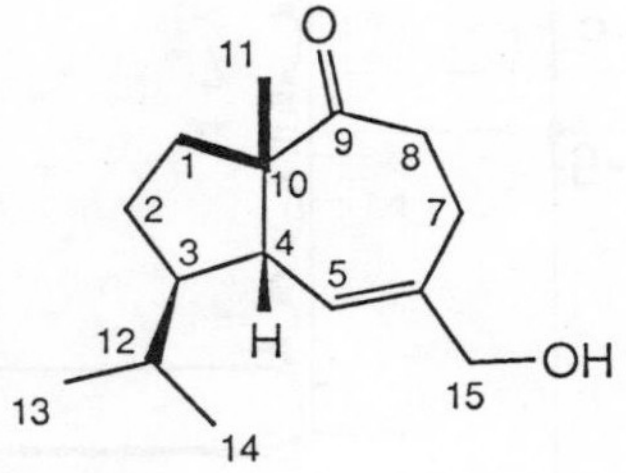

Delayed COSY (DCOSY)

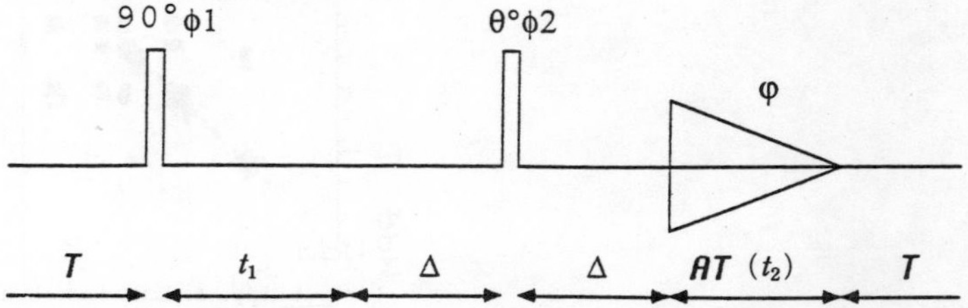

Δ depends on the J-value and transverse relaxation time. Usually, Δ is 60 to 100 ms. (Here, Δ is 400 ms.)

1) Ad Bax, *Two-Dimensional Nuclear Magnetic Resonance in Liquids*, Delft University Press, Delft, Holland (1982), pp.85-86

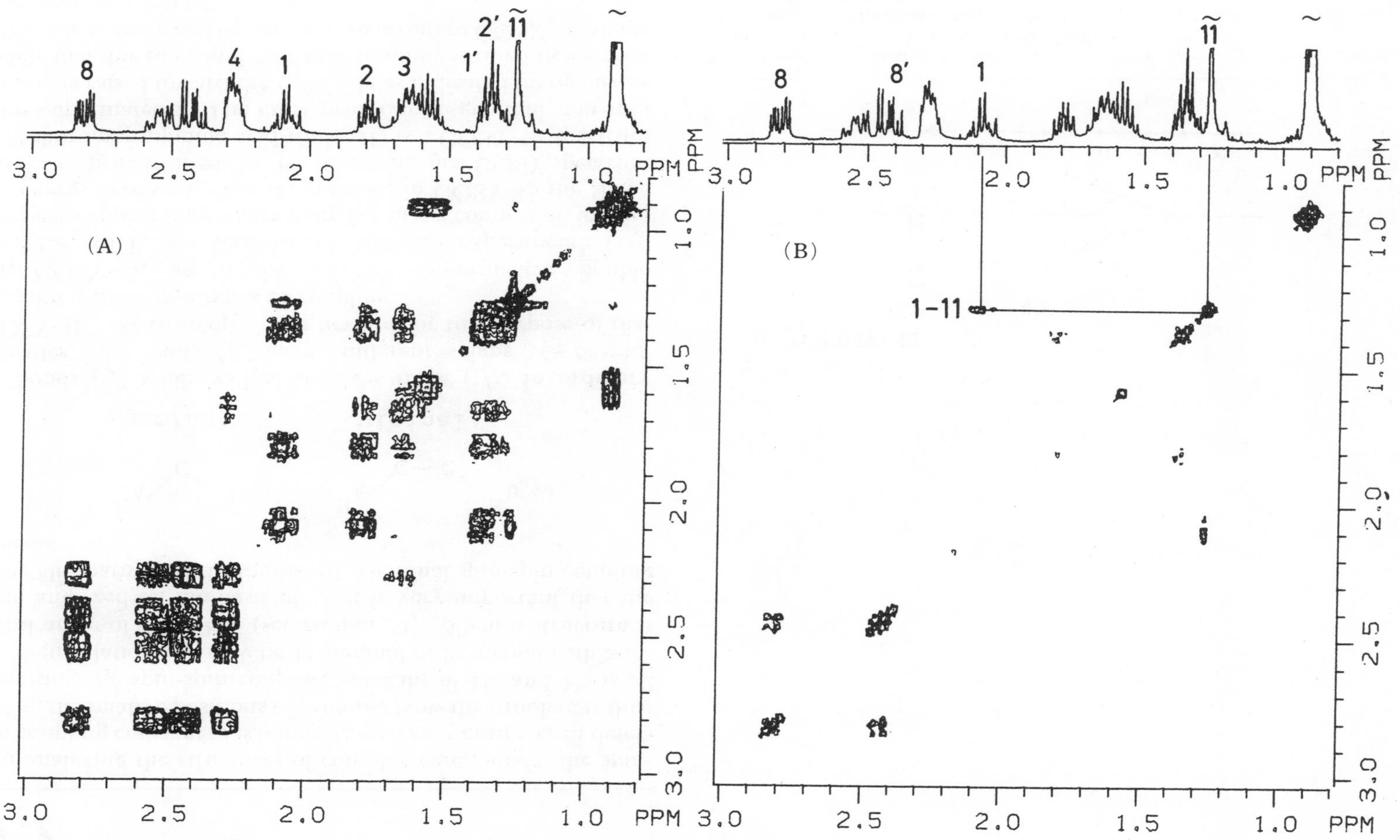

360MHz ^{1}H-NMR ($CDCl_3$) [TI]

49 COSY-45 Spectrum

In analyzing the structures of complex compounds, the spin-spin coupling constant J is a most important parameter in determining the relative positions of protons. Now the problem is this: Supposing the spin-spin coupling constant of H_A and H_B is 12 Hz, their relationship may be 1) geminal or 2) vicinal with a dihedral angle of 0 or 180° (see section 34). When a structure is being analyzed on the basis of J, it is very important that the above alternatives be distinguished. Geminal spin-spin coupling involves

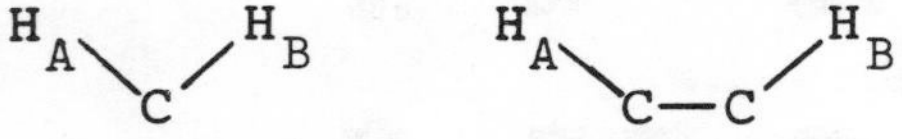

geminal vicinal

two bonds (^2J) while vicinal involves three (^3J). In ordinary molecules, ^3J and ^2J have different signs (+or−).[1] "COSY-45" spectroscopy[2] was devised for the purpose of distinguishing these two types of coupling.

In COSY-45, as in the INDOR (internuclear double resonance, SPT; see sections 14 and 15) experiment[3] (C), cross peaks appear only where there is a direct connection in Zeeman energy levels (A). For this reason, in COSY-45 the shape of the cross peak is tilted. In (B), (I) is a normal COSY spectrum of 1,2-dibromopropionic acid while (II) is a COSY-45 spectrum of the same material. The cross peak due to geminal spin-spin coupling is raised toward the right side as indicated by the arrow a, while that due to vicinal spin-spin coupling is tilted upward on the left side as indicated by arrow b. In ordinary COSY the shape of the peak is a square.

COSY-45 is used often since it is as simple to carry out as COSY. In certain compounds with specific structures, there are occasions when the signs of geminal and vicinal spin-spin couplings are the same (*e.g.* cyclopropanes).

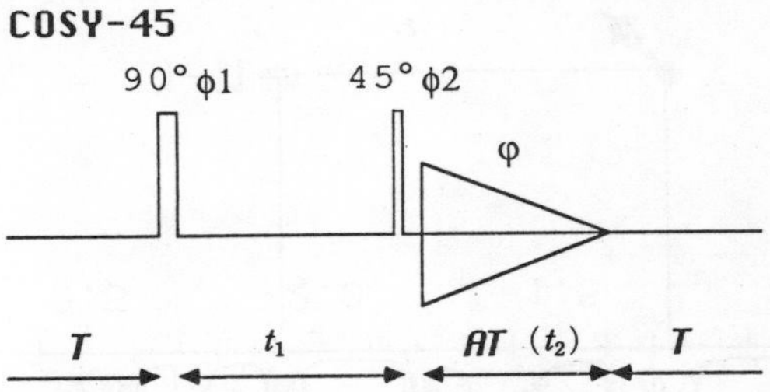

1) H. Günther (translated by R.W. Gleason), *NMR Spectroscopy—An Introduction*, John Wiley & Sons, Ltd., New York (1980), pp.99-120
2) Ad Bax, *Two-Dimensional Nuclear Magnetic Resonance in Liquids*, Delft University Press, Delft, Holland (1982), pp.78-84
3) H. Günther (translated by R.W. Gleason), *NMR Spectroscopy—An Introduction*, John Wiley & Sons, Ltd., New York (1980), pp.304-307

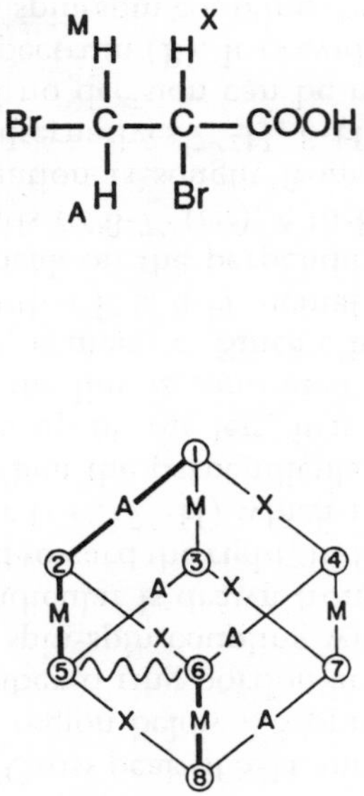

(Reproduced by permission from Bax, Ad and Freeman, Ray., *J. Magn. Reson.*, **44**, 554(1981))

(A)

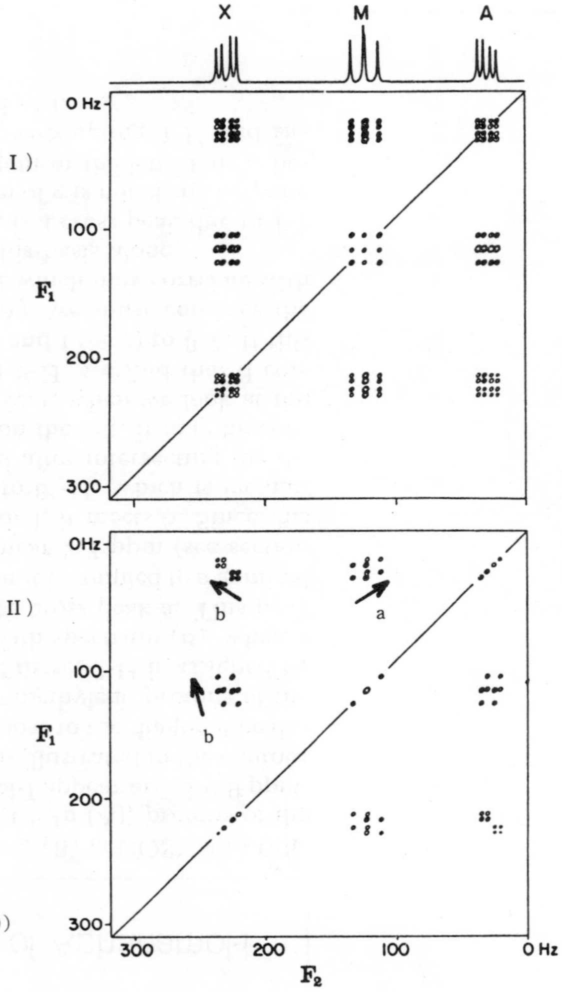

(B)

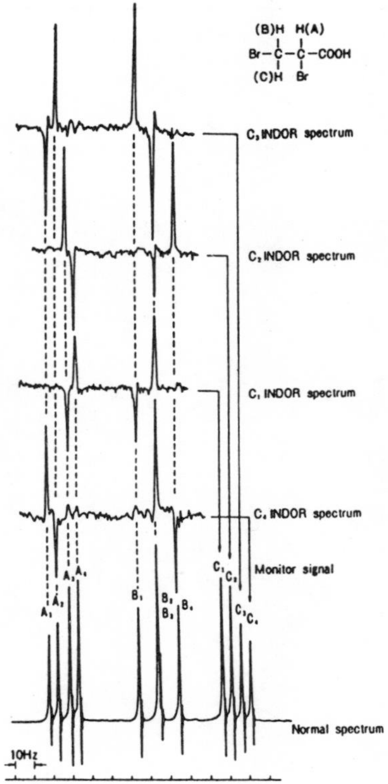

(Reproduced by permission from *The Hitachi Science Instrument News*, **17**, 37(1984))

(C)

50 COSY-45 Spectrum of Aphanamol-I

Spectrum (A) is COSY and spectrum (B) is COSY-45 of the aliphatic proton region of aphanamol-I.[1] In (A), protons of the 5-membered ring portion of aphanamol-I appear at 2.3-0.9 ppm. Their assignment may be carried out as illustrated in the contour map. (Cross peak of 3-H and 12-H is close to the diagonal peak.) In the region below 2.3 ppm the four-methylene protons of the 7-membered ring portion are seen. Of these, 7-H is assigned by allylic spin-spin coupling with 5-H. With spectrum (B), when a perpendicular is drawn from 7 it meets cross peak a. This peak is tilted toward the right, indicating that it is coupled to a geminal partner (*i.e.*, 7′-H) which is located near 2.3 ppm (see section 49). When the perpendicular is extended, it meets b. Since this is tilted up on the left, it is assigned to 8′-H, which is vicinal. When the line is continued downward after intersecting the diagonal, it meets c. Since c is also up on the left, it may be concluded that it is 8-H vicinal to 7-H. Next, when we look at the cross peak on the perpendicular from 8-H, we find that d corresponds to 8-7′ (*vic*), e to 8-8′ (*gem*) and f (or c) to 8-7. If this information is sought from COSY (A), we must consider the three alternatives 7′-H, 8-H and 8′-H which may correlate with 7, and no decision can be made on this basis alone.

In spectrum (B), it is evident that h is a cross peak due to 1-2 vicinal spin-spin coupling, but the form of g is not clear, *i.e.*, one cannot tell whether the tilt is on the right or the left. This is because g represents an overlap of the *gem*-coupling 1-1′ and *vic*-coupling 1-2′. The same may be said of i.

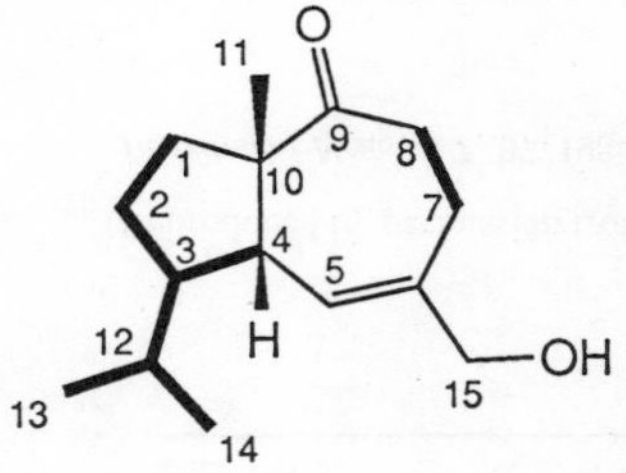

1) M. Nishizawa, A. Inoue, Y. Hayashi, S. Sastrapradja, S. Kosela, T. Iwashita, *J. Org. Chem.*, **49**, 3660 (1984)

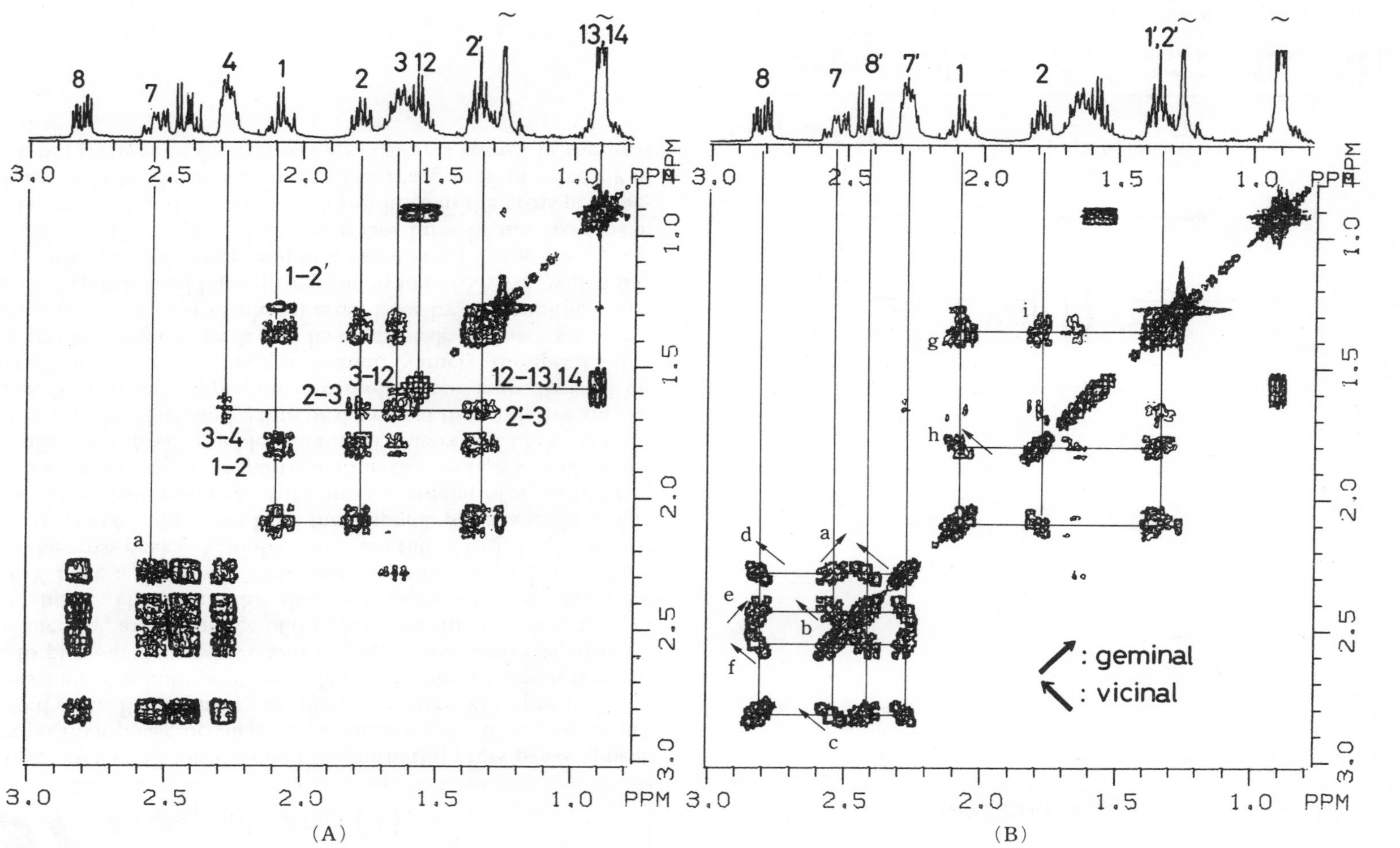

360MHz ^{1}H-NMR (CDCl$_3$) [TI]

51 Phase-sensitive DQF-COSY Spectrum of L-Aspartic Acid (I)

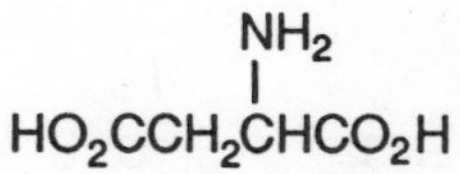

In ordinary 2D spectra described hitherto, cross peaks appear in a positive phase because the absolute values of the peaks are plotted out. By contrast, in phase-sensitive 2D spectra,[1] the peaks appear in both positive and negative phases, and it is possible to plot out the positive and negative cross peaks separately.

Generally, an advantage of the phase-sensitive method is good line shape. However, the diagonal peaks of phase-sensitive COSY have a dispersive character, and the long tail interferes with the cross peaks. A double quantum filter (DQF) can exclude this dispersive component from the diagonal peaks and the single quantum transitions such as the intense singlet of a methyl signal. The spectrum of L-aspartic acid in this section was obtained by phase-sensitive double quantum filtered COSY (DQF-COSY).[2] The black and white peaks reveal their relative sign of phase. The mosaic appearance in the contour map depends on the coupling pattern of the spin system. Namely, the specific line in a certain diagonal peak and its cross peaks forms a set of antiphase (*i.e.* 180° out of phase) cross lines by its coupling constant. L-Aspartic acid has a three-spin system (AMX), which has three different spin-spin coupling constants ($|J_{MX}| > |J_{AX}| > |J_{AM}|$). The most distant antiphase lines in the cross peak are between M and X, and the closest lines in the cross peak between A and M. These features can be recognized more easily in the cross sections of the contour spectrum as shown in the next section.

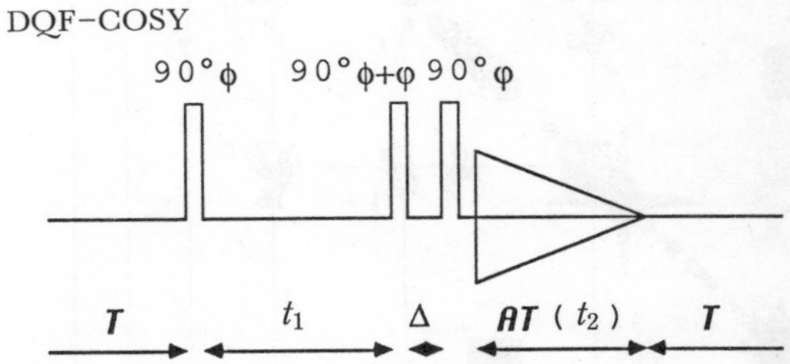

For a 2-quantum filter

$\phi = 0, \pi/2, \pi, 3\pi/2$; $\varphi = 0, \pi/2$

1) W.R. Croasmun and R.M.K. Carlson (Ed.) *Two-Dimensional NMR Spectroscopy—Applications for Chemists and Biochemists*, VCH Publishers, Inc., New York (1987), pp.109-121, 150-153

2) R.R. Ernst, G. Bodenhansen and A. Wokaun, *Principles of Nuclear Magnetic Resonance in One and Two Dimensions*, Oxford University Press, Oxford, (1987), pp.431-440

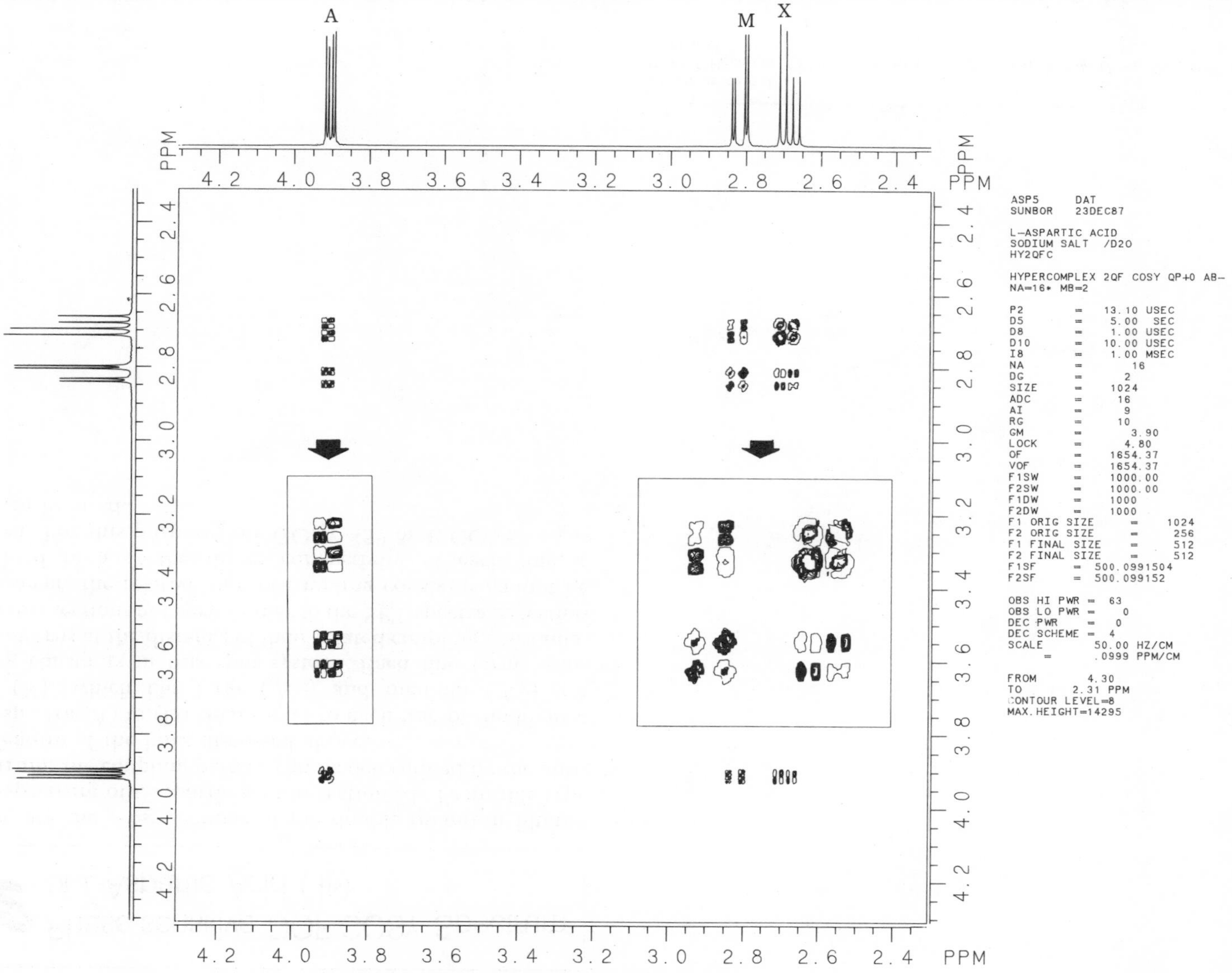

500 MHz ^{1}H-NMR (D_2O-NaOD) [TI]

52 Phase-sensitive DQF-COSY Spectrum of L-Aspartic Acid (II)

These are the cross sections of the double quantum filtered COSY spectrum of L-aspartic acid in section 51. From this type of spectrum, the coupling pattern can be determined by the anti-phase feature of the lines discussed above.

The spectra (A) to (D) correspond to each line of the highest proton (X), which has large (J_{MX}) and medium (J_{AX}) size coupling constants in this spin system. Each line forms anti-phase patterns at the distance of their related coupling constants. These cross sections are very similar to the SPT spectra in section 15. However, the relative sign of coupling constants cannot be determined because the direct connectivity of each line is unknown. For this purpose, the COSY-45[1] or E.COSY[2] technique can be used.

1) See section 49.
2) C. Griesinger, O.W. Sørensen and R.R. Ernst, *J. Am. Chem. Soc.*, **107**, 6394-6396 (1985)

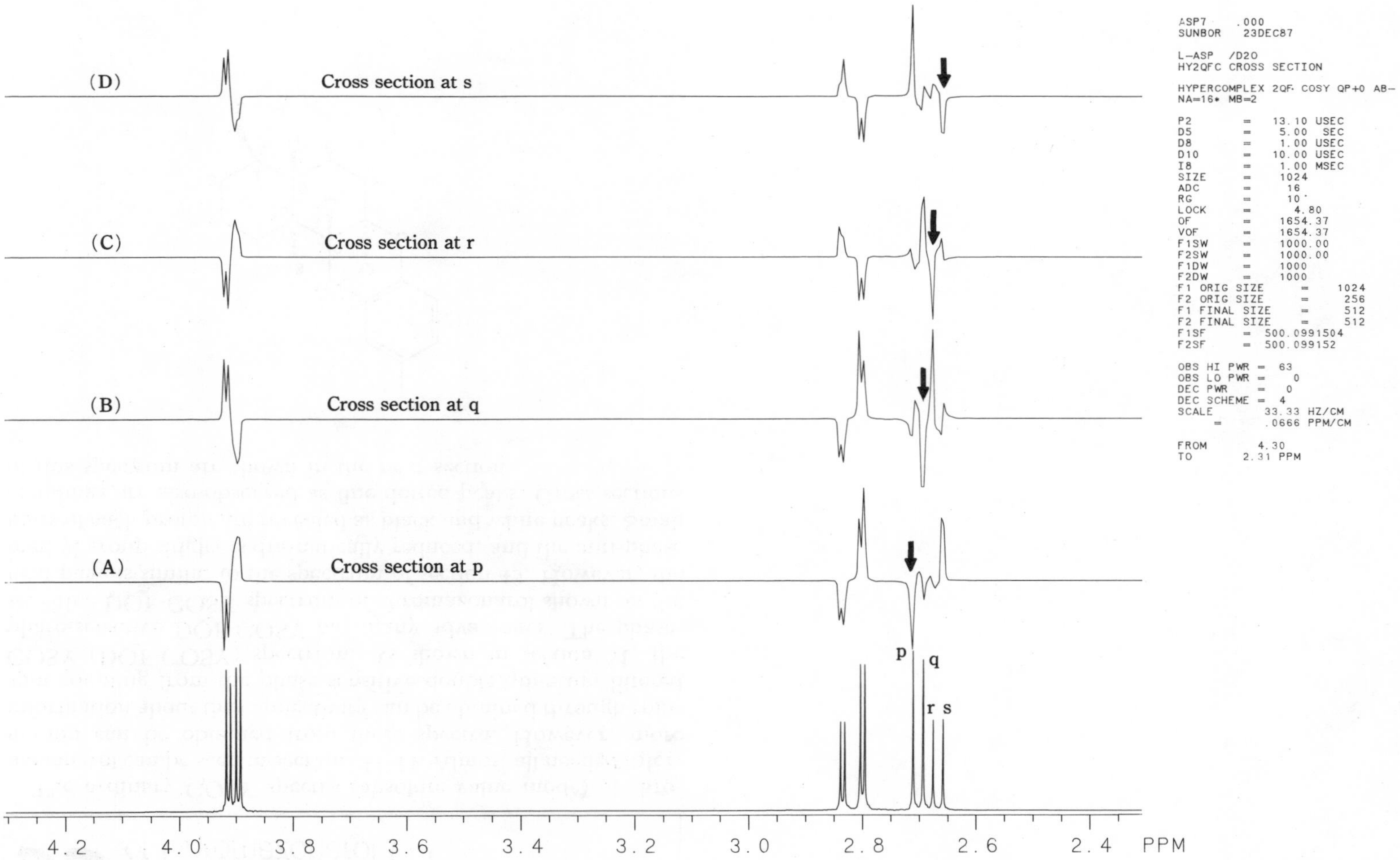

500 MHz ^{1}H-NMR (D_2O-NaOD) [TI]

53 Phase-sensitive DQF-COSY Spectrum of Chromazonarol (I)

The ordinary COSY spectra (absolute value mode) of chromazonarol can be seen in sections 41-44. Almost all needed information can be obtained from these spectra. However, more information about the connectivity can be obtained through spin-spin coupling from the phase-sensitive double quantum filtered COSY (DQF-COSY) spectrum. As shown in section 51, the phase-sensitive DQF-COSY has many advantages. The phase-sensitive DQF-COSY spectrum of chromazonarol shown on the next page is similar to the spectrum of section 43. However, the methyl group singlet is dramatically reduced, and the anti-phase pairs of each proton are revealed as black and white peaks. Small couplings are also observed as fine dotted peaks. Cross sections of this spectrum are shown in the next section.

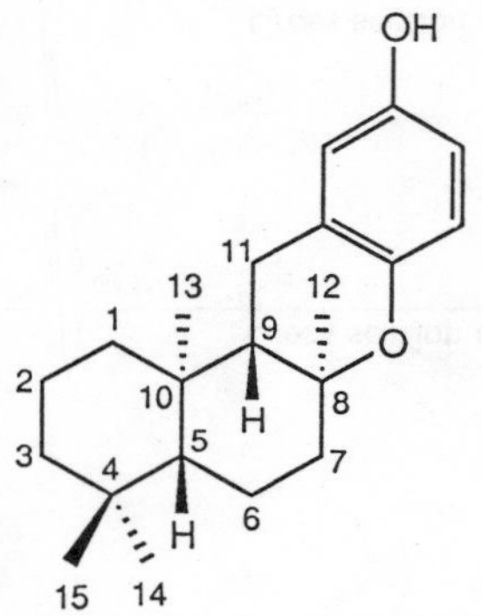

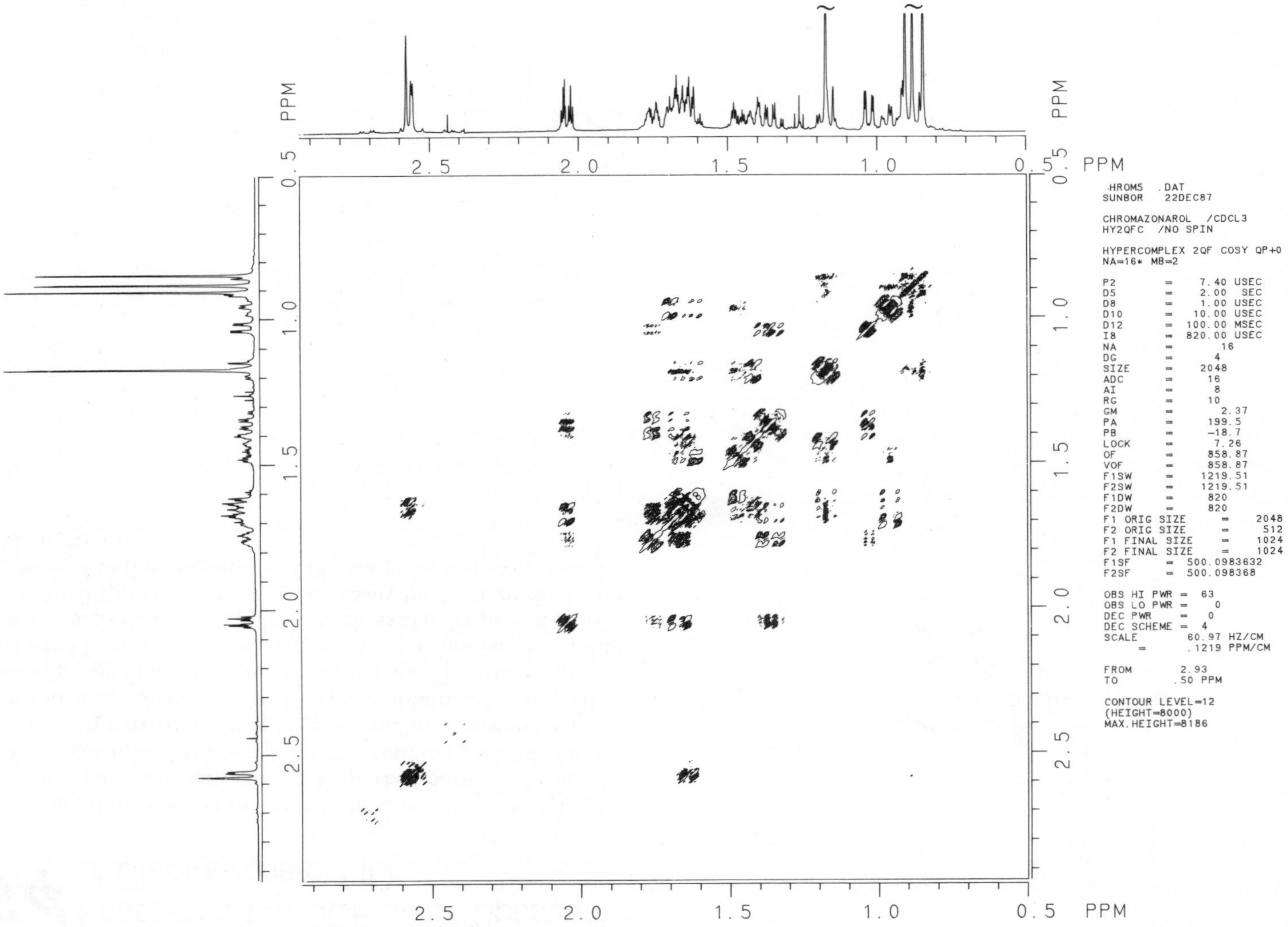

500 MHz ^{1}H-NMR ($CDCl_3O$) [TI]

54 Phase-sensitive DQF-COSY Spectrum of Chromazonarol (II)

These are the cross sections of the phase-sensitive DQF-COSY spectrum of chromazonarol at the ⬇ marked position. From (A), it can be determined that 7α proton is coupled with 7β, 6β and 6α proton. In particular the spin-spin coupling constant between 7α proton and 7β proton can be easily obtained because of its large antiphase doublet. Similarly in (B) and (D), the 6α proton is extracted from overlapped signals. A small spin-spin coupling yields a complex and fine pattern such as the 7α proton in (C). Thus a small spin-spin coupling can easily be distinguished from a large one and the magnitude of the coupling constant classified by this technique.

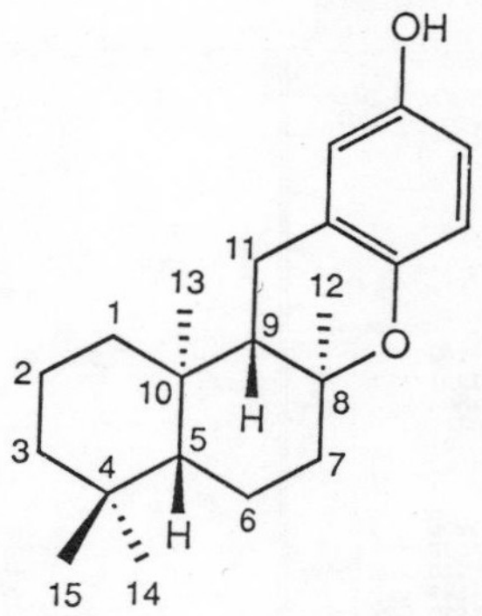

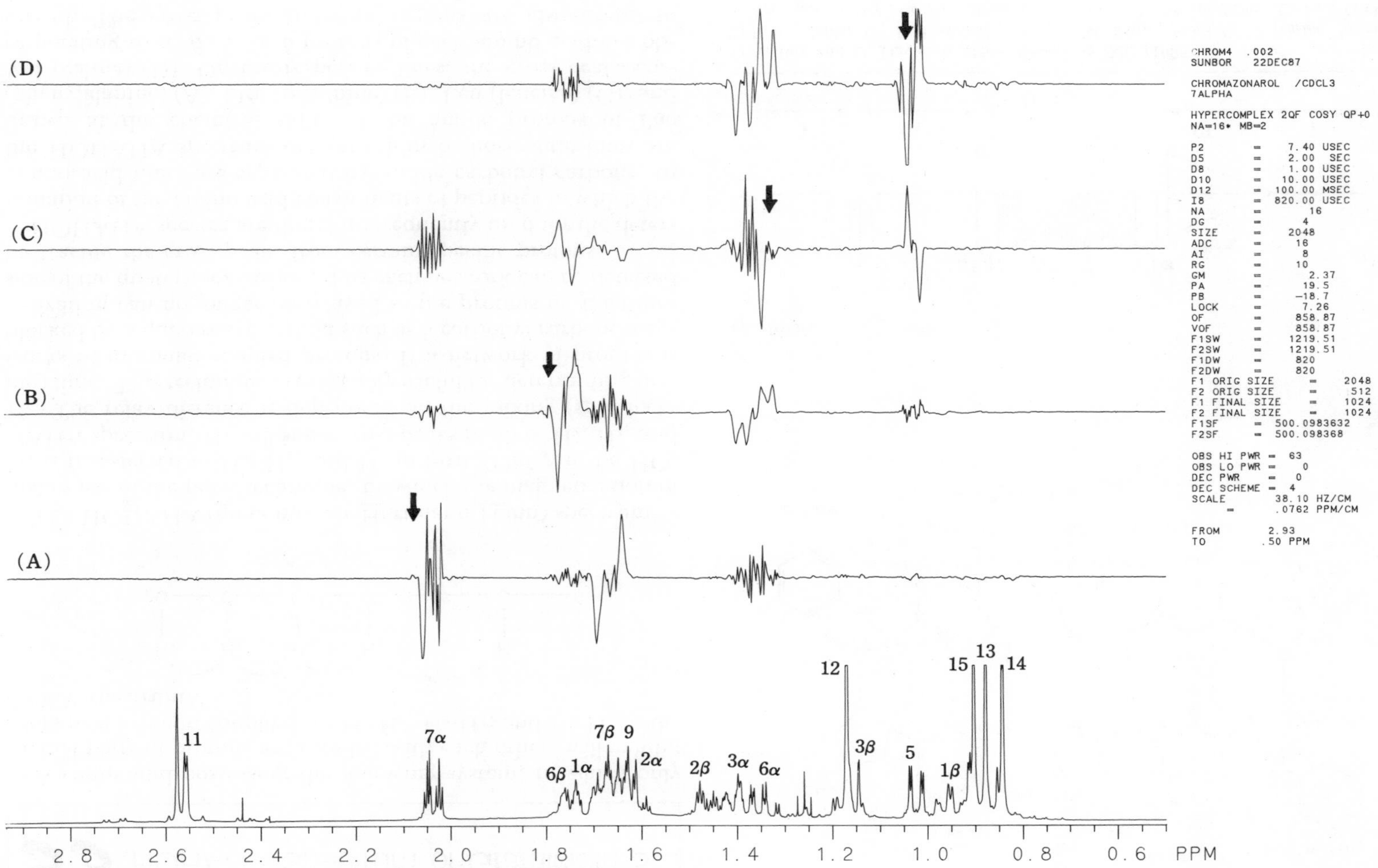

500 MHz ^{1}H-NMR ($CDCl_3$) [TI]

55 HOHAHA Spectrum of Gramicidin S

A compound possessing the following system, in which only vicinal pairs of protons are coupled with each other, will exhibit cross peaks due to couplings of H_1-H_2, H_2-H_3, and H_3-H_4 in the COSY spectrum.

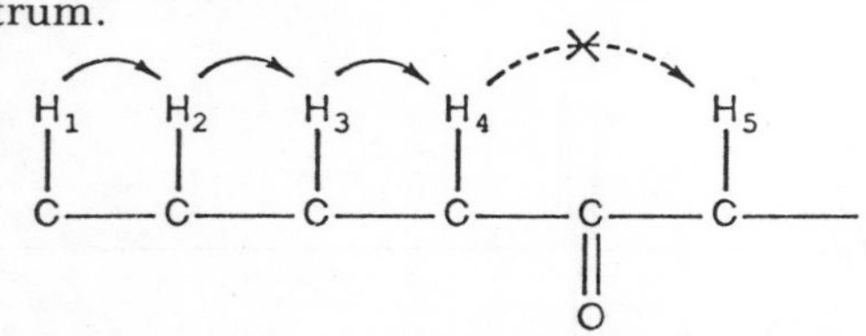

The HOHAHA (homonuclear Hartmann-Hahn) spectrum[1, 2] makes use of the pulse technique, by which the magnetization of H_1 is transferred to H_2, H_3, and H_4 in turn. Thus, in the HOHAHA spectrum, H_1 will show cross peaks to all of H_2, H_3, and H_4. The relay distance is dependent on the mixing (spin locking) time. This technique is especially useful for determining networks of mutually coupled protons. If a network of protons is blocked by a quaternary carbon such as a carbonyl carbon, magnetization can no longer be relayed to the protons on the other side of the quaternary carbon, thus each network can be detected by tracing the cross peaks from certain specific protons.

HOHAHA spectra are therefore frequently used for the determination of the amino acid constituents of peptides in which the amino acid units are separated by amide carbonyl carbons. In the HOHAHA spectrum of gramicidin S, horizontal lines are drawn at the chemical shifts of the amide protons of Phe (phenylalanine) (A), Orn (ornithine) (B), Leu (leucine) (C), and Val (valine) (D). On the respective lines, the cross peaks corresponding to α, β, γ, or δ protons of each amino acid are observed. The cross peaks in other regions are also useful to confirm the assignment.

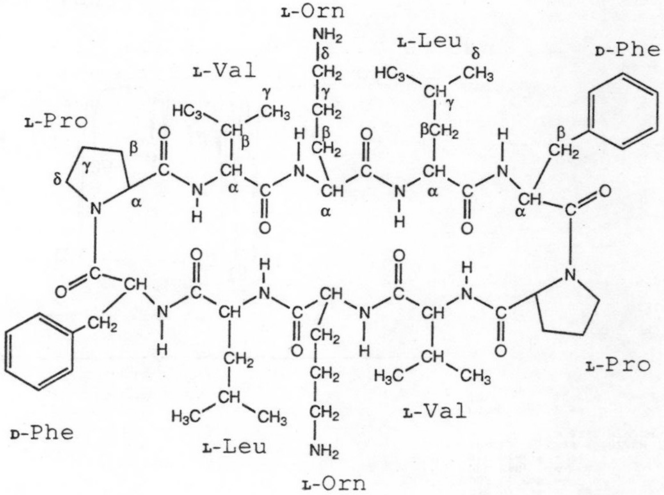

HOHAHA

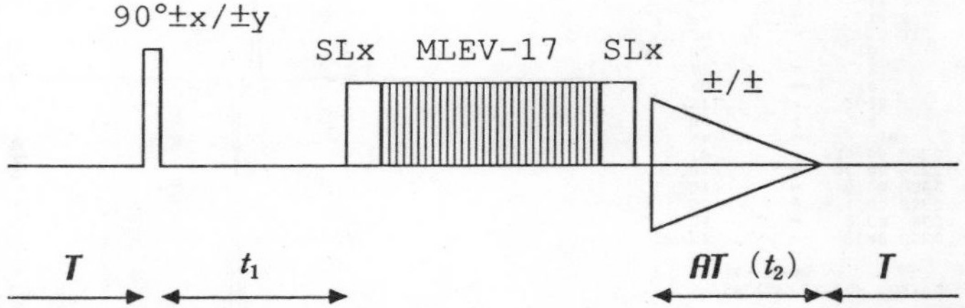

1) A. Bax and D. Davis, *J. Magn. Reson.*, **65**, 355 (1985)
2) R.R. Ernst, G. Bodenhansen and A. Wokaun, *Principles of Nuclear Magnetic Resonance in One and Two Dimensions*, Oxford University Press, Oxford (1987), pp.185-191, 444-448

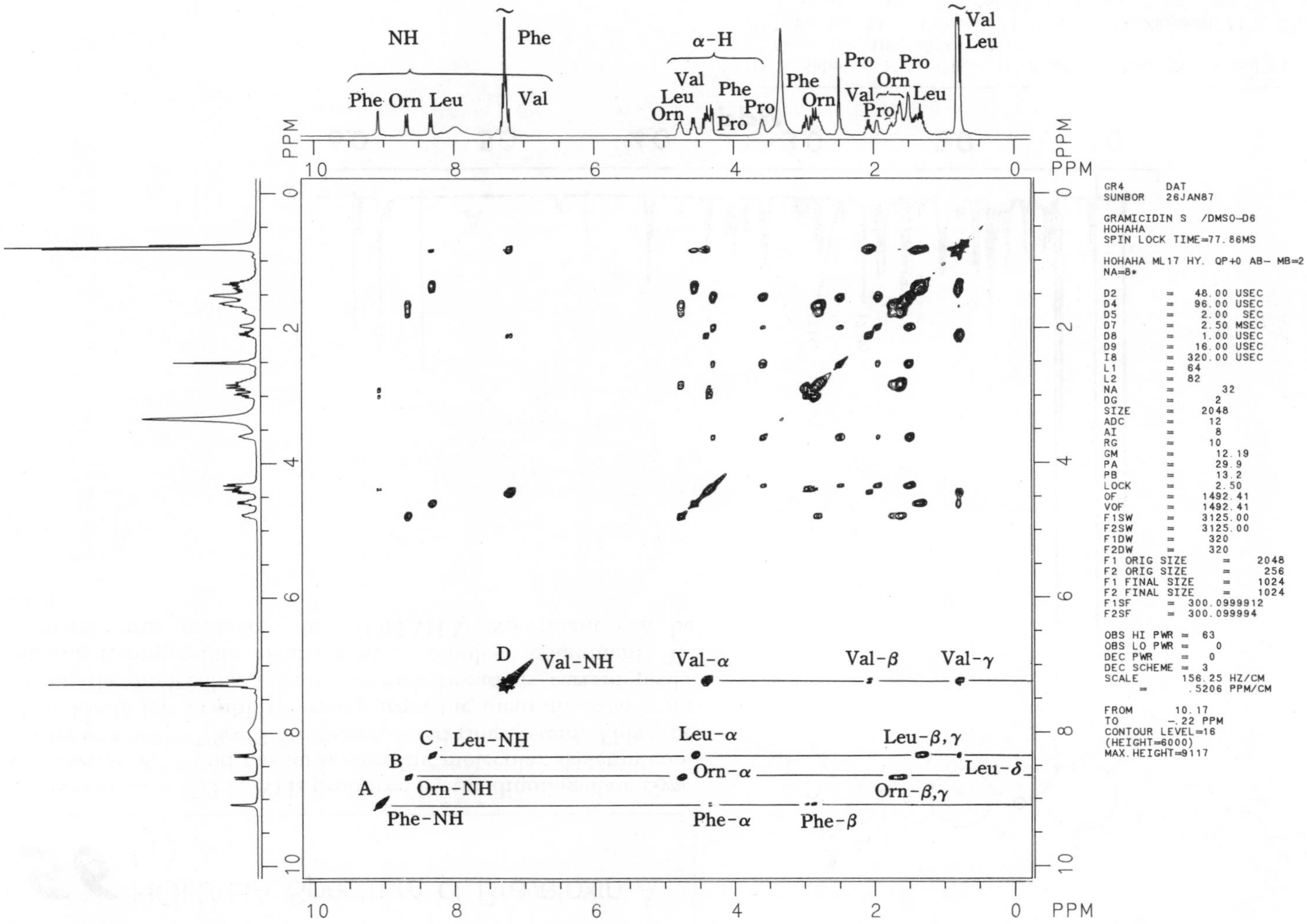

NH
Phe
Phe Orn Leu
Val
α-H
Val
Leu
Orn
Phe
Pro
Pro
Phe
Orn
Pro
Val
Pro
Orn
Pro
Leu
Val
Leu
PPM
GR4 DAT
SUNBOR 26JAN87
GRAMICIDIN S /DMSO-D6
HOHAHA
SPIN LOCK TIME=77.86MS
HOHAHA ML17 HY. QP+0 AB- MB=2
NA=8*
D2 = 48.00 USEC
D4 = 96.00 USEC
D5 = 2.00 SEC
D7 = 2.50 MSEC
D8 = 1.00 USEC
D9 = 16.00 USEC
I8 = 320.00 USEC
L1 = 64
L2 = 82
NA = 32
DG = 2
SIZE = 2048
ADC = 12
AI = 8
RG = 10
GM = 12.19
PA = 29.9
PB = 13.2
LOCK = 2.50
OF = 1492.41
VOF = 1492.41
F1SW = 3125.00
F2SW = 3125.00
F1DW = 320
F2DW = 320
F1 ORIG SIZE = 2048
F2 ORIG SIZE = 256
F1 FINAL SIZE = 1024
F2 FINAL SIZE = 1024
F1SF = 300.0999912
F2SF = 300.099994
OBS HI PWR = 63
OBS LO PWR = 0
DEC PWR = 0
DEC SCHEME = 3
SCALE 156.25 HZ/CM
= .5206 PPM/CM
FROM 10.17
TO -.22 PPM
CONTOUR LEVEL=16
(HEIGHT=6000)
MAX.HEIGHT=9117
D Val-NH
Val-α
Val-β
Val-γ
C Leu-NH
Leu-α
Leu-β,γ
B
Orn-α
Leu-δ
A Orn-NH
Orn-β,γ
Phe-NH
Phe-α
Phe-β

56 HOHAHA Spectrum of Brevetoxin A (1)

Brevetoxin A (BTX-A) is produced by the dinoflagellate *Gymnodinium breve*,[1, 2] and has an interesting molecular skeleton consisting of a 5/8/6/7/9/8/8/6/6/6-*trans*-fused ring system. This type of molecule has flexibility arising from the medium sized rings within the molecule, a feature which broadens certain peaks making it impossible to arrive at a complete assignment. To overcome this problem, the HOHAHA experiment can be used.[3)]

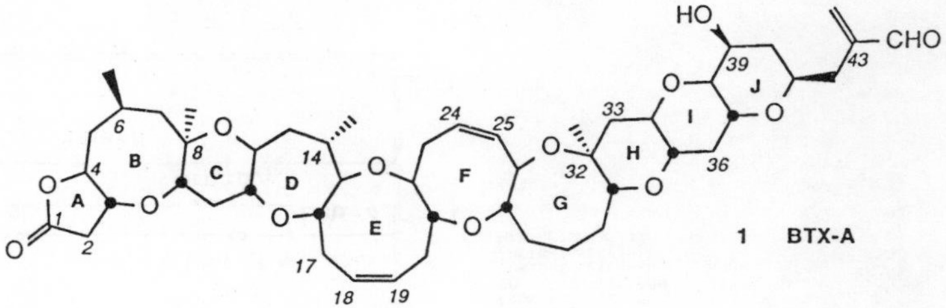

1 BTX-A

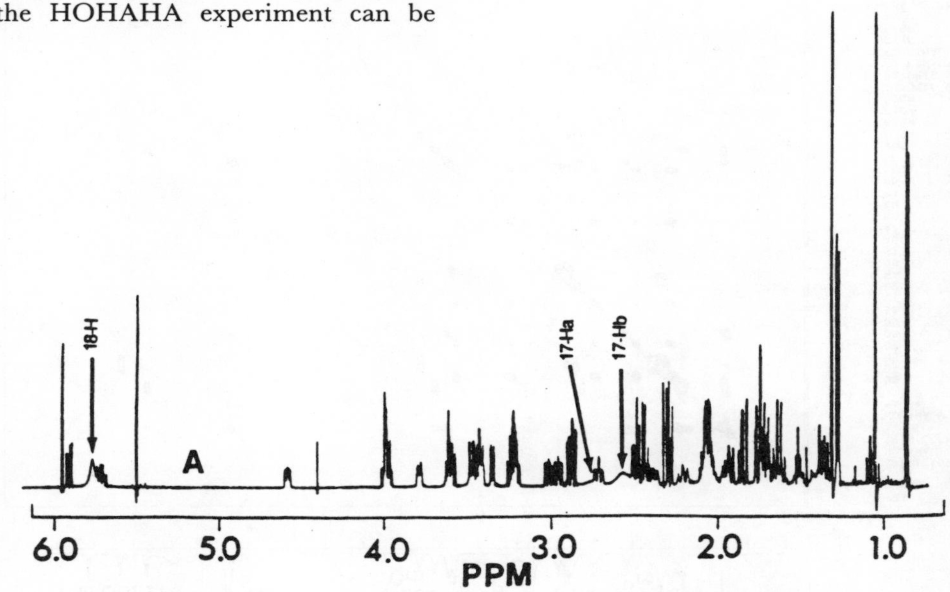

1) Y. Shimizu, H.N. Chou, H. Bando, G. Van Duyne and J.C. Clardy, *J. Am. Chem. Soc.*, **108**, 5614 (1986)
2) J. Pawlak, M.S. Tempesta, J. Golik, M.G. Zagorski, M.S. Lee, K. Nakanishi, T. Iwashita, M.L. Gross and K.B. Tomer, *J. Am. Chem. Soc.*, **109**, 1144 (1987)
3) M.S. Lee, K. Nakanishi and M.G. Zagorski, *New J. Chem.*, **11**, 753 (1987)

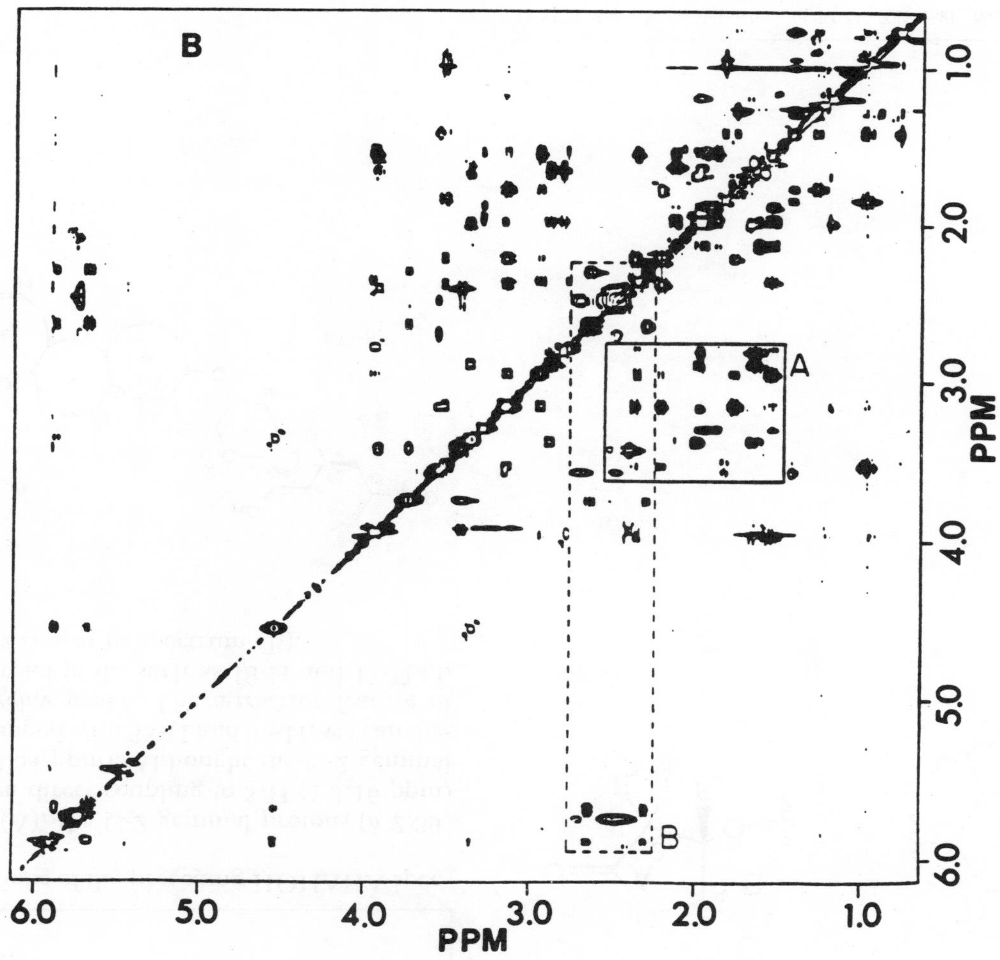

500 MHz ^{1}H-NMR [MZ][1)]

57 HOHAHA Spectrum of Brevetoxin A (II)

These are expanded regions of the preceding HOHAHA spectrum of brevetoxin A.

As shown in spectrum (A), the C-2 geminal protons (δ 2.39, 2.24 ppm) in ring A shown direct coupling to 3-H (δ 3.16 ppm) and also relay to 4-H (δ 3.54 ppm). Althought the C-2 geminal protons are severely overlapped with 33-H and 36-H, we can distinguish these peaks by relay peaks. The attractive feature of HOHAHA is that even broad peaks such as 18-H and 17-Ha,b reveal clear cross peaks as shown in spectrum (B).

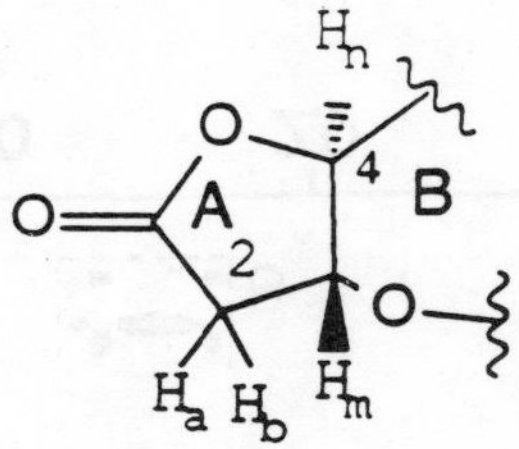

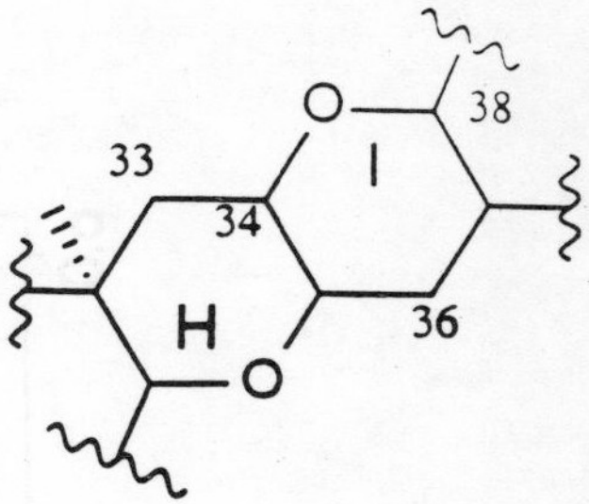

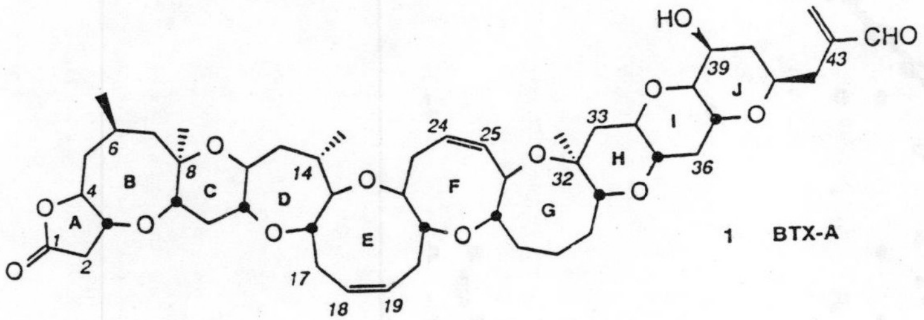

1 BTX-A

1) M.S. Lee, K. Nakanishi and M.G. Zagorski, *New Journal of Chemistry*, CNRS-Gauthier-Villans (1987), p.755

(A)

(B)

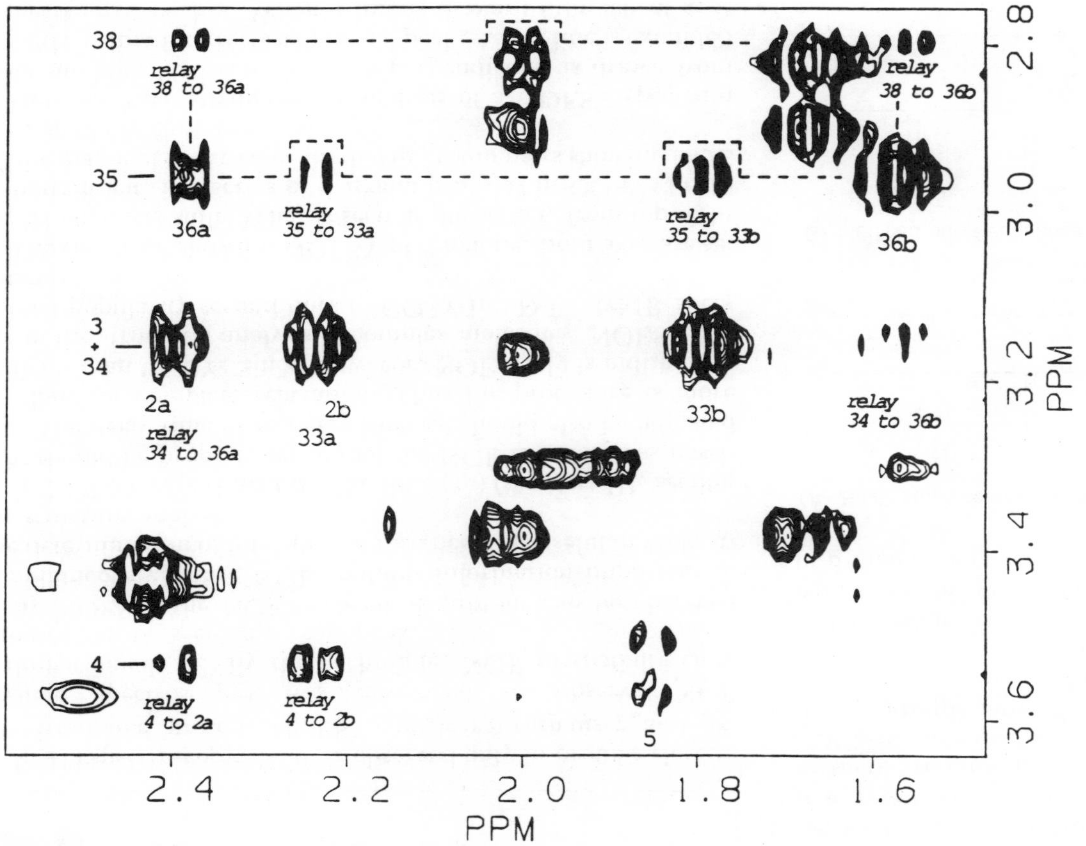

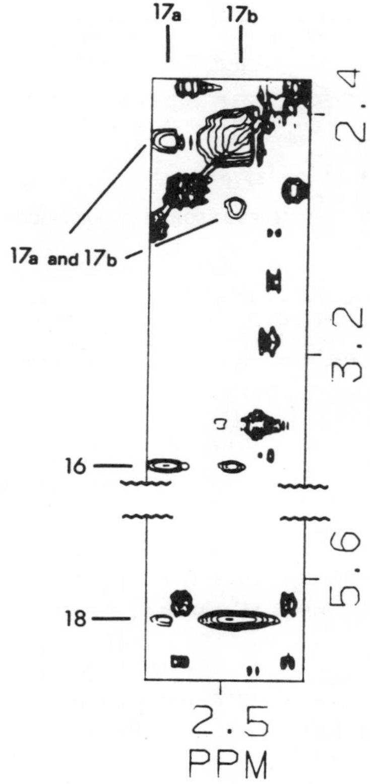

500 MHz ^{1}H-NMR [MZ][1)]

58 NOESY Spectrum of β-Ionone

In 1D spectroscopy, NOE is observed only in protons close to the irradiated proton. NOESY (nuclear overhauser and exchange spectroscopy) was developed to observe NOE 2-dimensionally.[1, 2] By this technique, NOE of spatially close protons can be seen as a cross peak.

By NOESY, the NOE between all protons can be observed simultaneously. Since NOE contains information important to the determination of inter-proton distance, it is useful in molecular structure analysis.

In NOESY experiment mixing time (τ_m) (see Part III, section 2.5) should be properly set so that the NOE is as large as possible. The delay time of recovery interval should also be adjusted to allow for complete relaxation. Thus the procedure is more difficult than COSY. Since, however, NOE study is indispensable in the structural analysis of complex molecules, NOESY has gained popularity second only to COSY in 2D-FT NMR spectroscopy.

The spectrum shown is NOESY of β-ionone. Both axes are set for ^1H chemical shift. What is seen at the center, from top right to bottom left, is a series of diagonal peaks as in COSY. Peaks off the diagonal are cross peaks due to proton pairs showing NOE (*i.e.*, spatially close protons).

Here we have attempted an analysis of a NOESY spectrum from the high field side. When a perpendicular is drawn from 1,1′-Me it first intersects a diagonal peak. As the line is extended it meets cross peak a. When a line is drawn to the left of a, it meets a diagonal peak. When a perpendicular is drawn upward from that point, it meets 3. This means that a is a cross peak which corresponds to the NOE between 1,1′-Me and 3-H. When the perpendicular from 1,1′-Me is further extended, it intersects cross peak b. When the same procedure as in a is applied, it is seen that b is a cross peak between 1,1′-Me and 8-H. Similarly c indicates an NOE between 1,1′-Me and 7-H. Thus by drawing perpendiculars down from the signals on the horizontal axis, it is seen that an NOE is obtained for the same number of signals as there are cross peaks.

When the same procedure is applied to 5-Me and 10, the methyl groups show NOE with the following protons:

5-Me: 4-H, 8-H, 7-H
10: 8-H, 7-H

The reason why olefinic protons 7-H and 8-H show an NOE with all methyl groups (1,1′-Me, 5-Me, 10) is that the side chain of β-ionone does not have a fixed conformation.

(A) Basic pulse sequence for NOESY

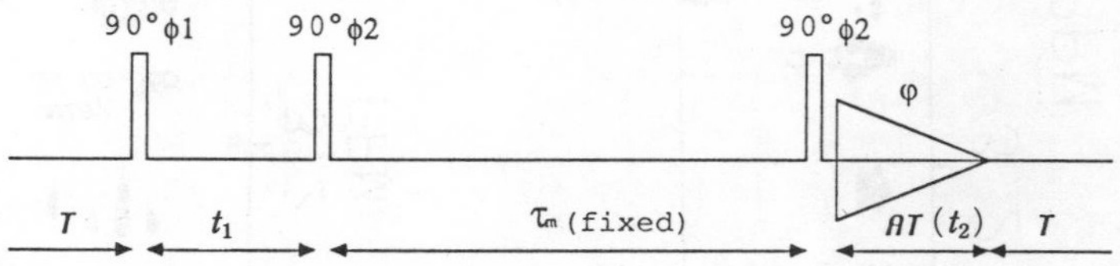

(B) Modified pulse sequence (suppression of *J* cross peaks)

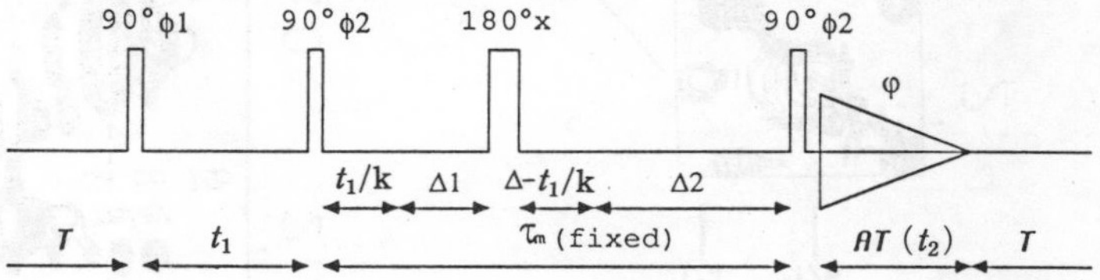

1) Ad Bax, *Two-Dimensional Nuclear Magnetic Resonance in Liquids*, Delft University Press, Delft, Holland (1982), pp.94-98

2) R.R. Ernst, G. Bodenhansen and A. Wokaun, *Principles of Nuclear Magnetic Resonance in One and Two Dimensions*, Oxford University Press, Oxford (1987), pp.490-538

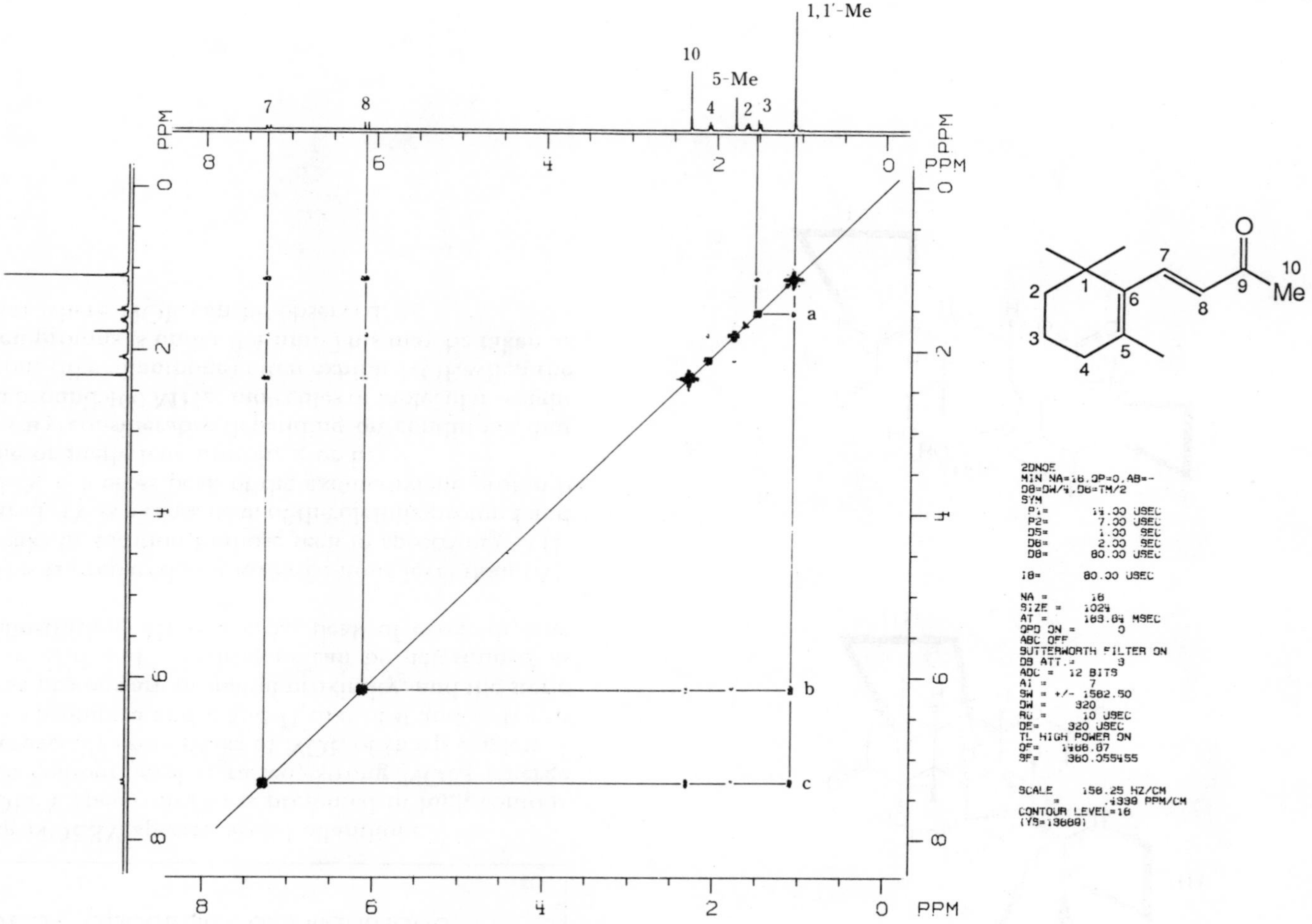

360MHz ¹H-NMR $(CDCl_3)$ [TI]

59 NOESY Spectrum of Ailanthone

The following NOESY spectra are of ailanthone.[1]

Here the NOESY spectrum (A) is presented in high contour level. When the contour level is raised, strong NOEs emerge preferentially, especially cross peaks of NOE of sharp singlets. I is a cross peak of protons a and c and II that of d and e. It can be seen that these protons are in spatial proximity, and the steric configurations of c, d and e hydrogens can be determined as shown in the illustration. III is a cross peak of exomethylene protons.

Spectrum (B) was prepared at a lower contour level than (A). Several cross peaks in addition to those seen in spectrum (A) (I-III) have appeared. IV is a cross peak of the olefinic proton f and the Me group i. V is a cross peak of the exomethylene proton b and the methine or methylene proton, g or h.

Spectra can vary considerably depending on conditions, but when studied at around 400 MHz, molecules of molecular weight of 200-400 daltons (like ailanthone) often exhibit NOE when the distance between protons is under 0.4 nm. This may be taken as a reference point where NOE can be observed.

(A)

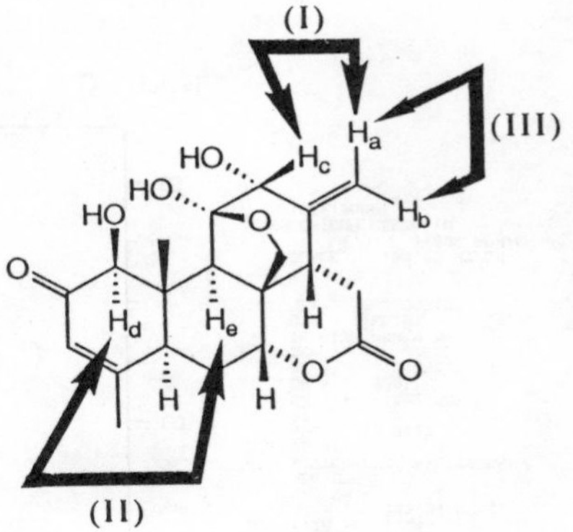

(B)

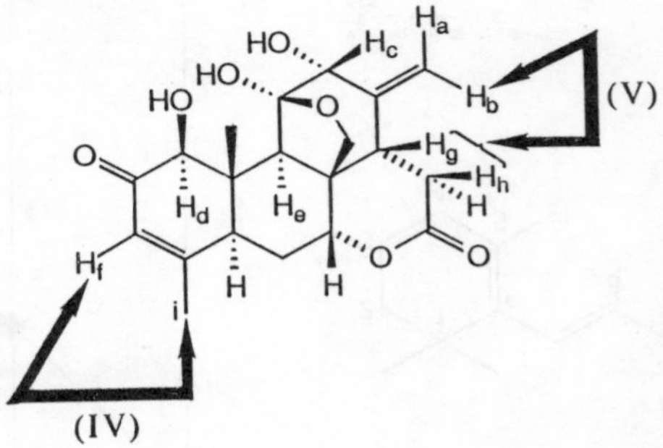

1) See section 12.

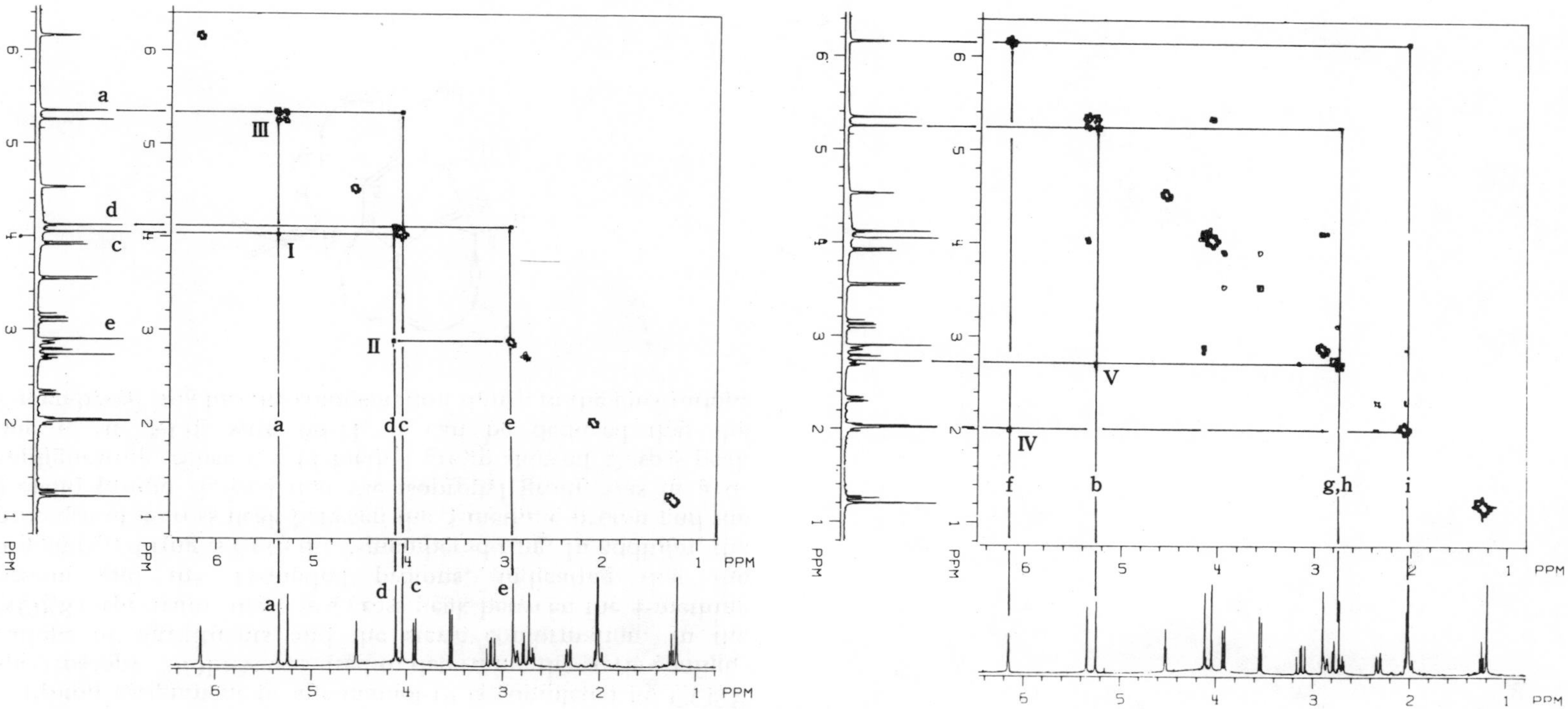

360 MHz ^{1}H-NMR ($CDCl_3$) [TI]

60 NOESY Spectrum of Aphanamol-I

Proton assignment in aphanamol-I[1] is completed by COSY spectroscopy. NOESY was used to determine the steric configurations of substituents and the steric conformation. In the NOESY spectrum, there is a cross peak between the 4-methine proton and the 11-methyl protons, indicating that the 5-membered ring is *cis* to the 7-membered ring. In addition, the presence of a cross peak between the 3-methine proton and the 5-olefin proton showed that the isopropyl group was in a β-configuration. Since the 11-methyl group showed a cross peak due to an NOE with 8β-H, it can be deduced that the 7-membered ring has the conformation shown in the illustration.

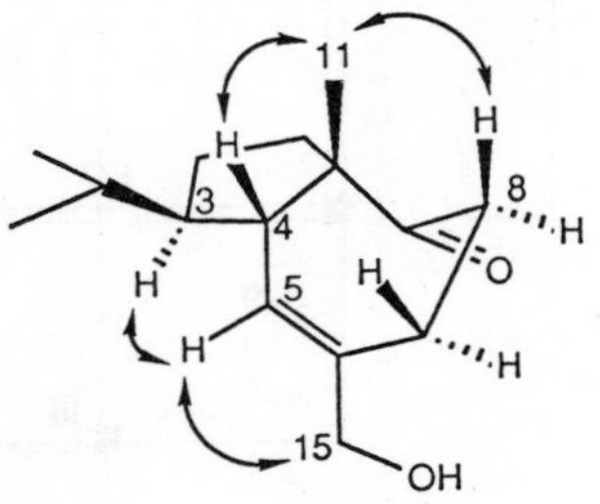

1) M. Nishizawa, A. Inoue, Y. Hayashi, S. Sastrapradja, S. Kosela, T. Iwashita, *J. Org. Chem.*, **49**, 3660 (1984)

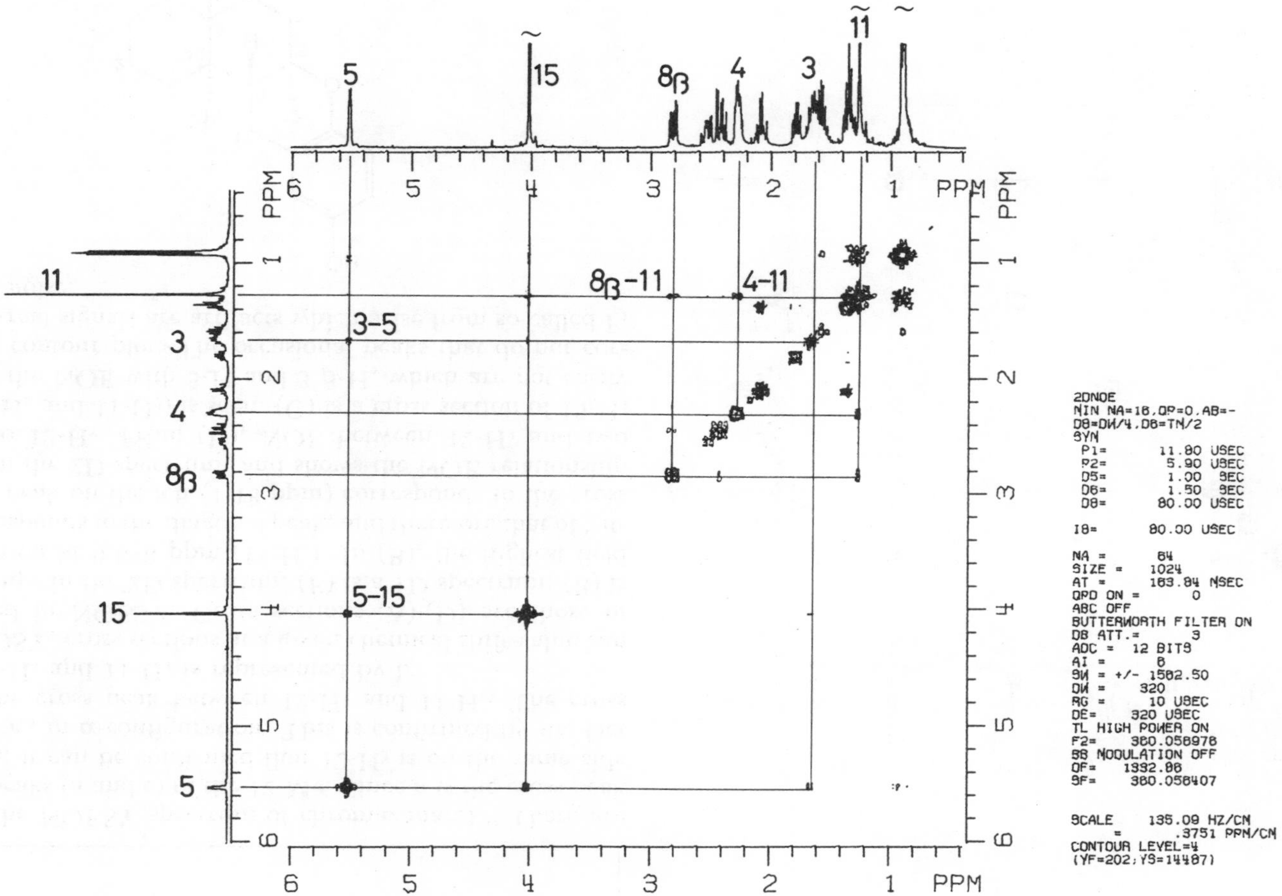

$\tau_m = 3$ sec

360MHz ^{1}H-NMR ($CDCl_3$) [TI]

61 NOESY Spectrum of Chromazonarol

This is the NOESY spectrum of chromazonarol.[1] There are two cross peaks (a and c) of the 12-Me. Since a is the cross peak with 13-H_3, it can be confirmed that 12-H_3 is on the same side as 13-H_3, *i.e.*, in α-configuration. This is confirmed by the fact that c is the cross peak between 12-H_3 and 11-H_2. The cross peak of 13-H_3 and 11-H_2 is represented by b.

As in COSY, cross sections at a given chemical shift value can be prepared in NOESY. Cross sections (A)-(D) are those of methyl groups in the 2D spectrum. (E) is a 1D spectrum. (B) is a cross section at 0.878 ppm (13-H_3). In (B), the highest field signal corresponds to the diagonal peak, and therefore that of "itself." The peak on the left (1.18 ppm) corresponds to the cross peak of a in the 2D spectrum, and shows the NOE relationship of 13-H_3 to 12-H_3. From (D), NOE between 12-H_3 and two others (13-H_3 and 11-H_2) is seen. (C) is a cross section of 15-H_3 and shows the NOE with 5-H and 3 β-H, which are not easily found in a contour plot. The occasional peaks that do not correspond to real signals are artifacts which arise from so-called F_1 noise or t_1 noise.

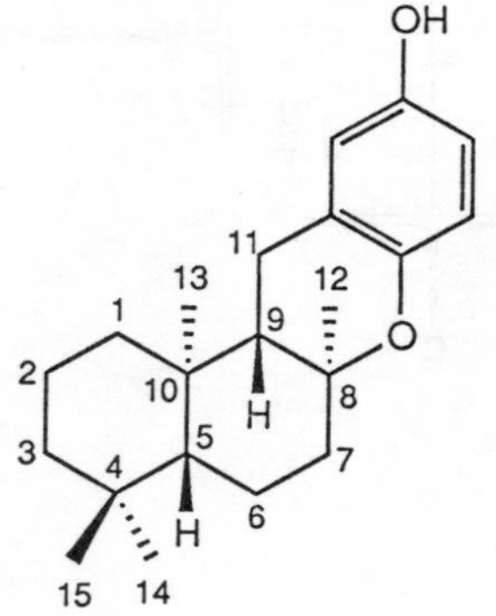

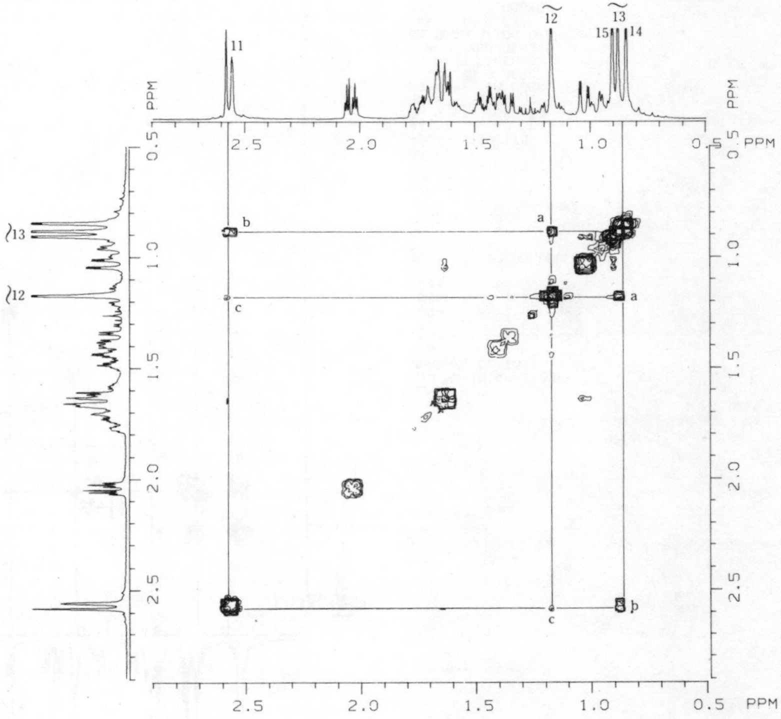

1) See section 41.

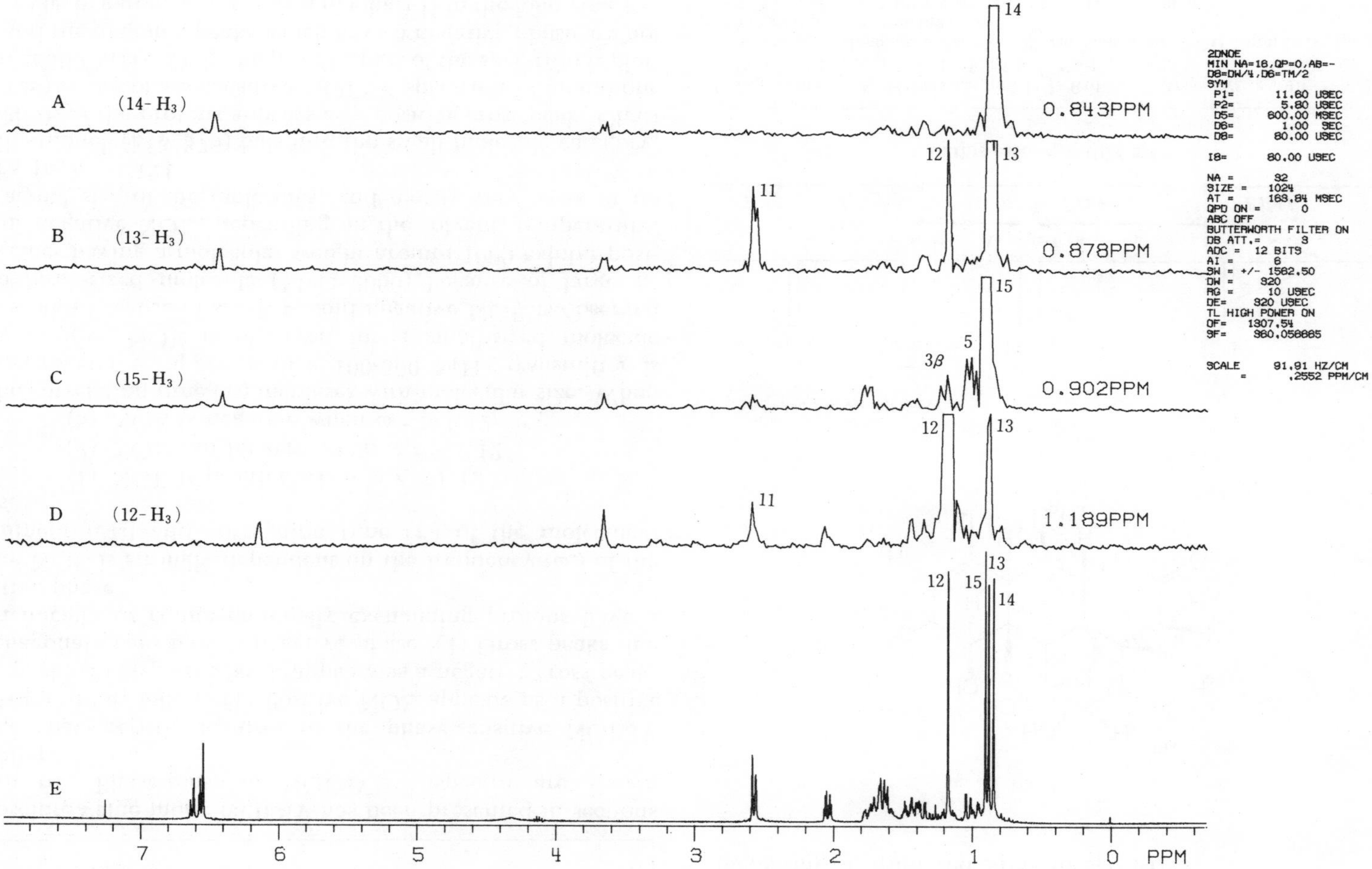

360MHz ^{1}H-NMR ($CDCl_3$) [TI]

62 Phase-sensitive NOESY Spectrum of Ailanthone (Ⅰ)

Absolute value mode NOESY has been presented in sections 58 to 61. Phase-sensitive NOESY[1, 2] spectra are shown hereafter.

The characteristic features of the phase-sensitive NOESY spectrum are as follows (1) Positive NOE appears as a positive cross peak. (2) Negative NOE appears as a negative cross peak. (3) Diagonal peaks have a negative phase. (4) Cross peaks due to chemically or conformationally exchanging protons have a negative phase.

The NOE is strongly dependent on the frequency (ω_o) of the instrument used and correlation time (τ_c) of the molecule.[3] Thus,

(1) NOE is positive when $\omega_o\tau_c < 1.12$
(2) NOE can be zero when $\omega_o\tau_c = 1.12$
(3) NOE is negative when $\omega_o\tau_c > 1.12$

The correlation time (τ_c) increases with molecular size. When a spectrometer equipped with a 100-500 MHz transmitter is used, positive NOE is observed for a small-sized molecule (Mw < 500) because of small τ_c, and negative NOE is observed for a large-sized molecule (Mw > 2000) because of large τ_c. Molecules having a molecular weight around 1000 exhibit positive or negative NOEs depending on the solvent, temperature, and actual size of the molecules, and mostly very weak or no NOEs $\{\omega_o\tau_c = 1.12\}$.

Ailanthone[4] (Mw 376) falls into the small molecule category, and NOE of the protons appears as a positive cross peak. Chart I (p.129) is the phase-sensitive NOESY spectrum of ailanthone taken at 300 MHz. Only the positive part of the spectrum is plotted, and the diagonal peaks which have a negative phase are not seen. (The negative part is shown in Chart II in the following sections.) A good number of cross peaks due to NOEs are observed.

The mark, a-b, means the NOE between proton a and b. When the spectrum was measured at 500 MHz, very few NOE peaks were present. This means that $\omega_o\tau_c$ becomes closer to 1.12 on increasing ω_o from 300 MHz to 500 MHz.

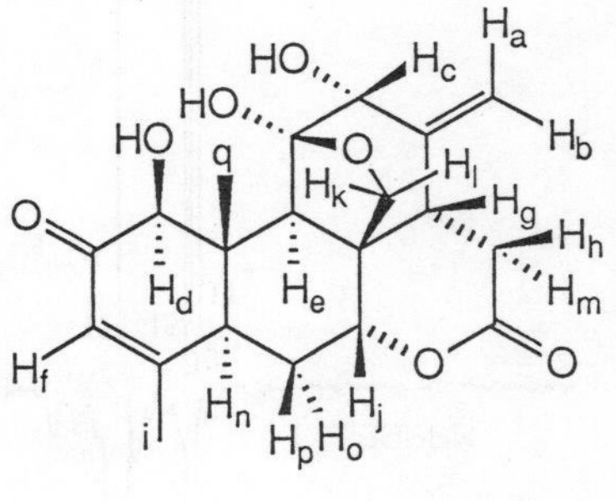

Phase-sensitive NOESY

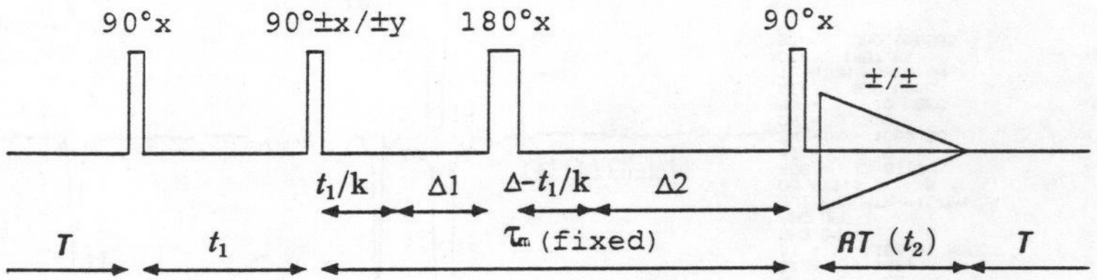

In this case, τ_m=1026 ms

1) D.J. State, R.A. Haberkorn and D.J. Ruben, *J. Magn. Reson.*, **48**, 286 (1982)
2) W.R. Croasmun and R.M.K. Carlson (Ed.), *Two-Dimensional NMR Spectroscopy—Applications For Chemists and Biochemists*, VCH Publishers, Inc., New York (1987), pp.153-162
3) S. Macura and R.R. Ernst, *Mol. Phys.*, **41**, 95 (1980)
4) See section 12.

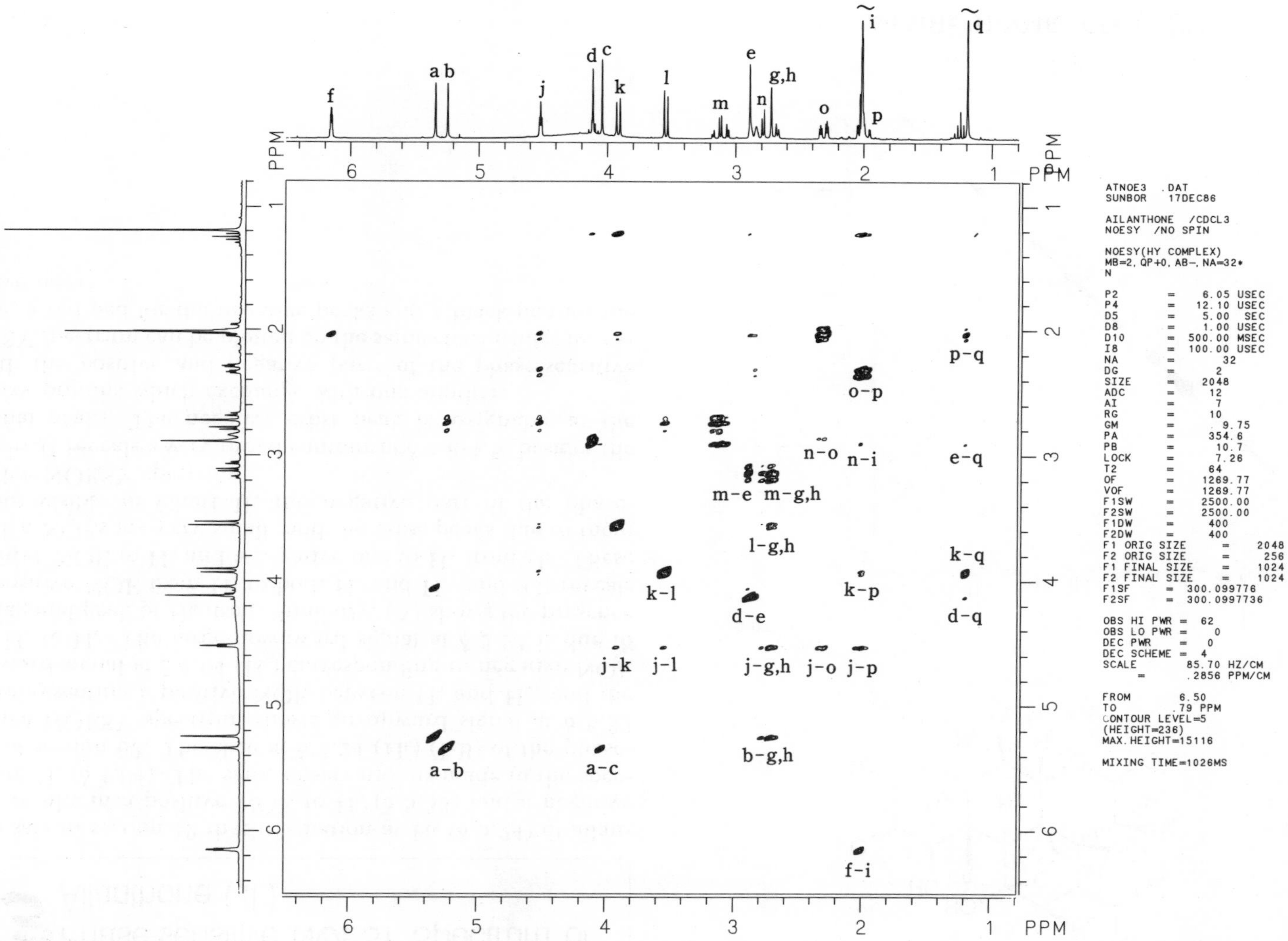

300 MHz ^{1}H-NMR ($CDCl_3$) [TI]

63 Phase-sensitive NOESY Spectrum of Ailanthone (II)

We saw in section 12 that irradiation at H_b (δ 5.24) of ailanthone results in a positive NOE to H_a (δ 5.33) and a negative NOE to H_c (δ 4.04). The same observation is made in the spectrum of section 62. The slice at δ 5.24 (H_b) (I-B) of the phase-sensitive NOESY spectrum shows an upward signal at δ 5.33 (H_a) representing a positive NOE between H_a and H_b, and the downward signal at δ 4.04 (H_c) corresponding to negative NOE from H_a to H_c. The large downward signal at δ 5.24 is due to the diagonal peak of H_b itself. Similarly, (A) shows the presence of a positive NOE from H_a to both H_b and H_c, and (C) reveals a positive NOE to H_a and a negative one to H_b from H_c. These negative NOEs are very small, and the cross peaks due to them are not visible in Chart II, the negative part of the phase-sensitive NOESY spectrum.

Chart II reveals a very broad contour at δ 3.8-4.5, besides the diagonal peaks. The negative cross peak is assignable as the hydroxy protons which exchange with one another.

Both the positive and negative parts of the phase-sensitive NOESY spectrum can be plotted on the same chart using, for example, a red pen for the negative peaks and a black pen for the positive ones.

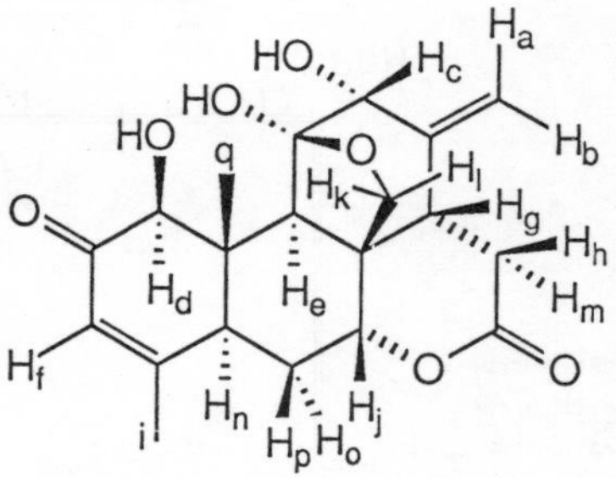

(Chart II)

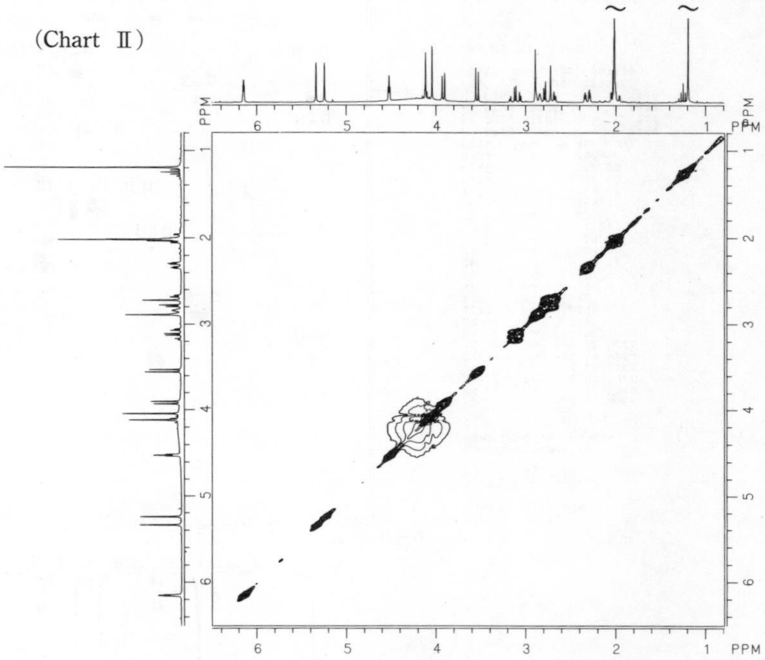

300 MHz ^1H-NMR ($CDCl_3$) [TI]

(Chart I)

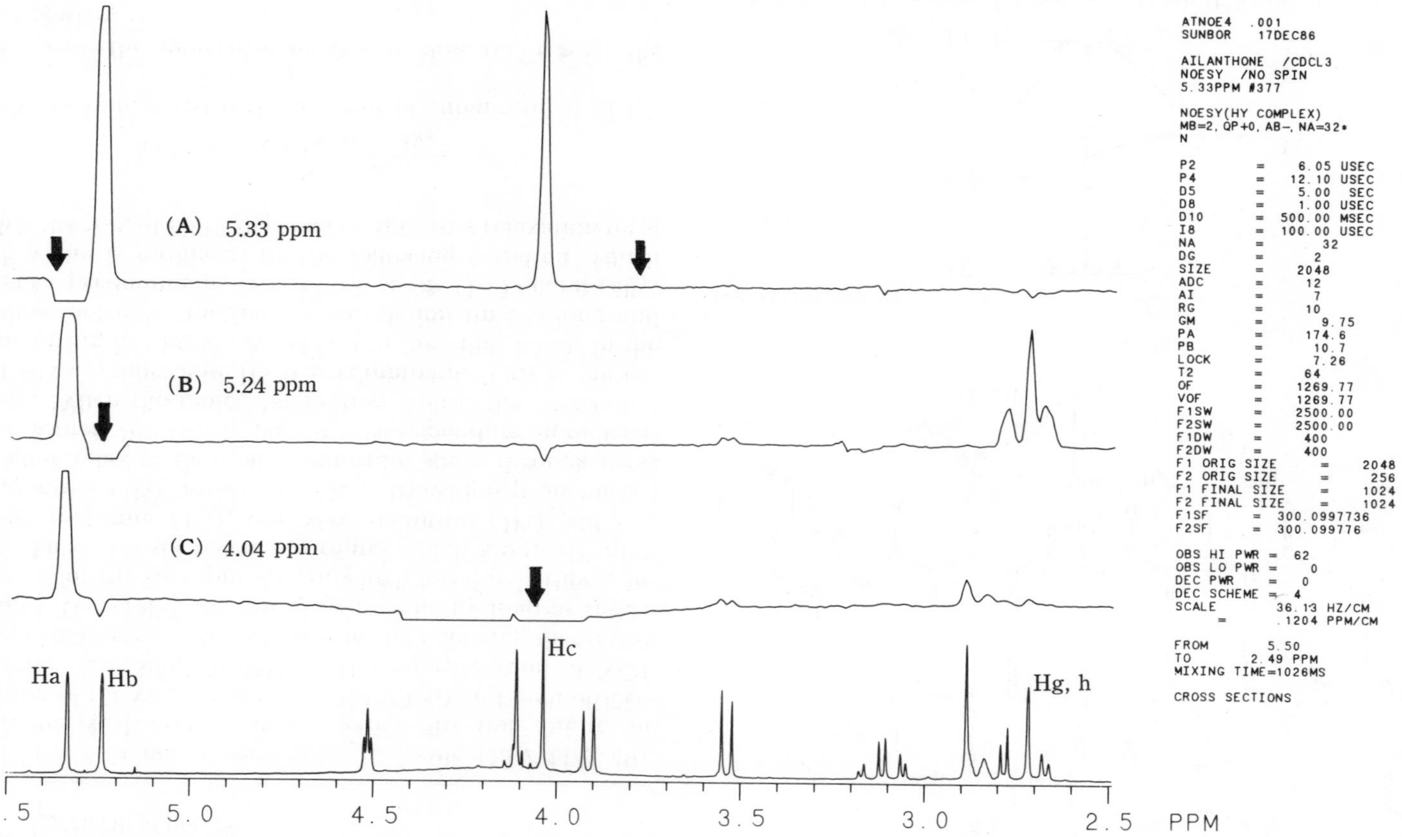

300 MHz ^{1}H-NMR ($CDCl_3$) [TI]

64 Phase-sensitive NOESY Spectrum of Gramicidin S

Gramicidin S is not a very large molecule (M_W 1141.49). However, the NOE appears in a negative direction due to the combination of the viscous solvent (DMSO-d_6) and high observing frequency (ω_o: 300-500 MHz). The interpretation of NOE which appears in section 62 is based on the following theoretical background. Generally, we can draw the energy map as shown in the figure for the two-spin system which has dipole-dipole interaction. There are six relaxation routes which are made up of four single quantum (W_1), one zero quantum (W_0) and one double quantum (W_2) transitions. (W is transition probability.) Excited spins relax to thermal equilibrium states through these relaxation routes. But two situations exist depending on molecular motion. When the molecular motion is fast, the correlation time (τ_c) is very short and W_2 is predominant. This is the extreme narrowing condition ($\omega_o\tau_c \ll 1$). On the other hand, in the case of slow molecular motion the correlation time is long and W_0 becomes predominant ($\omega_o\tau_c \approx 1$ or $\omega_o\tau_c \gg 1$). Theoretically, the NOE value is estimated by the following equation, which means that the sign of NOE depends on the cross relaxation term W_2-W_0.[1, 2]

$$f_I(S) = \frac{W_2 - W_0}{2W_1^I + W_0 + W_2}$$

(I and S are spin, f_I (S) is the amount of enhancement of I)

So in this case the molecular motion is slow ($\omega_o \tau_c \gg 1$); the NOE is negative.

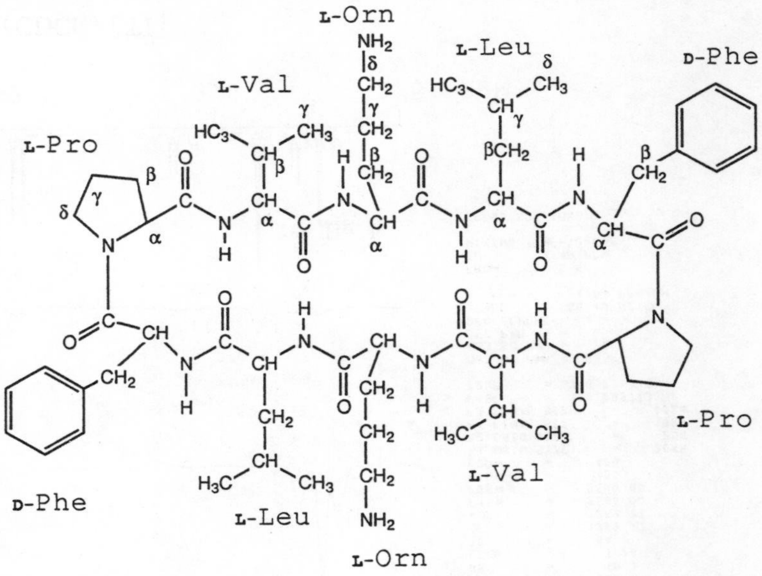

Relaxation routes

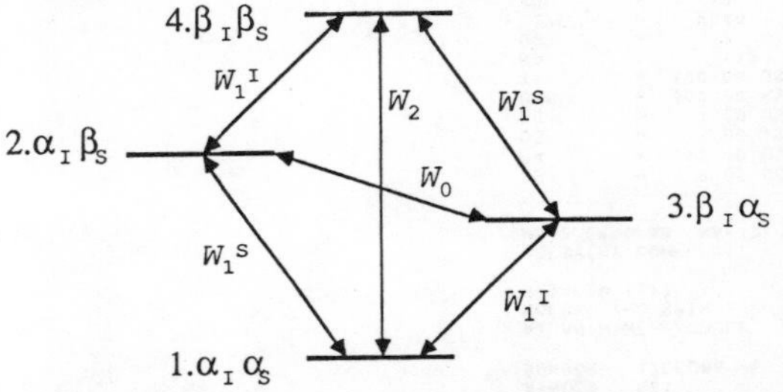

Energy level diagram for the two–spin system (I, S) which has dipole–dipole interaction. $\alpha_I\beta_S$ indicates that spin I is α and spin S is β.

1) J.H. Noggle and R.E. Schirmer, *The Nuclear Overhauser Effect Chemical Applications*, Academic Press, New York (1971), p.16

2) See Part III.

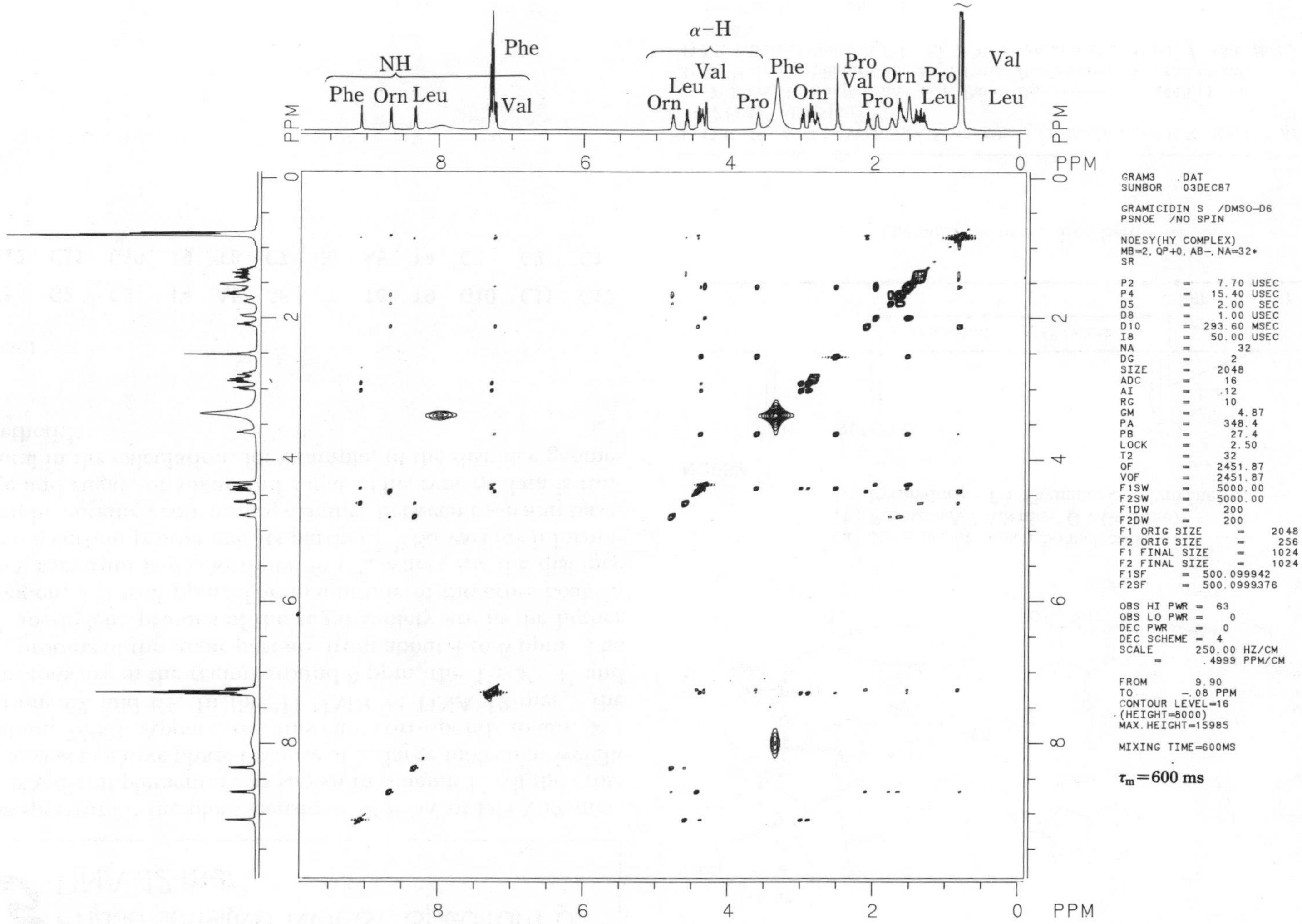

500 MHz ^{1}H-NMR (DMSO-d_6) [TI]

65 Phase-sensitive NOESY Spectrum of DNA 12-mer

This spectrum is the phase-sensitive NOESY of DNA 12-mer, which is self-complementary as shown in scheme 1. All the cross peaks have a negative phase because of its large molecular weight (M_W about 7400). Apparently, this case corresponds to $\omega_0\tau_c \gg 1$ in sections 62 and 64. In the ^{1}H-NMR of DNA 12-mer,[1] the base protons are in the region around 8 ppm, the 1′, 3′, 4′ and 5′, 5″ protons of the sugar part are from about 4 to 6 ppm. The 2′, 2″ methylene protons of the sugar moiety are in the higher field region, 1.5 to 3 ppm. The magnitude of the cross peak in NOESY spectrum is proportional to r^{-6}, where r is the distance between a certain proton and its partner.[2,3] So various information can be obtained concerning distance between base and base, or base and sugar, or sugar and sugar. This type of data is fundamental in the calculation, for example, of the distance geometry method.[4]

Scheme 1

C1	G2	C3	T4	A5	G6	C7	T8	T9	G10	C11	G12
G12	C11	G10	T9	T8	C7	G6	A5	T4	C3	G2	C1

Scheme 2

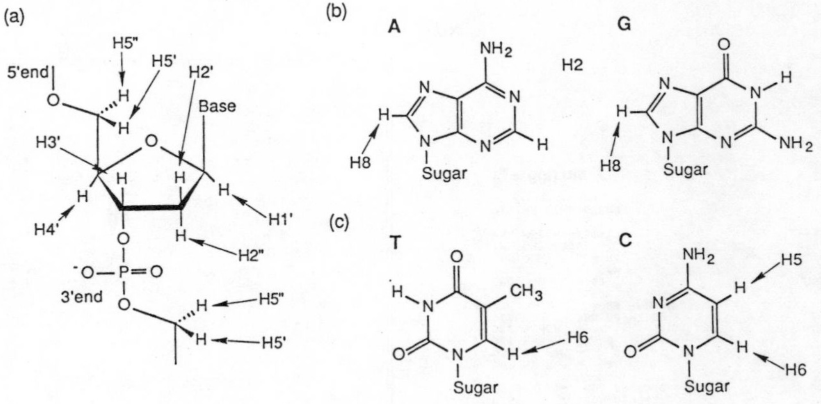

(a) Structure of deoxyribonucleotide
(b) Purines (A : Adenine, G : Guanine)
(c) Pyrimidines (T : Thymine, C : Cytosine)

NOESY

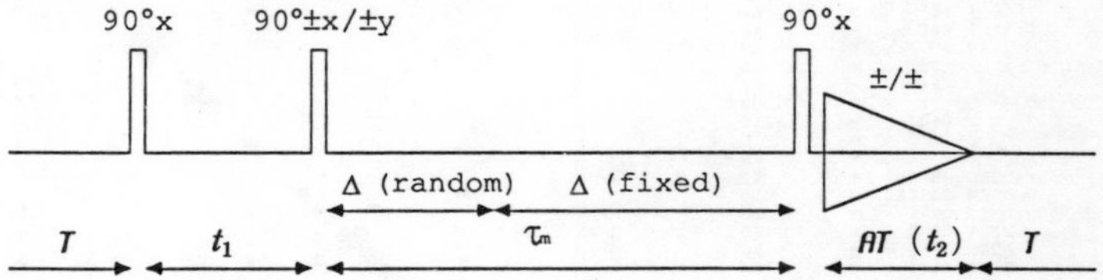

The pulse sequence used here[5]

1) D.R. Hare, D.E. Wemmer, S.-H. Chou, G. Drobny and B.R. Reid, *J. Mol. Biol.*, **171**, 319 (1983)
2) D. Hare, L. Shapiro and D.J. Patel, *Biochemistry*, **25**, 7445 (1986)
3) D. Hare, L. Shapiro and D.J. Patel, *Biochemistry*, **25**, 7456 (1986)
4) M. Kouchakdjian, B.F.L. Li, P.F. Swan and D.J. Patel, *J. Mol. Biol.*, **20**, 139 (1988)
5) See Part III, 2.5 and 2.6.

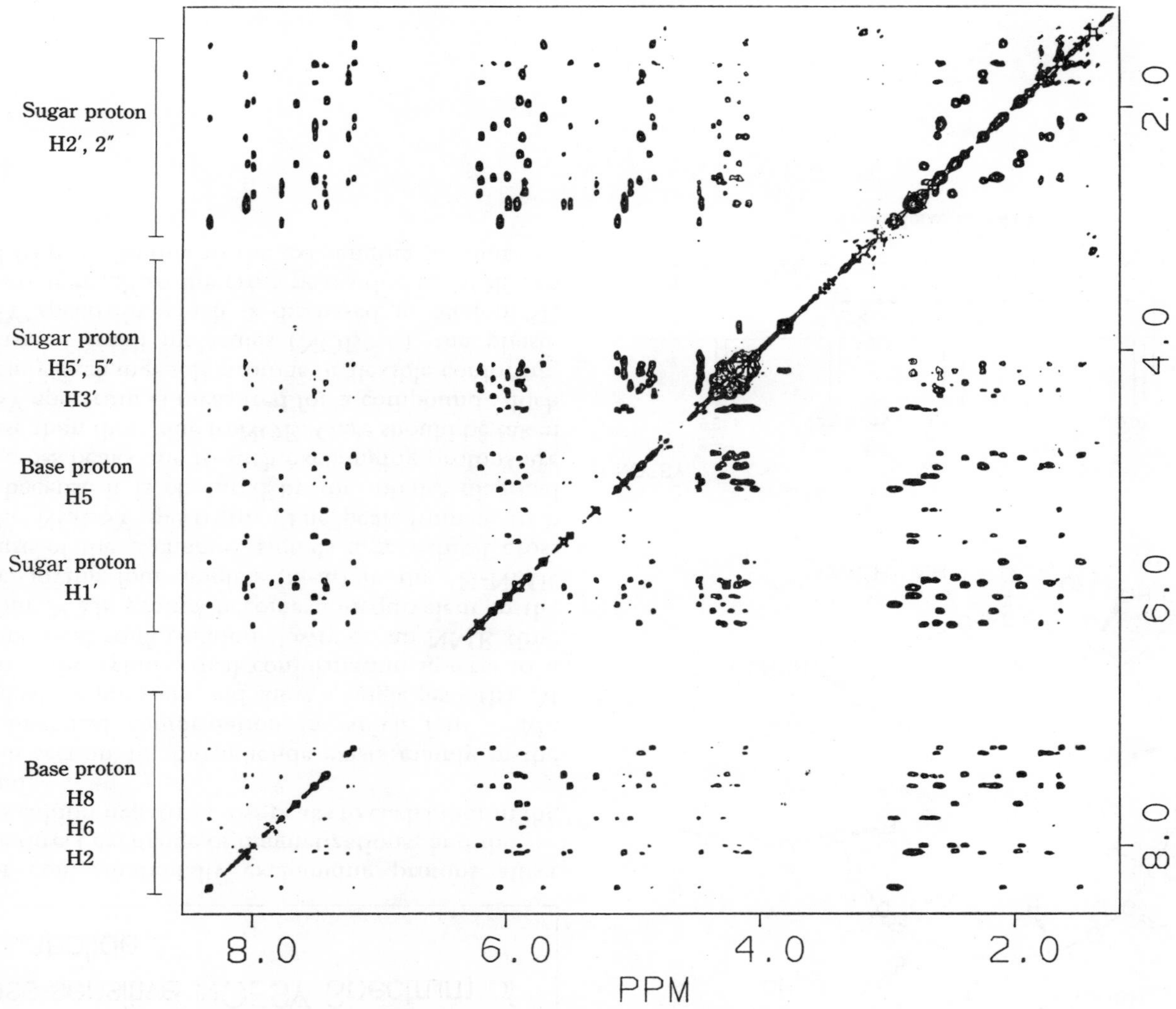

500 MHz ^{1}H-NMR (2.1 mM in D_2O Mixing time=250 ms) [MZ]

66 Phase-sensitive NOESY Spectrum of Bassianolide

Chemically or conformationally exchanging protons affect each other by the direct exchange of magnetizations, and the exchanging protons exhibit negative cross peaks to each other in the NOESY spectrum.

As discussed in section 13, bassianolide exists mainly in the most stable symmetrical conformation in which four *N*-Me groups are chemically equivalent and show a single peak (b). At room temperature, the symmetrical conformation inverts to a less stable asymmetrical conformation slowly on an NMR time scale, and the four *N*-Me groups become nonequivalent in the latter conformer, giving four singlets (a_1-a_4) in the ^{1}H-NMR spectrum. Because of the exchange, signals a_1-a_4 exhibit cross peaks to b in the NOESY spectrum. (The peak from a_3 to b cannot be seen because it is obscured by the intense diagonal peak of b.) The cross peaks due to such exchanging protons are often more intense than those due to NOE. Care should be taken when the NOESY spectrum is measured for a compound which possesses chemically exchangeable protons or flexible conformations. In the case of small molecules (NOE>0), the phase-sensitive NOESY spectrum which is discussed in section 61, gives a more clearcut result so the cross peaks due to NOE can be distinguished from those due to the exchanging protons.

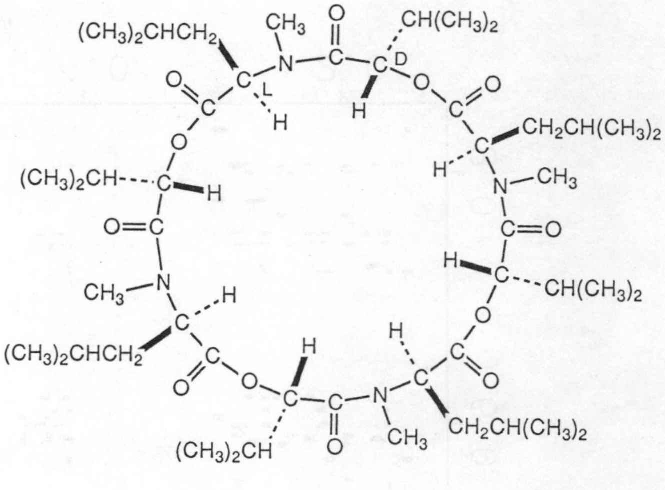

NOESY

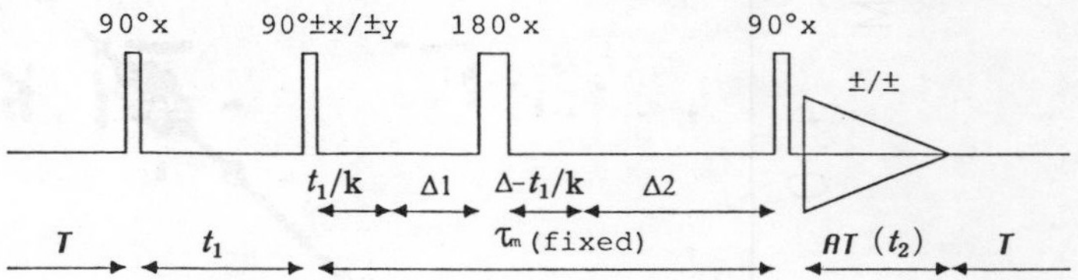

Here, τ_m=414 ms

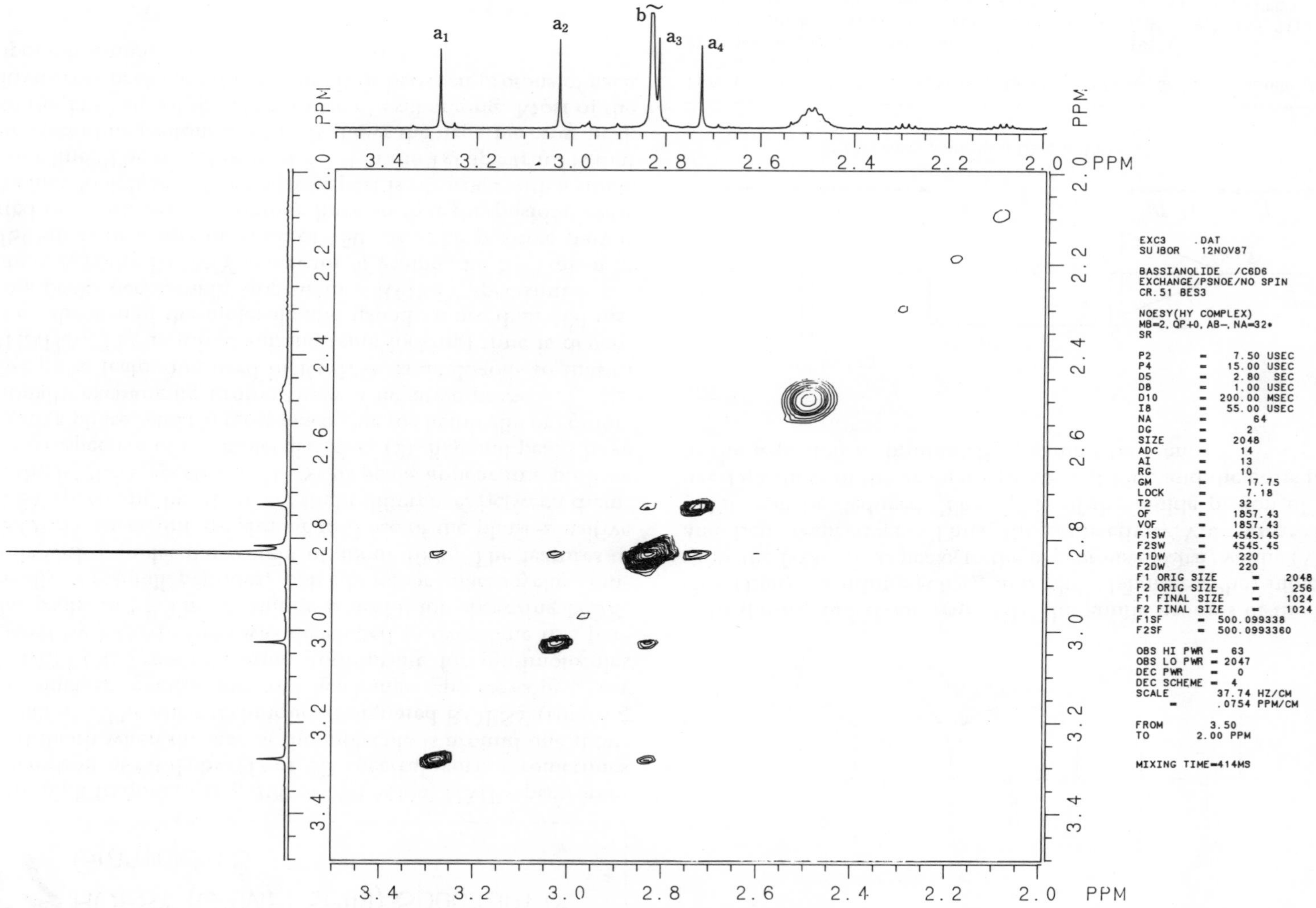

500 MHz ^{1}H-NMR (C_6D_6) [TI]

67 ROESY (CAMELSPIN) Spectrum of Gramicidin S

With high frequency (*e.g.* 400 or 500 MHz) NMR spectrometers detection of NOE by 1D or 2D spectral work is sometimes quite difficult when the size of the molecule is around one thousand daltons. The pulse technique, designated ROESY (rotating frame nuclear overhauser and exchange spectroscopy), (or CAMELSPIN: cross-relaxation appropriate for minimolecules emulated by locked spins) was developed to overcome this particular problem.[1, 2] The technique is useful for observing NOE, especially for small peptides, polyethers, or macrocyclic compounds with a molecular weight of about 1000. The features of the ROESY spectrum are similar to those of the phase-sensitive NOESY spectrum, but there are slight differences between them.

In the ROESY spectrum, (1) NOE peaks appear in a positive phase irrespective of the molecular size, (2) diagonal peaks have a negative phase, and (3) cross peaks due to chemically or conformationally exchanging protons have a negative phase.

The pulse technique used in ROESY is analogous to that of HOHAHA. The required mixing (spin-locking) time is dependent on the size of the molecule and usually more than 100 ms. *J* cross peaks occasionally appear in a ROESY spectrum.

Chart A is the ROESY spectrum of gramicidin S[3, 4] taken in DMSO-d_6 with a mixing time of 150 ms. The positive part is plotted out with several contour lines so that the positive cross peaks may look black. The negative part is recorded with a single contour line. The broad peak at δ 7.9 in the 1D-spectrum is that of the guanidine protons of Orn. It shows the negative cross peak (a) to the H_2O signal (δ 3.3) because of exchanging. Most of the positive cross peaks are due to the NOE between protons of each component amino acid.

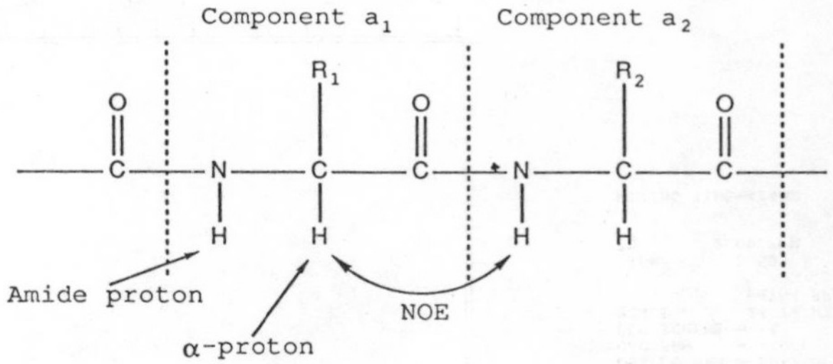

In the expanded spectrum (B), the amide protons of the leucine (Leu), ornithine (Orn), and phenylalanine (Phe) moieties show the NOE cross peaks to the α-protons of Orn, valine (Val), and Leu, respectively. Thus, the connectivity Val→Orn→Leu →Phe can be deduced. The signal of the amide proton of Val overlaps those of the aromatic protons of Phe, and the cross peak to the α-proton of proline (Pro) cannot be seen.

ROESY

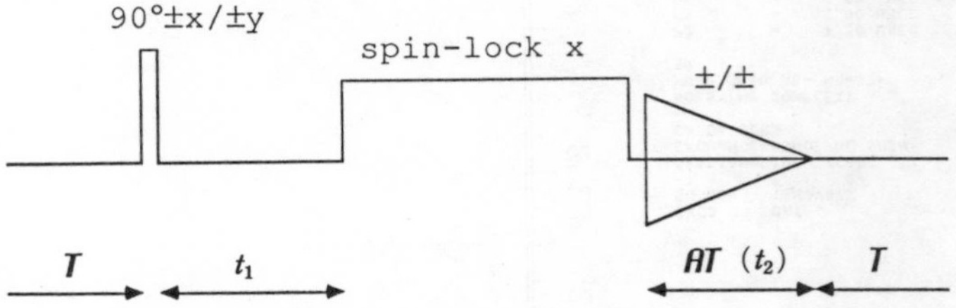

In this case, spin-lock time is 150 ms.

1) A.A. Bothner-By, R.L. Stephens, J. Lee, C.D. Warren and R.W. Jeanlo, *J. Am. Chem. Soc.*, **106**, 81 (1984)
2) A. Bax and D.G. Davis, *J. Magn. Reson.*, **63**, 207 (1985)
3) K. Wüthrich, G. Wider, G. Wagner and W. Braun, *J. Mol. Biol.*, **155**, 311 (1982)
4) M. Billeter, W. Braun and K. Wüthrich, *J. Mol. Biol.*, **155**, 321 (1982)

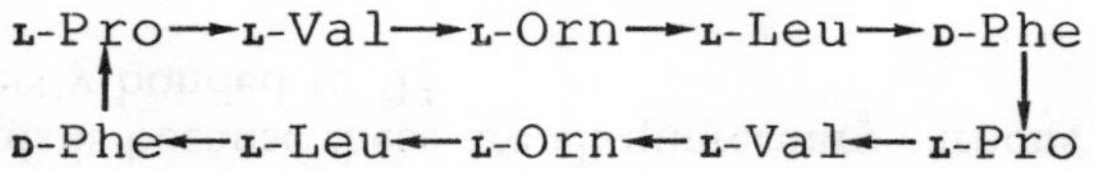

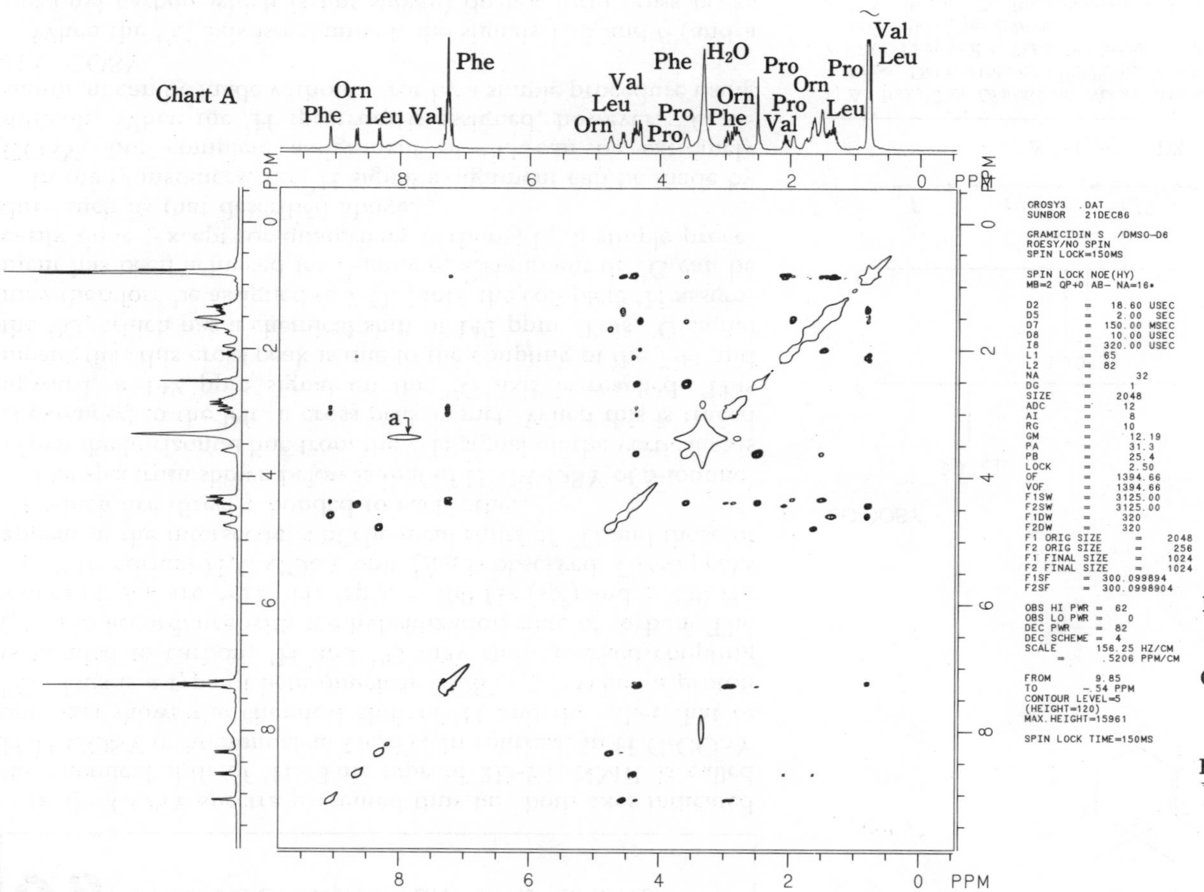

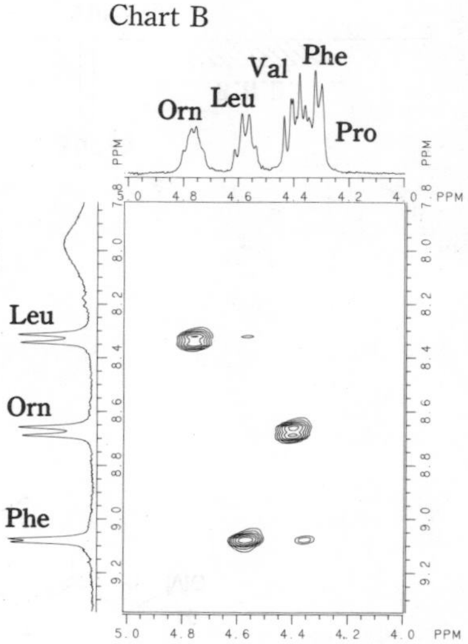

300 MHz ¹H-NMR (DMSO − d_b) [KM]

(DMSO

68 H,C-COSY Spectrum of β-Ionone

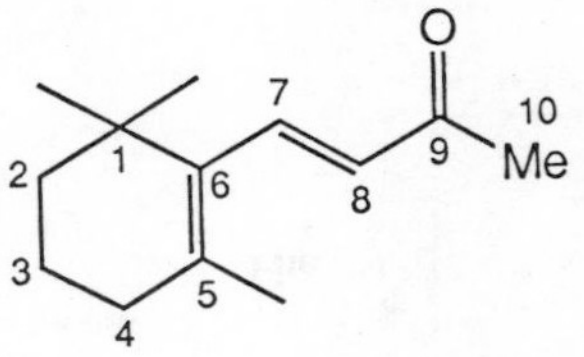

In the COSY spectra presented thus far, both axes indicated the chemical shift of ^{1}H. This type of 2D-FT NMR is called H,H-COSY or homonuclear COSY. In contrast, in H,C-COSY one axis shows the chemical shift of ^{1}H and the other that of ^{13}C. This is a type of heteronuclear COSY. [1–3] When a proton is bonded to carbon, ^{1}H and ^{13}C may show marked coupling ($^{1}J_{CH}$) in accordance with the hybridization state of carbon. The values of $^{1}J_{CH}$ are ~125 Hz (sp^3), ~150 Hz (sp^2) and ~250 Hz (sp).[4] In normal H,C-COSY only $^{1}J_{CH}$ is observed. Cross peaks appear at the intersections of chemical shifts of ^{13}C and those of ^{1}H which are directly bonded to each other.

The spectrum shown below is that of H,C-COSY of β-ionone. When the horizontal line from the 7-H signal on the vertical axis is extended to the left, a cross peak is met. When this is traced upward, a 142 ppm signal on the ^{13}C axis is reached. This means that this cross peak is due to the coupling of the 7-H and the ^{13}C, which has a chemical shift of 142 ppm. This ^{13}C signal may therefore be assigned to 7-C. Since the complete ^{1}H assignment has been achieved for β-ionone, assignment of ^{13}C can be easily done (except for quaternary carbons) by a simple procedure such as that described above.

In many instances, the ^{1}H signal assignment can be made by COSY, but complete assignment of ^{13}C can be extremely difficult. When the ^{1}H spectrum is assigned, however, ^{13}C assignment can be made without error by a simple procedure using H,C-COSY.

When the ^{13}C axis is examined, the signals 1, 5 and 6 (and a carbonyl carbon which is not shown) do not form cross peaks with ^{1}H. This is because these are all quaternary carbons which are not directly bonded to ^{1}H.

H, C-COSY

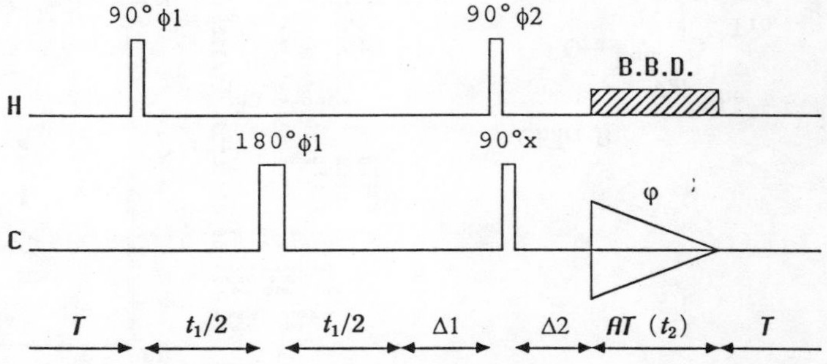

$\Delta_1 = 1/2\,^{1}J_{CH}$ (D3), $\Delta_2 = 1/3\,^{1}J_{CH}$ (D4)

1) A. Bax, *Two-Dimensional Nuclear Magnetic Resonance in Liquids*, Delft University Press, Delft, Holland (1982), pp.50-64
2) G.C. Levy (Ed.), *Topics in Carbon-13 NMR Spectroscopy*, vol. 4, John Wiley & Sons, Inc. (1984), pp.200-205
3) R.R. Ernst, G. Bodenhansen and A. Wokaun, *Principles of Nuclear Magnetic Resonance in One and Two Dimensions*, Oxford University Press, Oxford (1987), pp.471-479
4) E. Breitmaier and W. Voelter, *^{13}C NMR Spectroscopy—Methods and Applications in Organic Chemistry*, Verlag Chemie, Weinheim, New York (1978), pp.93-98

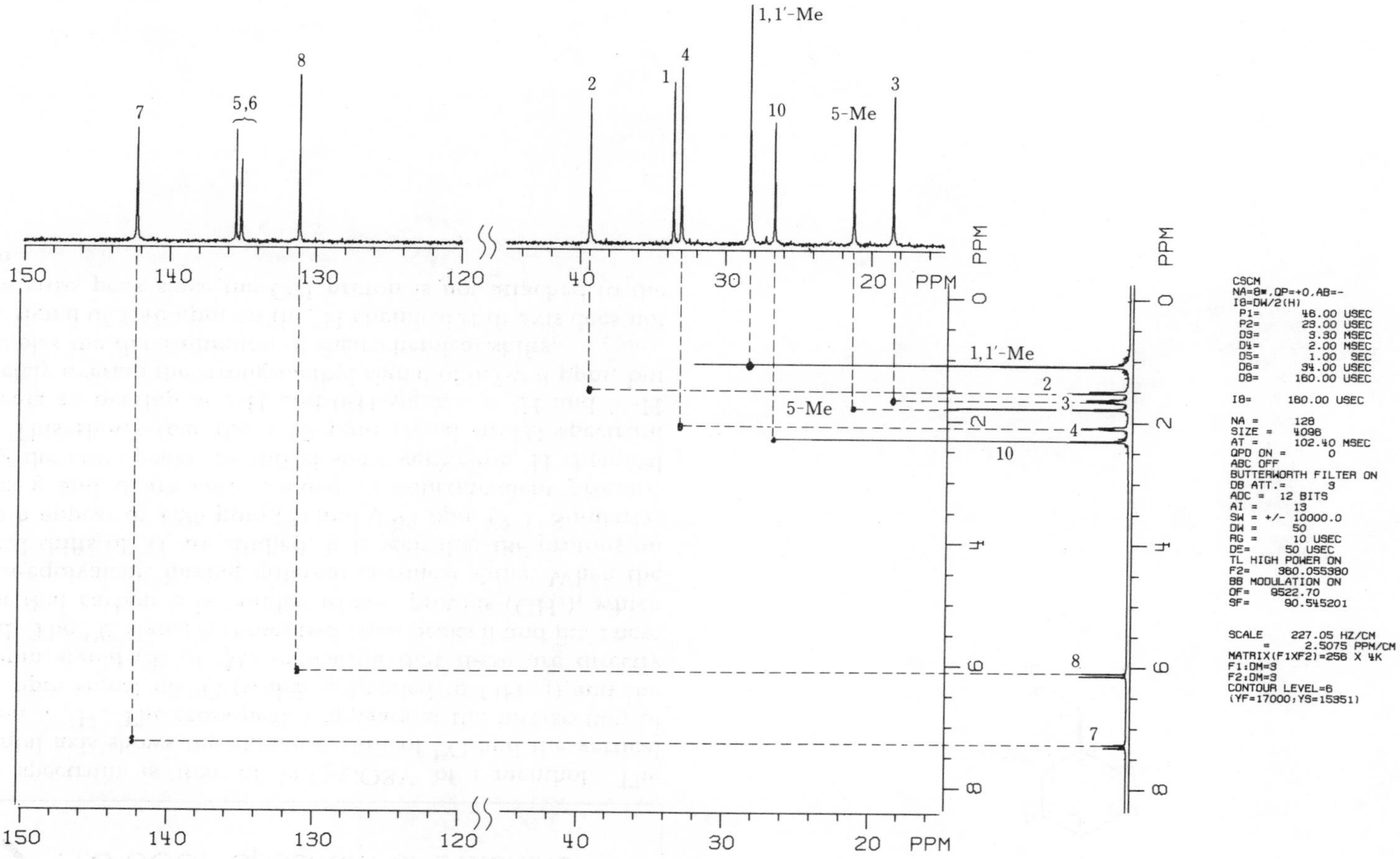

90MHz ^{13}C-NMR ($CDCl_3$) [TI]

69 H,C-COSY Spectrum of L-Menthol

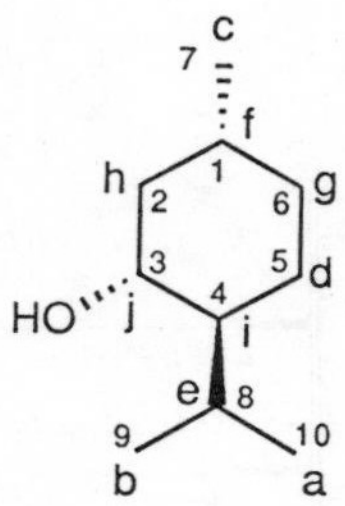

The spectrum is that of H,C-COSY of L-menthol. The horizontal axis shows the chemical shift of ^{13}C and the vertical axis that of ^{1}H. The cross peak i appears at the intersection of the 72 ppm signal of ^{13}C (which is bonded to OH; j) and the 3.28 ppm signal (3) of ^{1}H, indicating that these are directly bonded. The ^{13}C signal h shows two cross peaks ii and iii. These suggest that carbon h is bonded to two protons (CH_2), which are non-equivalent, having different chemical shifts. When the chemical shifts of ^{1}H are studied, it is seen that the protons on carbon h appear at 1.95 ppm (2) and 0.93 ppm (2′). Similarly, carbons g and d are each bonded to nonequivalent protons. Among the cross peaks, iv and vi show very close ^{1}H chemical shifts. This shows that the 1.55 ppm signal in 1D spectrum represents an overlap of 5-H and 6-H signals. 5′-H and 6′-H completely overlap the strong methyl signal of 0.7-0.8 ppm, but 2D enables the determination of their chemical shifts.

The signal of 4.46 ppm on the ^{1}H chemical shift axis does not show a cross peak since the OH proton is not attached to the carbon.

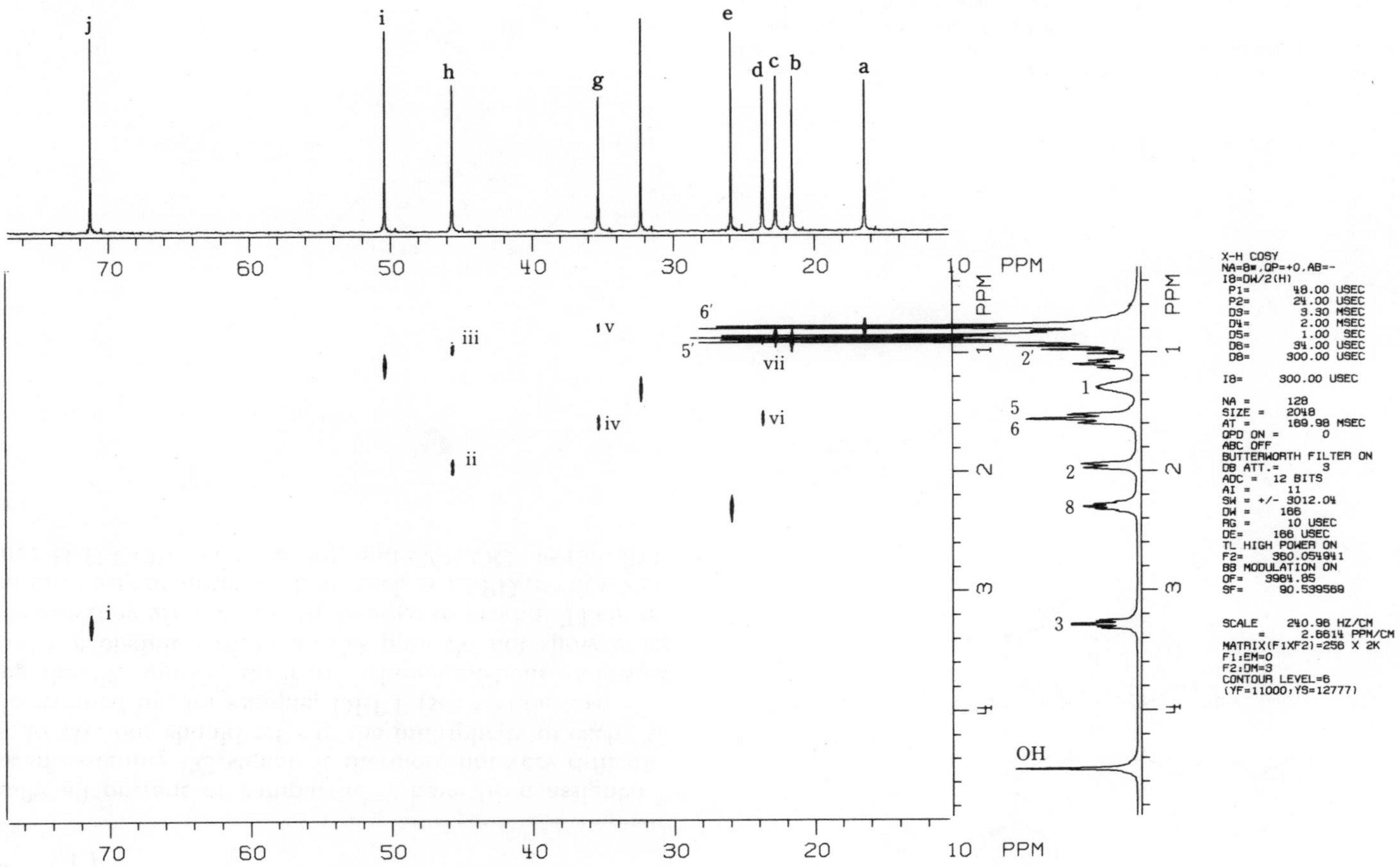

90MHz ^{13}C-NMR (C_6D_6) [TI]

70 H,C-COSY Spectrum of Compactin (I)

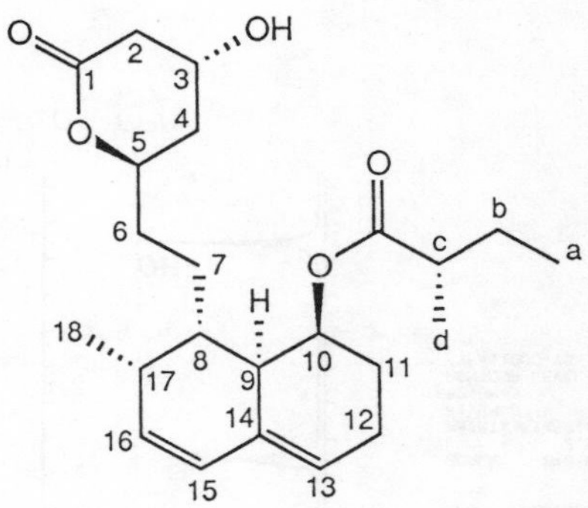

Virtually all protons of compactin[1–3] have been assigned.[3] The task of assigning ^{13}C signals is therefore not very difficult. Needless to say, one should refer to the multiplicity of each ^{13}C signal determined by, for exmple, DEPT (see section 26).

Among the ^{13}C signals, the two carbonyl carbons at lowest field[4] and the olefinic carbon at 134 ppm do not show cross peaks because they are not directly bonded to proton. Their assignment must rely on other methods such as LSPD (section 22), long-range H,C-COSY (section 80), and COLOC (section 81).

1) A.G. Brown, T.C. Smale, T.J. King, R. Hasenkamp and R.H. Thompson, *J. Chem. Soc., Perkin Trans. 1.*, **1976**, 1165
2) A. Endo, M. Kuroda and Y. Tsujita, *J. Antibiot.*, **29**, 1346 (1976)
3) M. Hirama and M. Uei, *J. Am. Chem. Soc.*, **104**, 4251 (1982); *Idem, Tetrahedron Lett.*, **23**, 5307 (1982)
4) See sections 39 and 40.
5) See section 82.

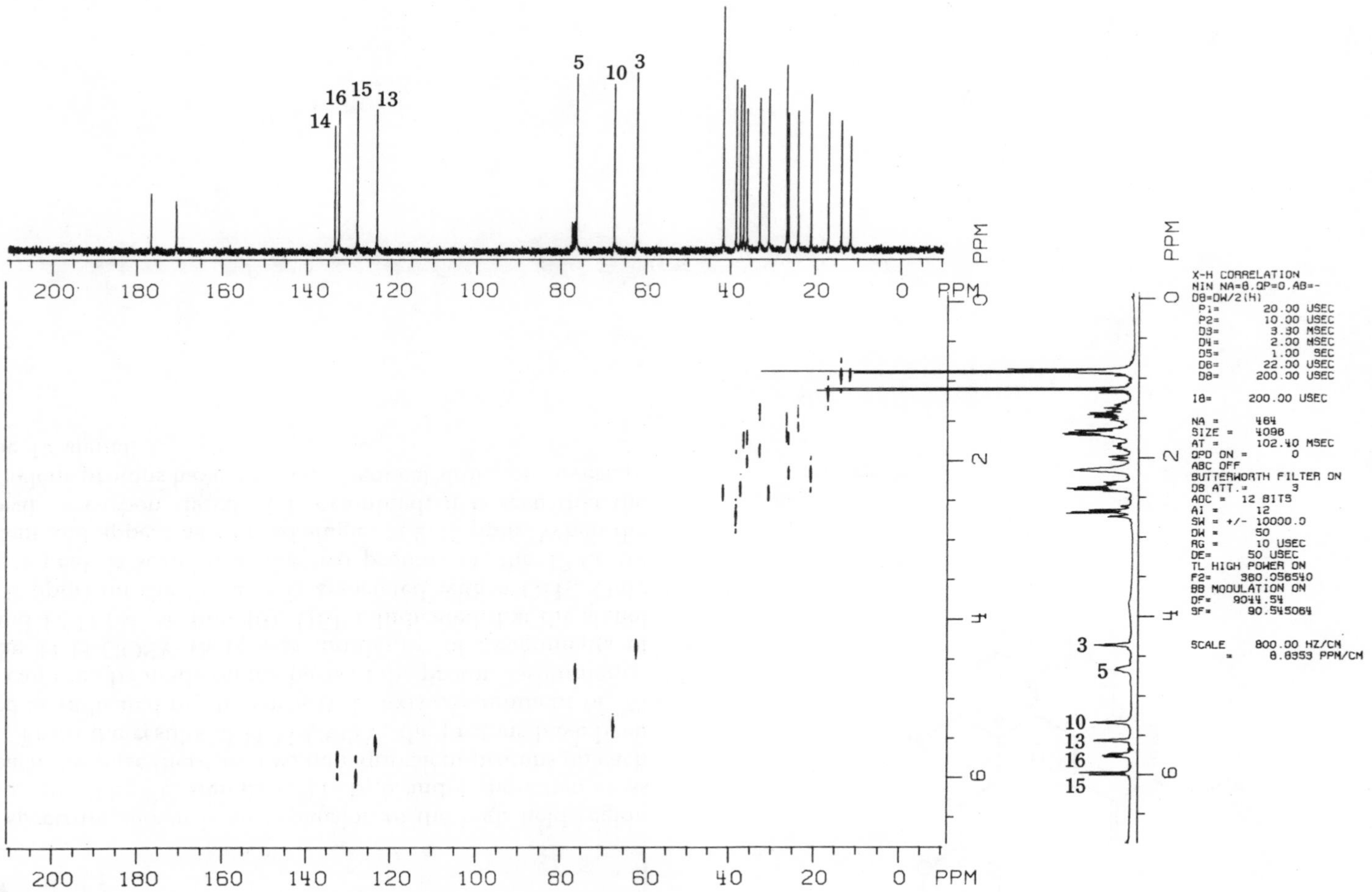

90MHz ^{13}C-NMR ($CDCl_3$) [TI][1)]

71 H,C-COSY Spectrum of Compactin (II)

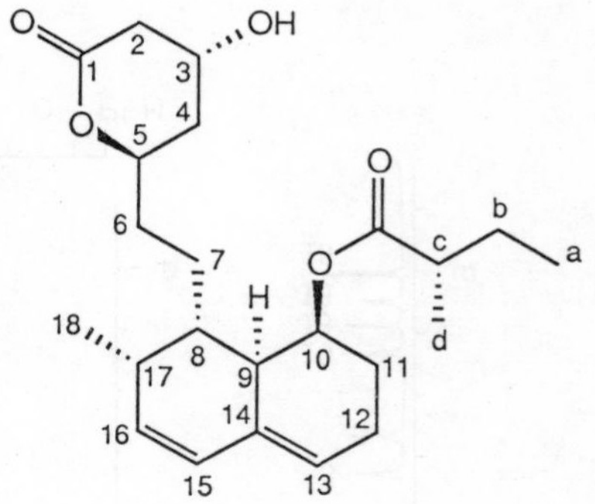

The spectrum shown is an expansion of the high field region of compactin. The ^{13}C signals 7, 11, b, 6 and 4 show two cross peaks each, because there are two nonequivalent protons on each carbon. From the results of H,H-COSY, the protons have been assigned as indicated on the vertical ^1H axis. Assignment of ^{13}C signals can then be made on the basis of the proton assignments.

In the H,H-COSY there was ambiguity of assignments of 11-H and 12-H (see section 40). DEPT indicated that the signal 12 (20.8 ppm) on the ^{13}C axis is associated with a CH_2. Only one cross peak is seen since the two protons on the 12-C are equivalent and appear as a broad singlet at 2.15 ppm. When the cross peak of carbon signal 11 is examined, it is seen that the 11-methylene protons have different chemical shifts, one overlapping the 12 signal.

1) A. Endo *et al.*, *J. Antibiol.*, **38**, 444 (1985)

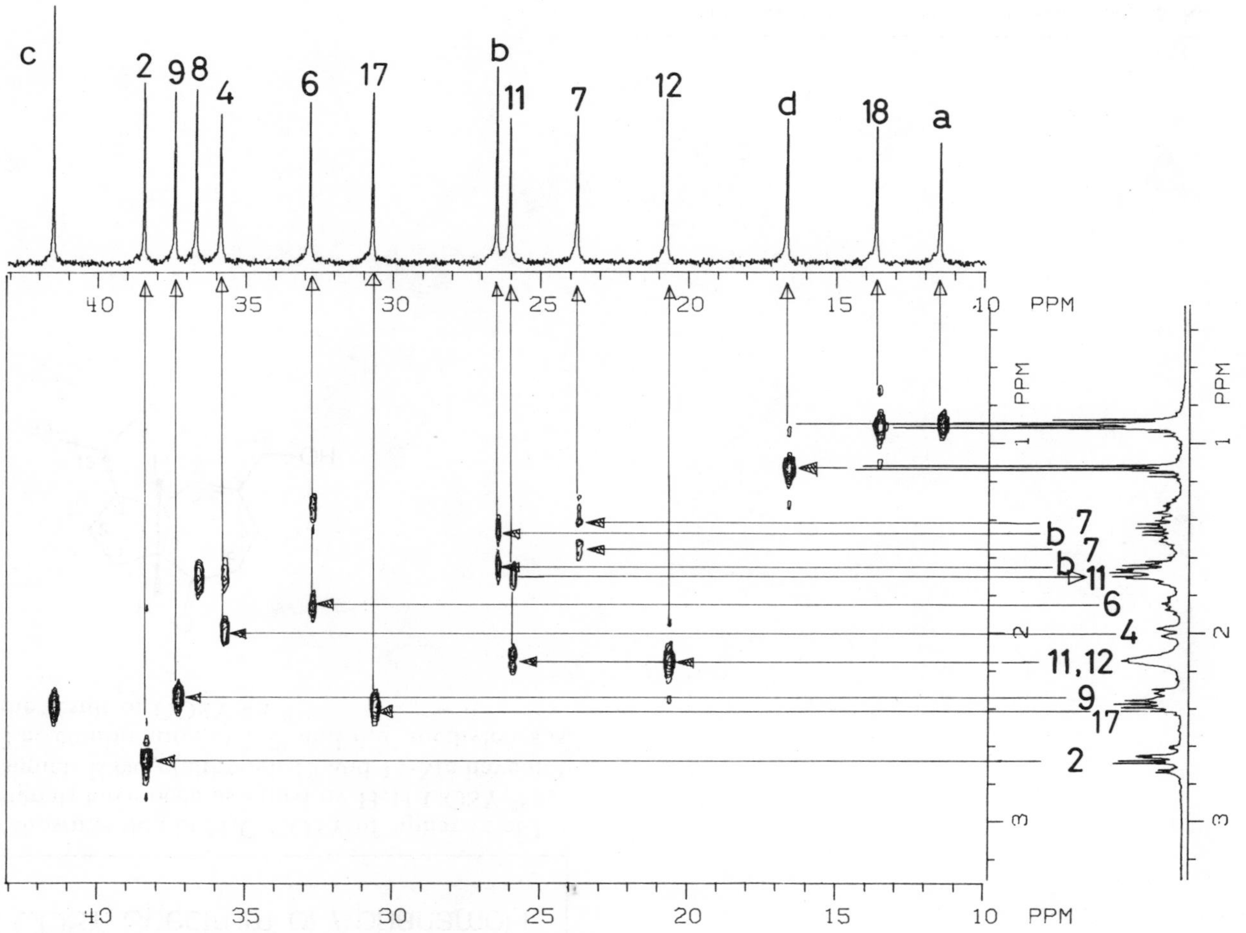

90MHz ^{13}C-NMR ($CDCl_3$) [TI][1)]

72 H,C-COSY Spectrum of Aphanamol-I

The spectrum shown is that of H,C-COSY of aphanamol-I.[1)]
Since proton signals have been assigned by H,H-COSY,[2)] assignment of ^{13}C signals is easy (although 13 and 14-Me have not been assigned). The combination of 7,7′ and 8,8′ methylenes is consistent with the result of COSY 45.[3)]

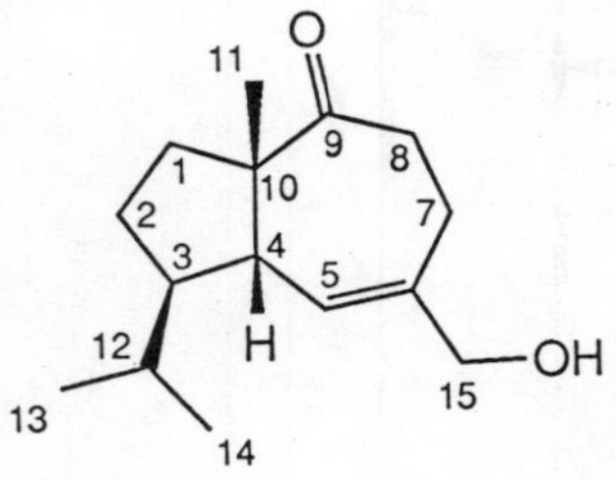

1) M. Nishizawa, A. Inoue, Y. Hayashi, S. Sastrapradja, S. Kosela, T. Iwashita, *J. Org. Chem.*, **49**, 3660 (1984)
2) See sections 47 and 48.
3) See section 50.

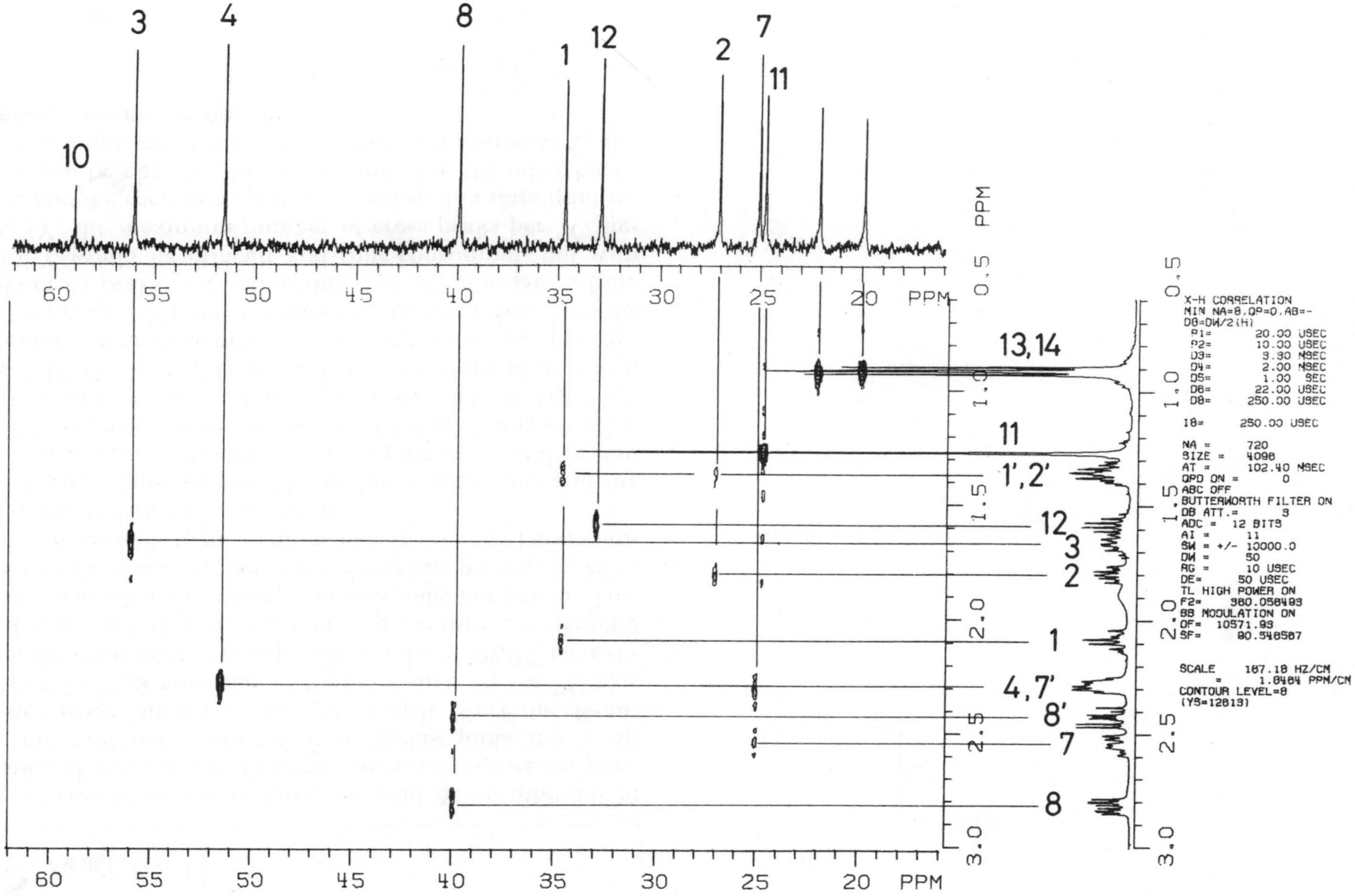

90MHz ^{13}C-NMR ($CDCl_3$) [TI]

73 H,C-COSY Spectrum of Metasequoic Acid A (I)

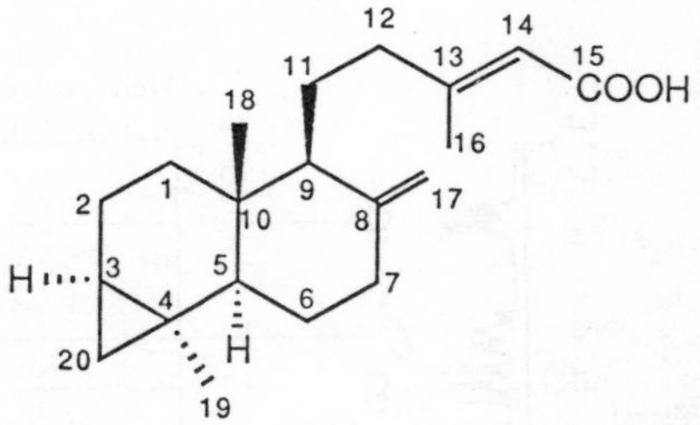

The spectrum is that of metasequoic acid A, an inhibitor of germination of the spores of the organisms which causes rice blast disease, the acid being isolated from metasequoia tree.[1] All methylene carbons show two cross peaks each. Since the chemical shifts of the 17.9 ppm and 18.0 ppm signals on the ^{13}C axis are close, the cross peaks overlap. The results of DEPT have established that 3 is CH and 20 is CH_2. Of the three overlapping cross peaks, a and b are contour peaks of equally low height. This indicates that these correspond to the methylene protons on 20-C and that c corresponds to the methine proton on 3-C. This is supported by the H,H-COSY spectrum.

The 1.6 ppm signal on the ^1H axis has a large area and the pattern is complex, representing overlapping of several proton signals. When a line is drawn to the left of 1.6 ppm, it is seen that there are a total of six overlapping protons: 2H, 11-H_2, 1-H, 5-H and 9-H. Of these, 2-H and 1-H represent one proton each of the 2- and 1-methylene protons. Their respective partners appear at 2.0 ppm (2′-H) and 1.0 ppm (1′-H), as evident from the 2-C, 1-C cross peaks. To obtain this type of information from H,H-COSY would be difficult and time-consuming, but with H,C-COSY, the maximum number of cross peaks per ^{13}C signal is two and the spectrum itself is simple, so that important information can be obtained in a short time. On the other hand, since it is a simple spectrum, information concerning C-C and H-H linkages cannot be obtained.

1) F. Sakan, T. Iwashita and N. Hamanaka, *Chem. Lett.*, **1988**, 123

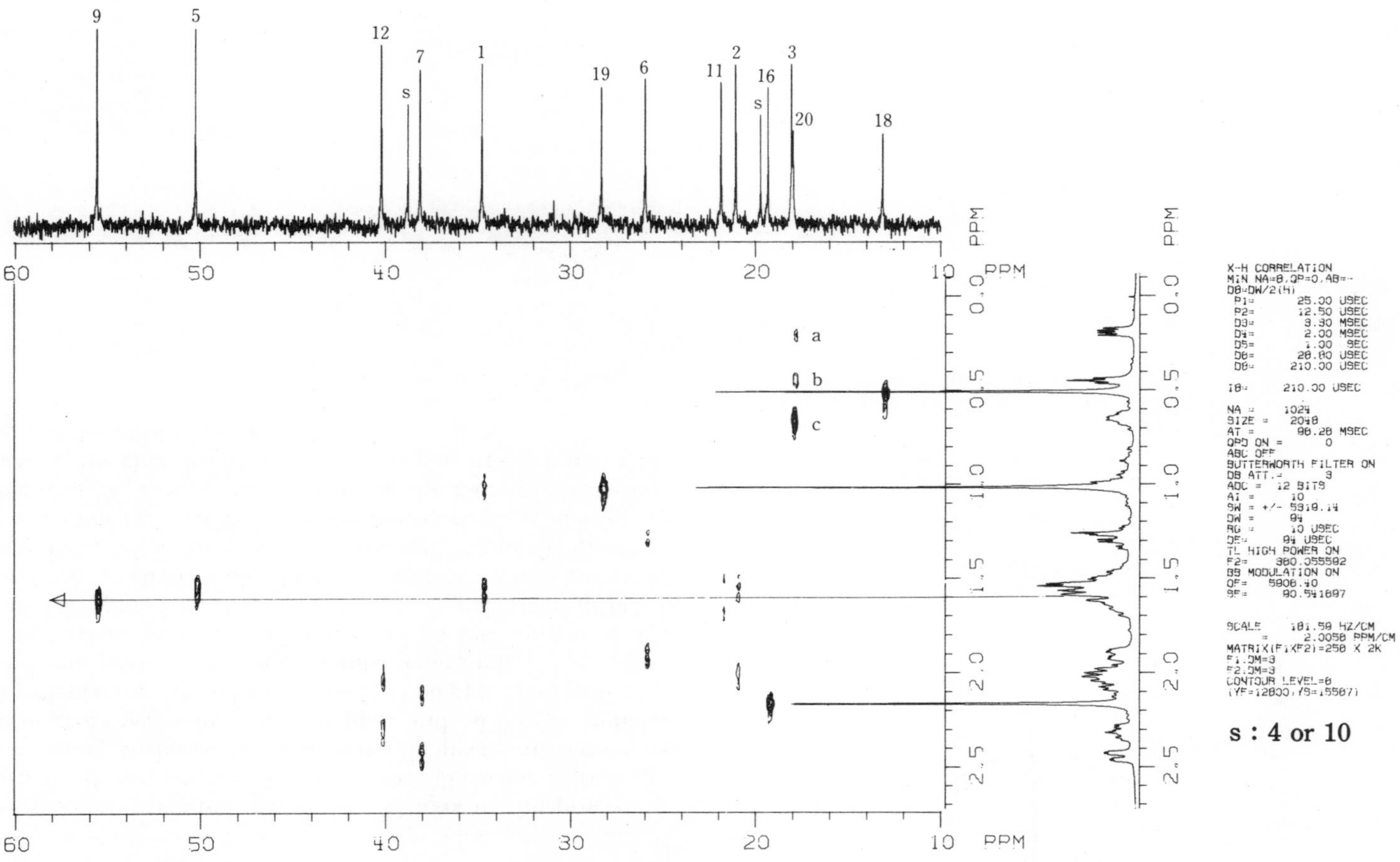

s : 4 or 10

90 MHz ^{13}C-NMR ($CDCl_3$) [TI]

74 H,C-COSY Spectrum of Metasequoic Acid A (II)

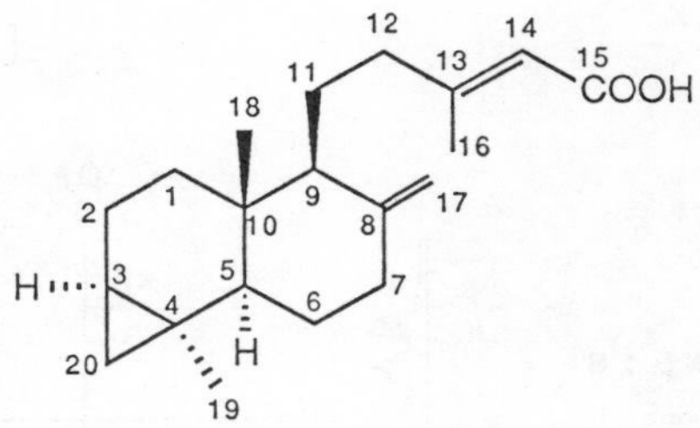

Cross sections prepared along the ^{13}C axis in the preceding spectrum are shown below. They represent all carbon signals except quaternary carbons. In the preceding spectrum, there was overlap of cross peaks for the 17.9 ppm and 18.0 ppm signals. Here, by displacing the cross sections by 0.08 ppm, two peaks for 20-C and one peak for 3-C are cleanly separated.

The methylene protons 1-H and 1′-H on 1-C appear at 1.5 ppm and 1.0 ppm. In 1D spectrum these signals overlap other peaks and are not discernible. The cross sections show, however, that these have been completely "extracted" from the cluster.

The merits of H,C-COSY may be summarized as follows: 1) assignment of ^{13}C can be made easily on the basis of ^{1}H assignment, and 2) the chemical shift of a proton pair in a geminal relationship can be easily determined.

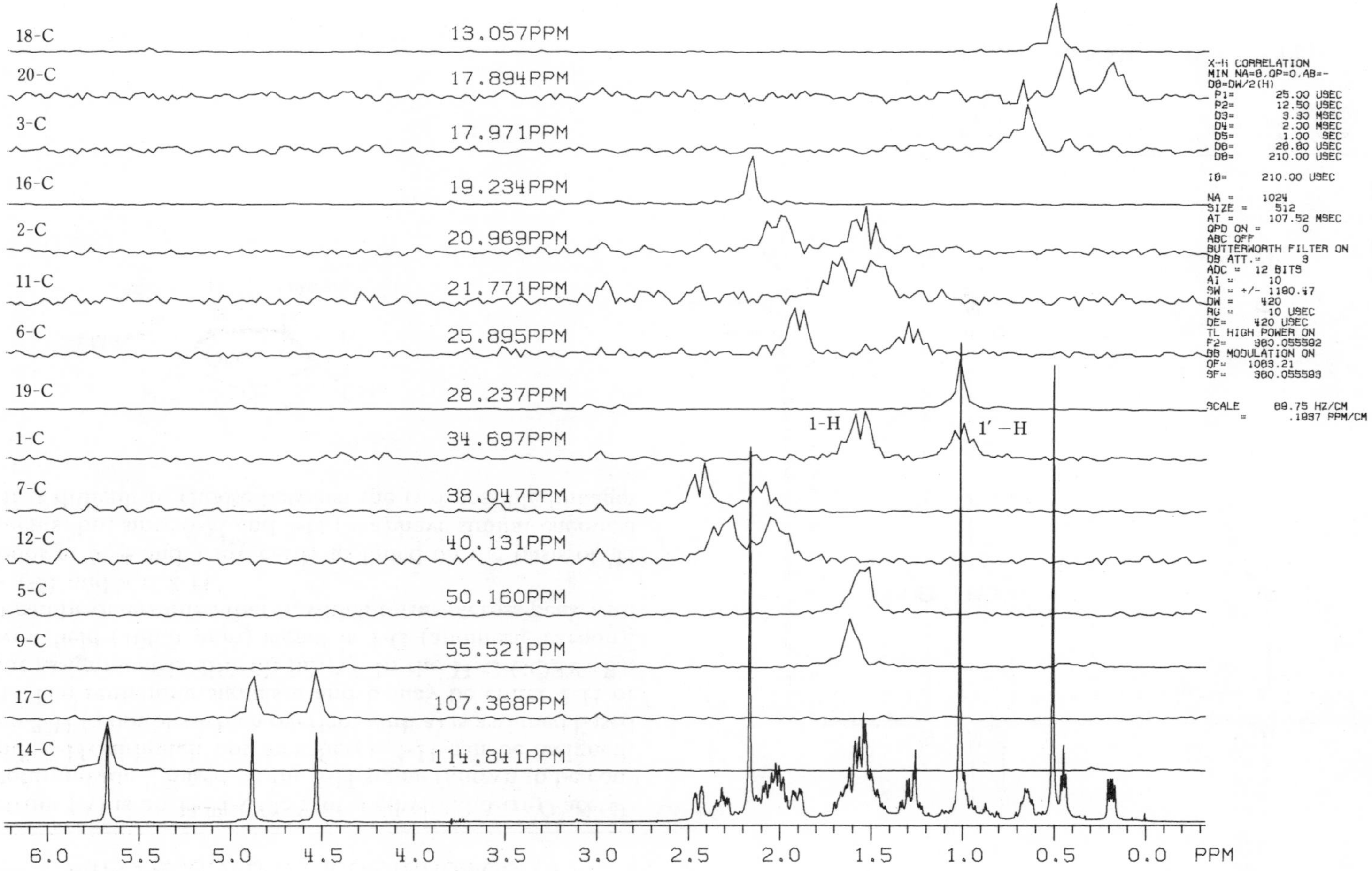

90MHz ^{13}C-NMR ($CDCl_3$) [TI]

75 H,C-COSY Spectrum of Methyl-2,3,5-tri-*O*-acetyl-β-D-fucofuranoside (I)

Spectrum (A) is an H,H-COSY of methyl-2,3,5-tri-*O*-acetyl-β-D-fucofuranoside.[1] Based on the 5-H signal (known to be coupled with 6-H_3 although not seen here), 4-H can be assigned, and then 3-H (a doublet which overlaps with a) is assigned based on 4-H. The remaining signals a and b may be either 2-H or 1-H, but assignment is difficult in (A). In the H,C-COSY (B), the lowest field (106.5 ppm) signal is 1-C (anomeric carbon). Thus it can be determined that b, which forms a cross peak with 1-C, is 1-H and a is 2-H.

Carbons at 5, 4 and 1 are easily assigned on the basis of ^{1}H assignments, but since 3-H and 2-H (= a) have similar chemical shifts, it is difficult to choose between the two possible linkages p and q.

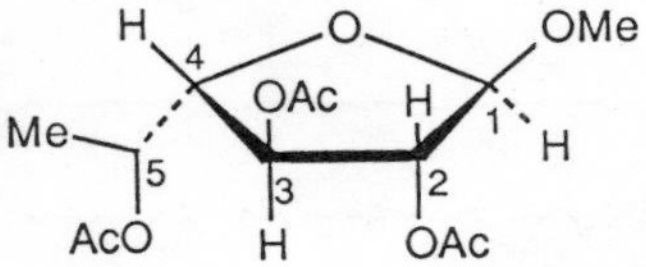

(A)

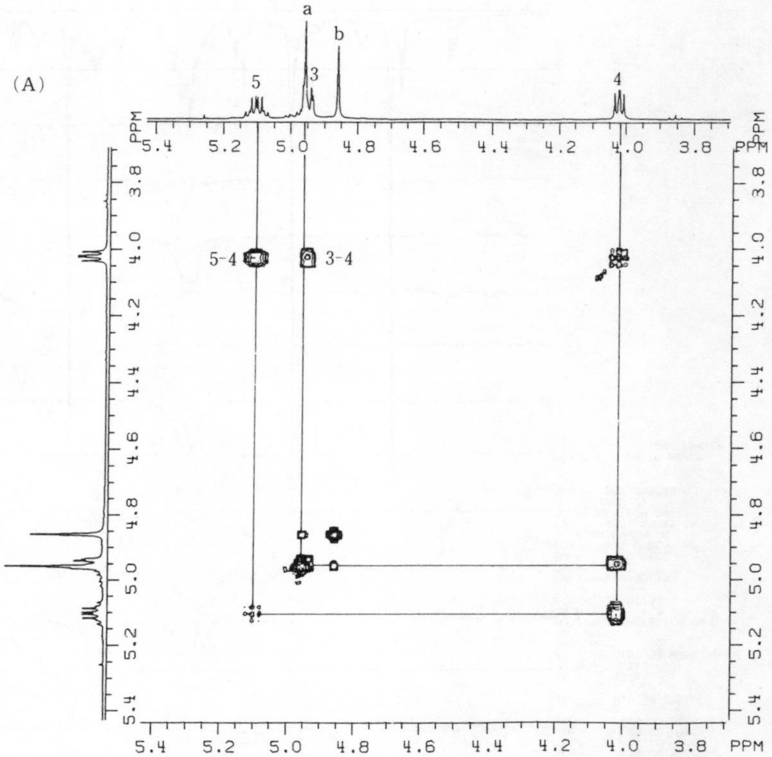

(A) 360MHz ^{1}H-NMR ($CDCl_3$) [TI]

1) T. Kinoshita, T. Miwa and J. Clardy, *Carbohydr. Res.*, **143**, 249 (1985)

(B)

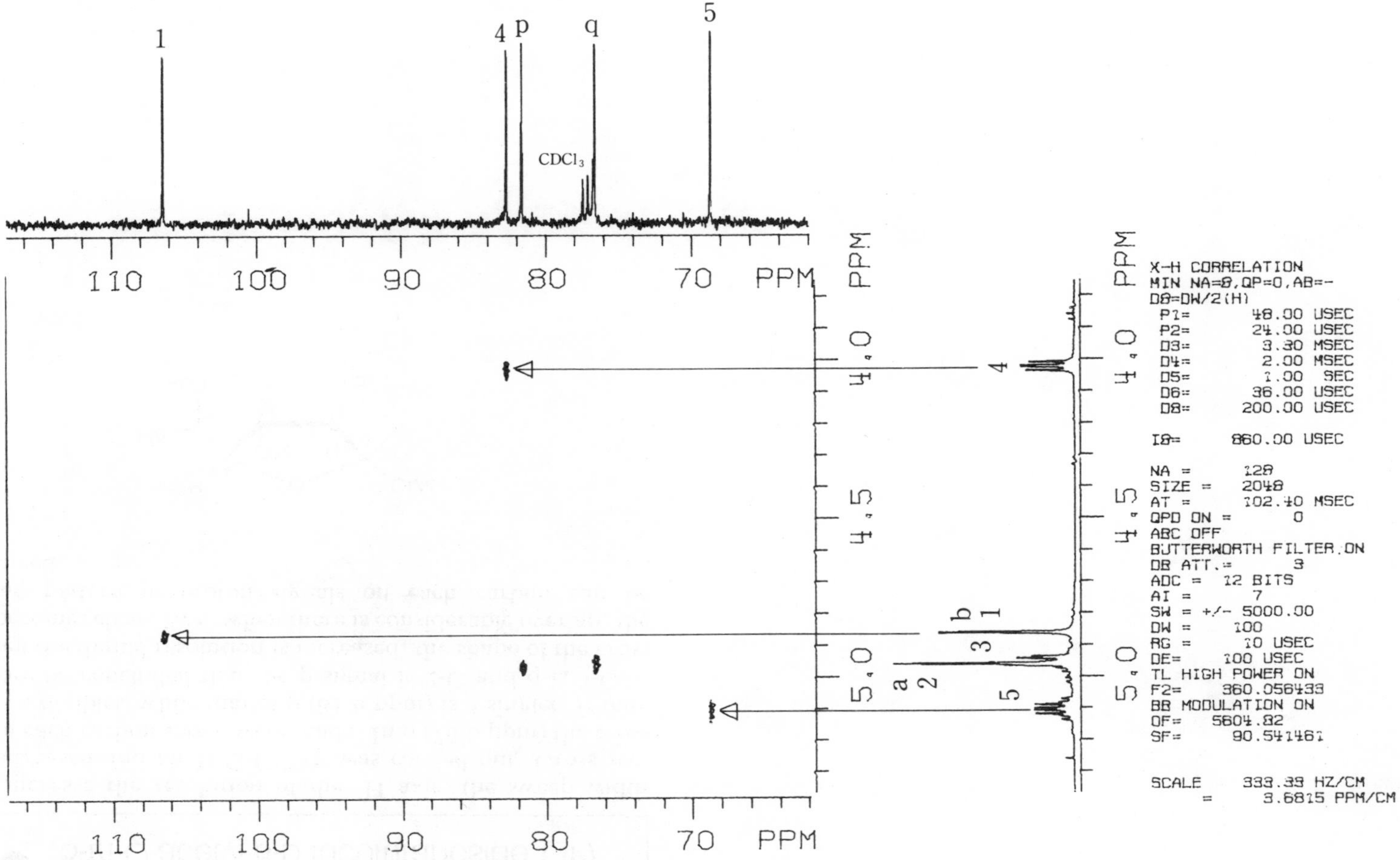

(B) 90MHz ^{13}C-NMR ($CDCl_3$) [TI]

76 H,C-COSY Spectrum of Methyl-2,3,5-tri-*O*-acetyl-β-D-fucofuranoside (II)

To increase the resolution of the ^{1}H axis, the sweep width was decreased and an H,C-COSY was carried out. Cross sections of each carbon signal were made. In q (76.5 ppm) the cross peak is a doublet, while that of p (81.6 ppm) is a singlet. It may therefore be concluded that the p signal is 2-C and q is 3-C.

When the digital resolution is increased, the shape of the cross peak becomes clear. Even when there is considerable overlap, the splitting pattern of proton signals on each carbon can be recognized.

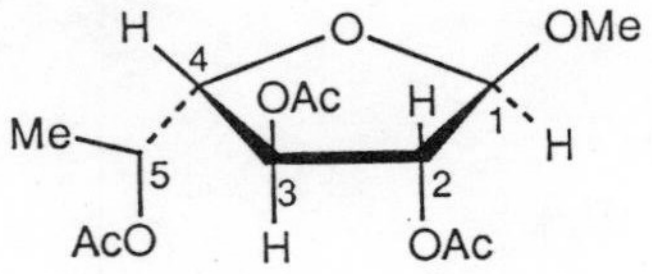

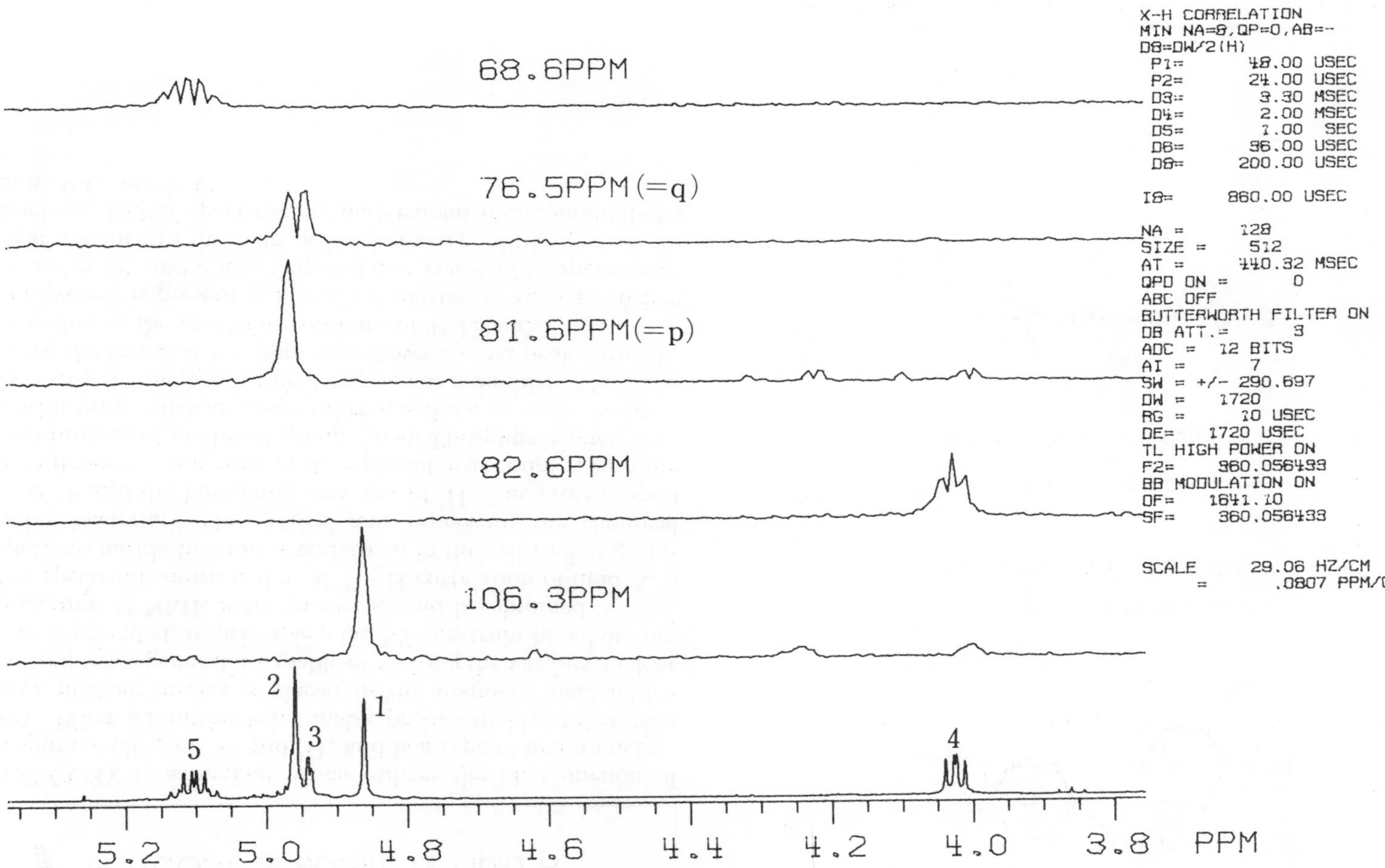

90MHz ^{13}C-NMR ($CDCl_3$) [TI]

77 H,P-COSY Spectrum of Lipid A

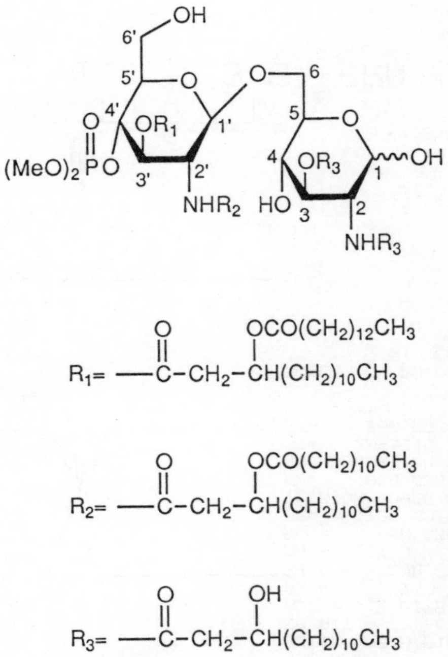

H,C-COSY is a method which utilizes the phenomenon of spin-spin coupling of ^{13}C and ^{1}H, and is a type of heteronuclear COSY. When a multinuclear tunable probe capable of detecting various nuclear species is placed in the magnetic field and a radio-frequency generator capable of exciting the various nuclear species is provided, in principle a COSY spectrum based on any combination of NMR active nuclei should be obtained.

The spectrum shown is that of ^{31}P-^{1}H correlation of lipid A, a lipopolysaccharide fraction of endotoxin in the cell wall of gram-negative bacteria.[1)] The vertical axis represents the chemical shift of ^{31}P and the horizontal axis that of ^{1}H. The cross shaped peak at the center is a cross peak of phosphorus coupled with the methyl protons of methoxyl group ($^{3}J_{HP}$). Phosphorus often couples with other nuclear species through three or more bonds.[2)] This is why the methoxyl signal appears as a doublet. Along the ^{1}H axis, the signal at 4.4 ppm also shows a cross peak with ^{31}P. This is due to the spin-spin coupling of 4′-H and P.

Phosphorus is present in biological substances such as nucleic acids and ATP, and it is anticipated that H,P-COSY spectroscopy will advance in the field of biochemistry. With progress in technology, COSY spectroscopy may extend into combinations such as P-C and N-C.

1) M. Imoto, S. Kusumoto, T. Shiba, H. Naoki, T. Iwashita, E. Th. Rietschel, H. W. Wollenweber, C. Galanos and O. Lüderitz, *Tetrahedron Lett.*, **24**, 4017 (1983)
2) W. Brügel, *Handbook of NMR Spectral Parameters*, vol. 2, Heydon & Son Ltd., London (1979) pp.515-516; vol. 3, pp.810-832

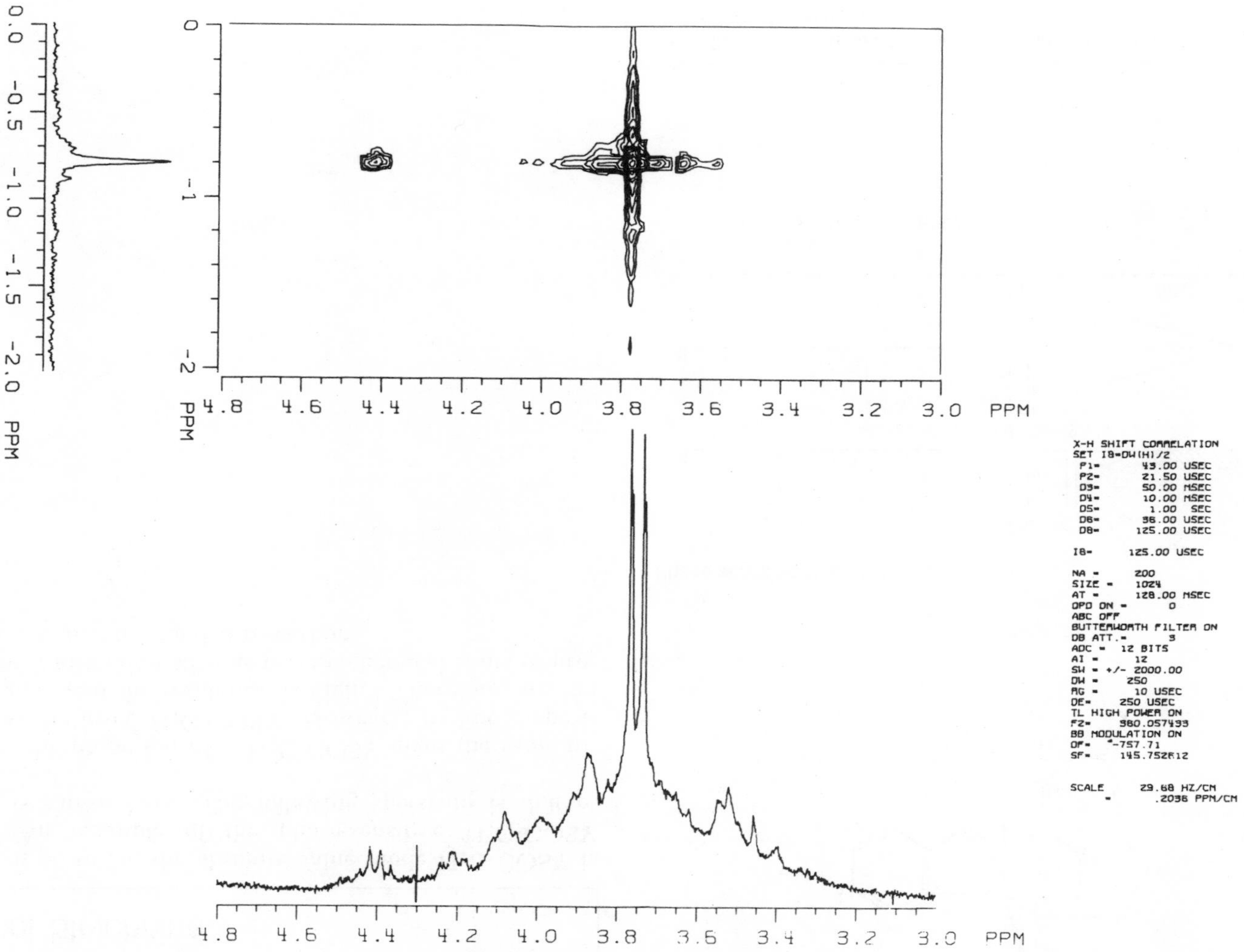

145.8MHz ^{31}P-NMR ($CDCl_3$) [TI]

78 Phase-sensitive H,C-COSY Spectrum of Strychnine

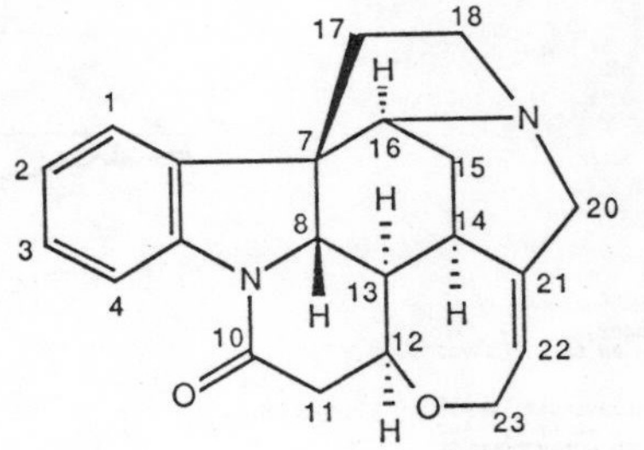

In sections 68 to 76, the absolute value mode H,C-COSY is presented. An example of the phase-sensitive H,C-COSY spectrum[1] is shown here. The following spectrum is that of strychnine.

Basically, the phase-sensitive H,C-COSY gives the same information as the usual H,C-COSY. However, its line shape is extremely good and the resolution is high.[2] Therefore, we can get excellent information such as precise chemical shifts of protons which are directly bonded to carbon.

Phase-sensitive H, C-COSY

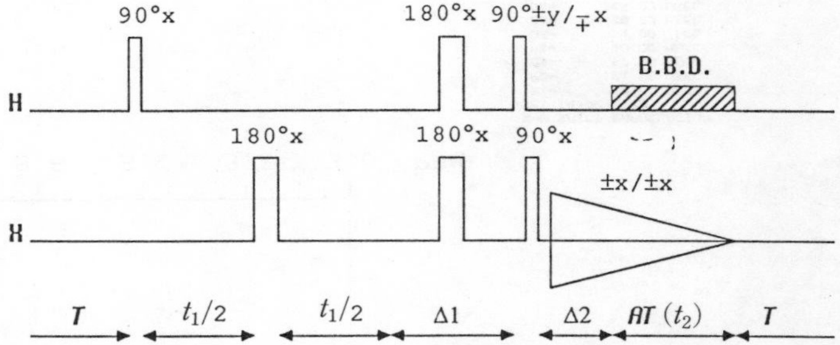

$\Delta_1=1/2^1J_{CH}$, $\Delta_2=1/3^1J_{CH}$

1) A. Bax and S. Sarkar, *J. Magn. Reson.*, **60**, 170 (1984)
2) See Part III, section 2.4.

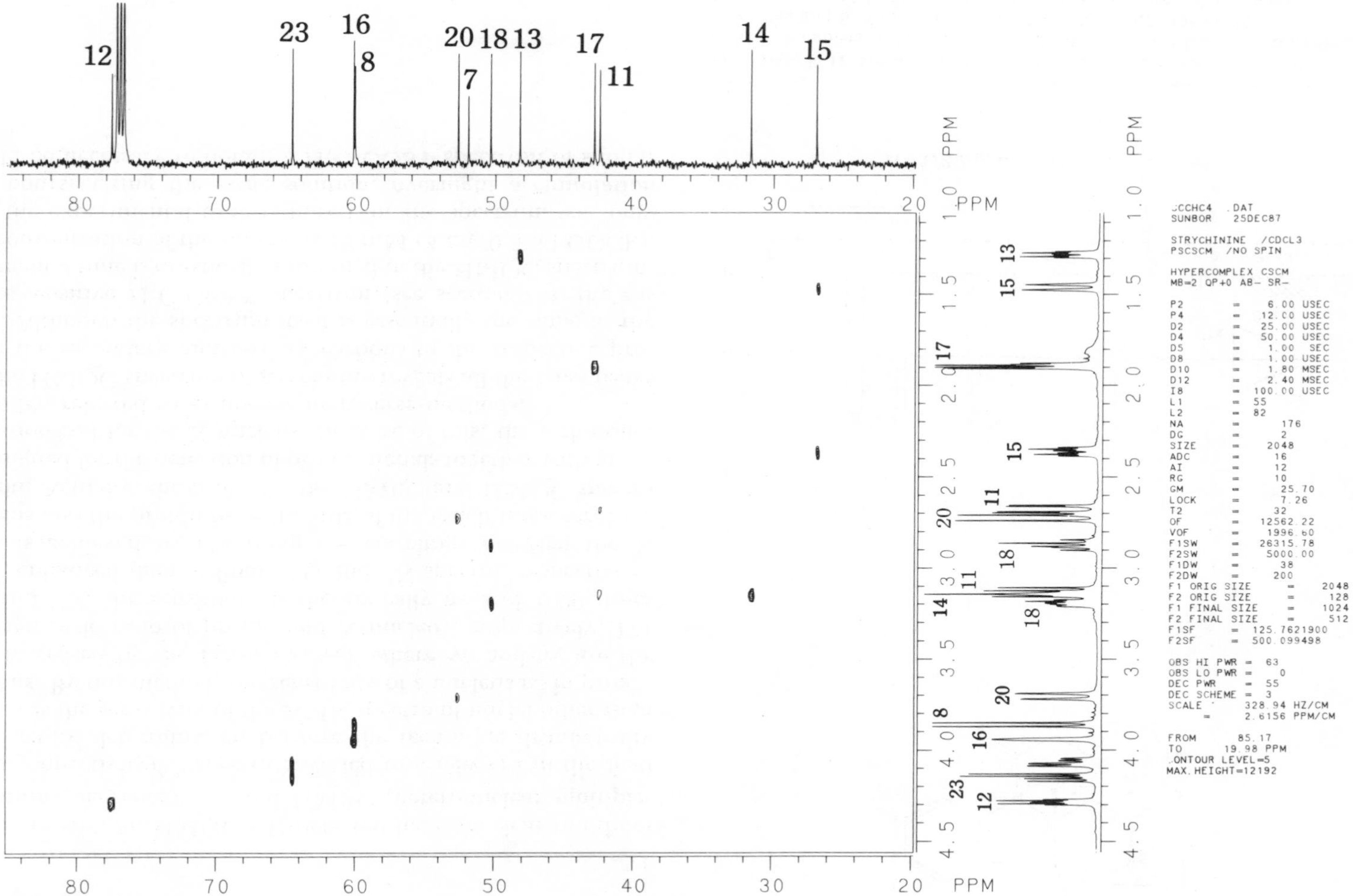

125 MHz ^{13}C-NMR ($CDCl_3$) [TI]

79 HMQC Spectrum of Strychnine

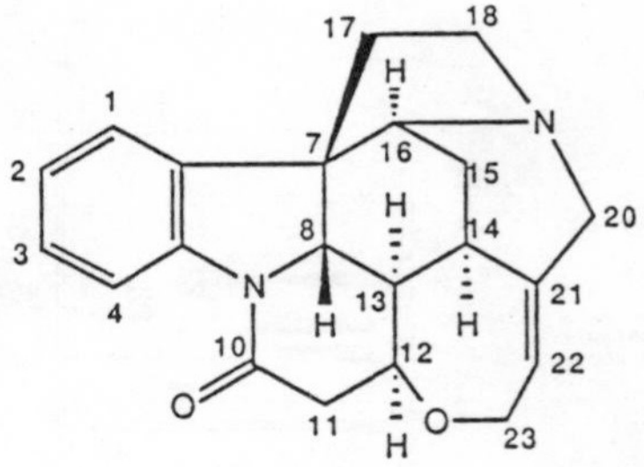

The reports on HMQC (^{1}H-detected heteronuclear multiple-quantum coherence)[1, 2, 3] and HMBC (heteronuclear multiple-bond connectivity)[2, 4] spectra have led to a new era in the field of structure determination because the technique dramatically enhances the sensitivity of the NMR spectra of nuclei other than protons. By this method, the sensitivity of a nucleus is, in principle, increased by the factor $(\gamma_H/\gamma_X)^3$ where γ_H and γ_X are the magnetogyric ratio of proton and X nucleus, respectively. For ^{13}C and ^{15}N, the sensitivity is theoretically 64 and 1000 times more enhanced than ordinary ^{13}C and ^{15}N spectra, respectively. This is achieved by observing the couplings between the X nucleus and the proton from the side of the much more sensitive proton. Actually, the probe for the HMBC and HMQC spectra is designed for the detection of proton signals together with an irradiation coil for the X nucleus. Because of this, the techniques are often referred to as inverse or reverse methods.

The HMQC spectrum of strychnine reveals all the cross peaks from the secondary and tertiary carbons to the respective protons. Although the spectrum itself is essentially the same as the phase-sensitive H,C-COSY spectrum (see section 78), the experimental time is drastically shortened in the HMQC spectrum; the concentration of the sample is 18 mM (3 mg/0.5 ml $CDCl_3$), and the experimental time required for the spectrum was only two hours. Using the same solution, overnight accumulation would be necessary to obtain a H,C-COSY spectrum of similar quality.

HMQC

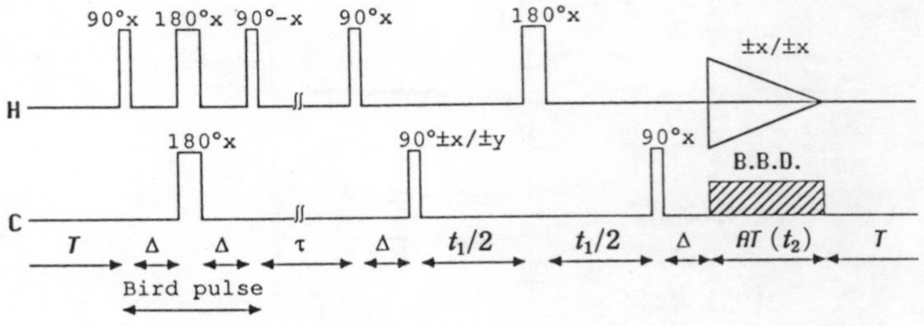

$\Delta = 1/2\,^{1}J_{CH}$, In this case, τ is 200 ms.

1) A. Bax, R.H. Griffey and B.L. Hawhins, *J. Magn. Reson.*, **55**, 301 (1983)
2) M.F. Summers, L.G. Marzilli and A. Bax, *J. Am. Chem. Soc.*, **108**, 4285 (1986)
3) A. Bax and S. Subramanian, *J. Magn. Reson.*, **67**, 565 (1986)
4) A. Bax and M.F. Summers, *J. Am. Chem. Soc.*, **108**, 2093 (1986)

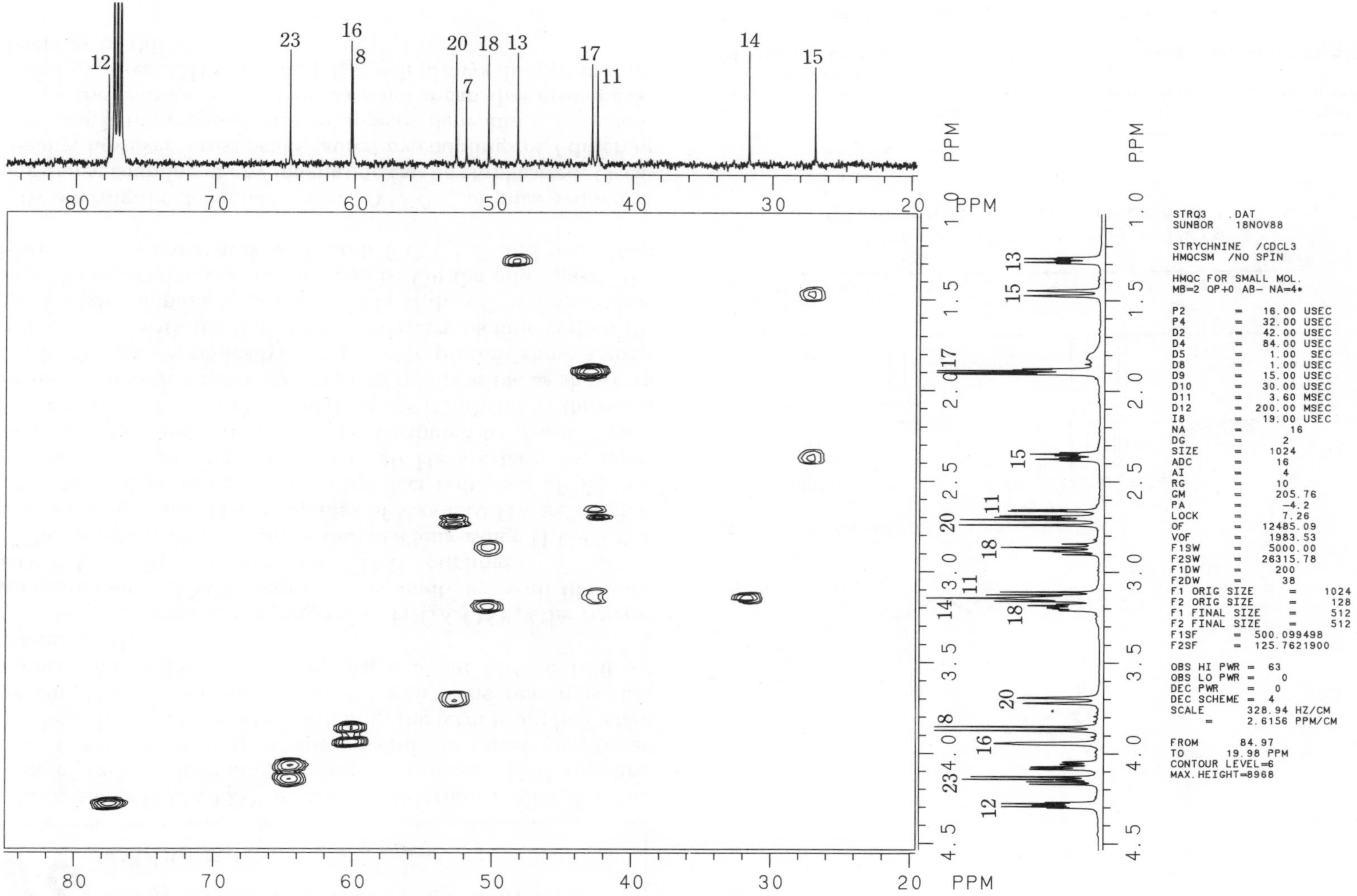

500 MHz ^{1}H-NMR ($CDCl_3$) [TI]

80 Long-range H,C-COSY Spectrum of β-Ionone

Long-range H,C-COSY spectrum yields information about all kinds of carbons including quaternary carbons. H-H coupling over a distance of four or more bonds is called long-range coupling. In the case of H-C coupling, the term is applied when the coupling is over two or more bonds. The reason is that whereas $^{1}J_{CH}$ is 250 to 120 Hz, $^{2}J_{CH}$ is about 1/10 to 1/20, or around 10 Hz.[1)]

In the pulse sequence of long-range H,C-COSY,[2)] the reverse discrimination TANGO sequence[3)] is combined with the ordinary H,C-COSY to observe small H-C couplings.

The spectrum shown below is that of a long-range H,C-COSY in which long-range H-C couplings of about 10 Hz are emphasized. Several peaks due to the imperfect reduction of $^{1}J_{CH}$ are present. Spin coupling constants of 10 Hz are large for long-range coupling and may usually be attributed to $^{2}J_{CH}$ or $^{3}J_{CH}$.

Assignment of 5-C and 6-C of β-ionone is difficult by the usual method, but with long-range coupling it is possible as shown in the illustration. Specifically, the 1,1′-Me protons show a cross peak (a) only with the high field quaternary olefinic carbon (6) ($^{3}J_{CH}$) while coupling of about 10 Hz with 5-C is improbable since the separation is over four bonds. On the other hand, the 5-Me shows a cross peak with both 5-C ($^{2}J_{CH}$) and 6-C ($^{3}J_{CH}$) (b).

By altering the delay parameter (Δ1, Δ2) in pulse sequence, long-range coupling of any magnitude can be emphasized. Occasionally, however, cross peaks caused by couplings of J different from that being emphasized can appear, depending, for example, on the contour level. This does not mean that cross peaks caused by direct CH coupling ($^{1}J_{CH}$) will always disappear completely as in this spectrum.

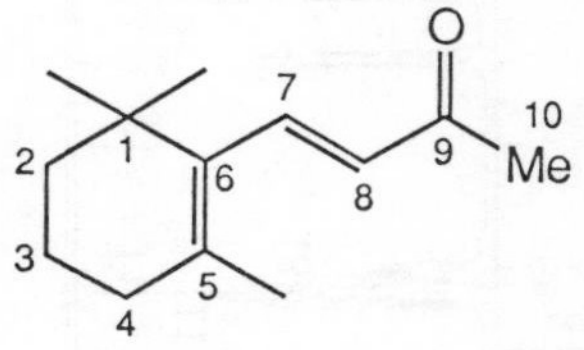

Long-range H, C-COSY

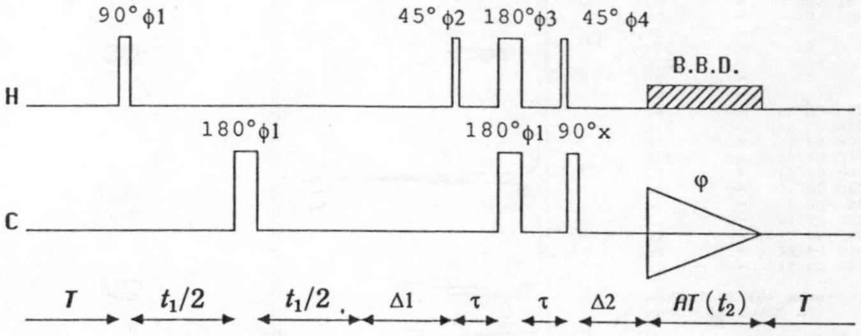

$\Delta=1/2^{lr}J_{CH}$, $\Delta_2=1/2^{lr}J_{CH} \sim 1/3^{lr}J_{CH}$, $\tau=1/2^{1}J_{CH}$
($^{lr}J_{CH}$=long range spin-spin coupling constant)

1) J.L. Marshall, *Carbon-Carbon and Carbon-Proton NMR Couplings: Application to Organic Stereochemistry and Conformational Analysis*, Verlag Chemie International, Inc., Deerfield Beach, Florida (1983), pp.11-64
2) M.J. Quast, A.S. Zektzer, G.E. Martin and R.N. Castle, *J. Magn. Reson.*, **71**, 554 (1987)
3) S. Wimperis and R. Freeman, *J. Magn. Reson.*, **58**, 348 (1984)

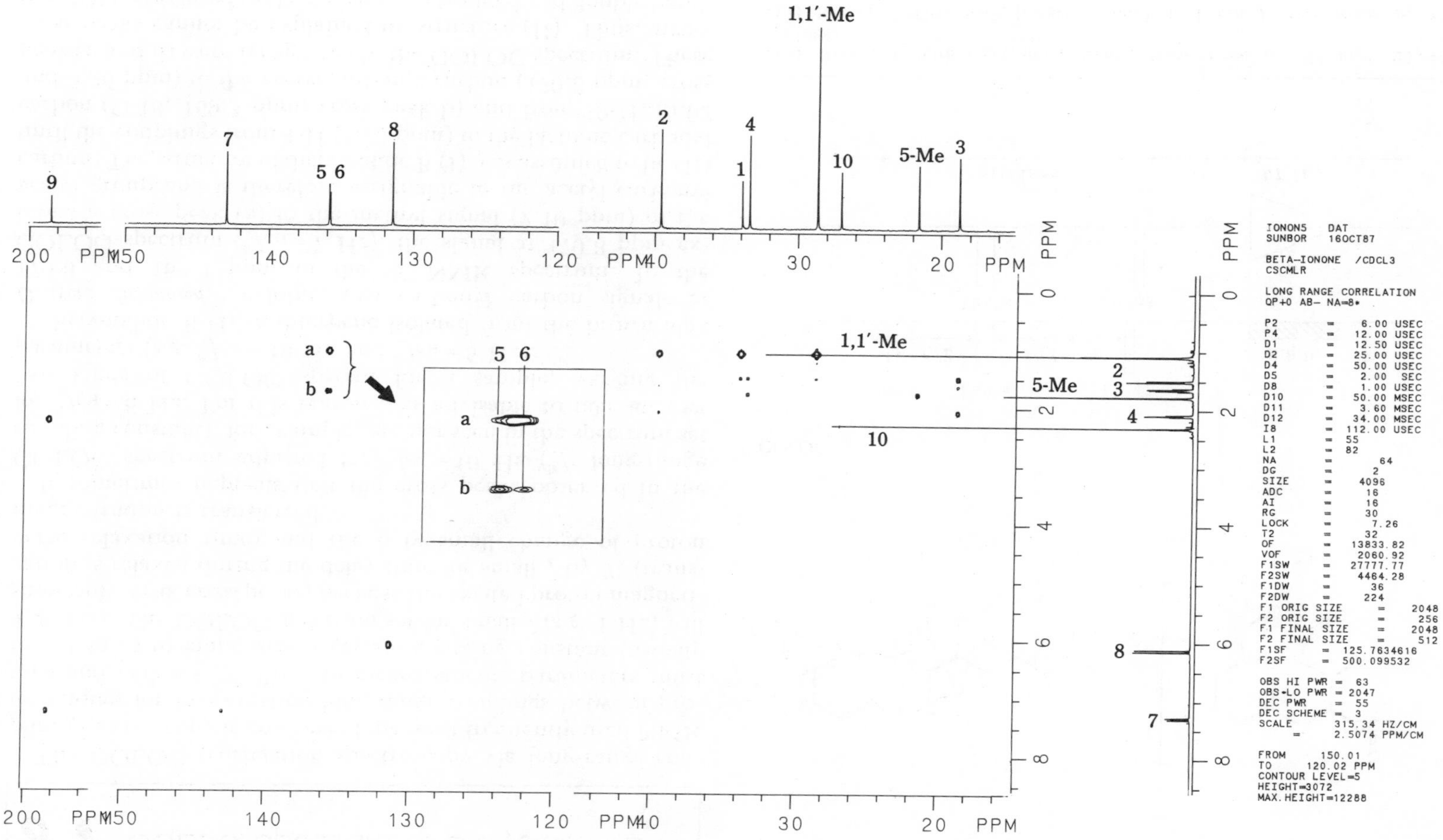

125 MHz ^{13}C-NMR ($CDCl_3$) [TI]

81 COLOC Spectrum of Dictyotalide B

The COLOC (correlation spectroscopy via long-range coupling) spectroscopy is now one of the most frequently used NMR techniques for investigating long-range couplings between protons and carbons.[1–3] Prior to measurement, parameters must be set so as to emphasize a certain coupling constant (usually 4-20 Hz). The COLOC spectrum set for small J (*e.g.* 1 Hz) will show only weak cross peaks, because the excited proton magnetization is relaxed during the delay time for small J by T_2 (transverse relaxation time) and the only small change of proton magnetization is transferred.

It sometimes happens that the cross peaks observed in the COLOC spectrum adjusted for ${}^{lr}J_{CH} = 10$ Hz (${}^{lr}J$: long-range coupling constant), for example, are not seen in the spectrum set for ${}^{lr}J_{CH} = 6$ Hz. For this reason it is advisable to take at least two kinds of COLOC spectra for a sample, varying the parameters (*e.g.* ${}^{lr}J_{CH} = 10$ Hz and ${}^{lr}J_{CH} = 5$ Hz).

Dictyotalide B (I), a diterpene isolated from the brown alga *Dictyota dichotoma*,[4] exhibits two carbonyl carbon signals at 170.8 and 169.3 ppm in the ^{13}C-NMR spectrum. In the COLOC spectrum (${}^{lr}J_{CH} = 7$ Hz), the signal at 170.8 ppm exhibits a cross peak (a) to the methyl signal (2.10 ppm) of the acetyl group and is therefore assignable to the acetyl carbonyl carbon. The structure of dictyotalide B (I) was assumed to be (II) until the couplings from 4-H (4.73 ppm) to the lactonic carbonyl carbon (C-18, 169.3 ppm; cross peak b) and from 19-H_2 (5.02 and 5.26 ppm) to the acetyl carbonyl carbon (170.8 ppm; cross peaks c and d) were recognized in the COLOC spectrum. These cross peaks cannot be explained by structure (II). Thus, structure (II) was revised to (I), which has a bridge-head double bond. Note that the 4-H and 19-H_2 are coupled to the carbonyl carbons through an oxygen atom.

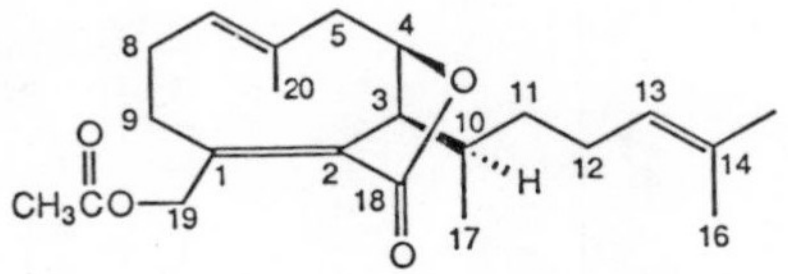

I

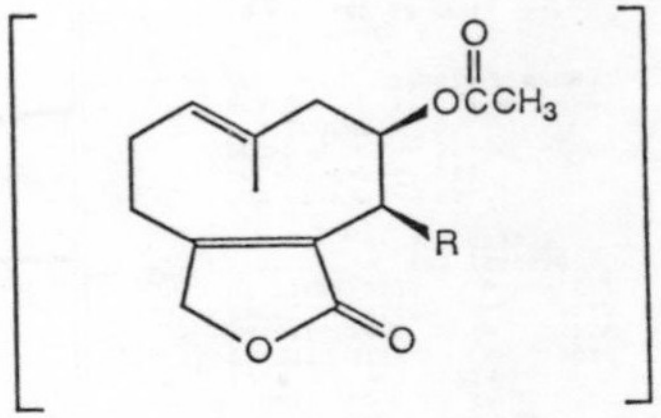

II

COLOC

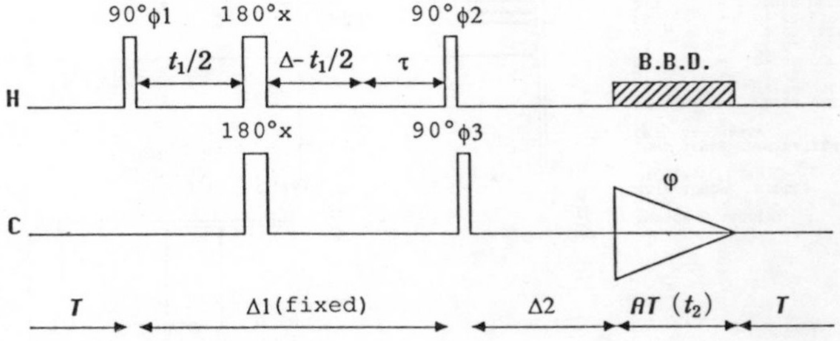

1) H. Kessler, C. Griesinger and J. Lautz, *Angew. Chem. Int. Ed. Engl.*, **23**, 444 (1984)
2) H. Kessler, C. Griesinger, J. Zarbock and H.R. Loosli, *J. Magn. Reson.*, **57**, 331 (1984)
3) H. Kessler, W. Bermel and C. Griesinger, *J. Am. Chem. Soc.*, **107**, 1083 (1985)
4) M.O. Ishitsuka, T. Kusumi and H. Kakisawa, *J. Org. Chem.*, **53**, 5010 (1988)

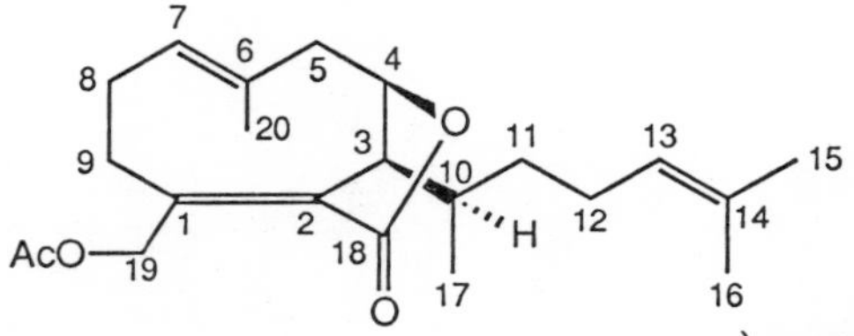

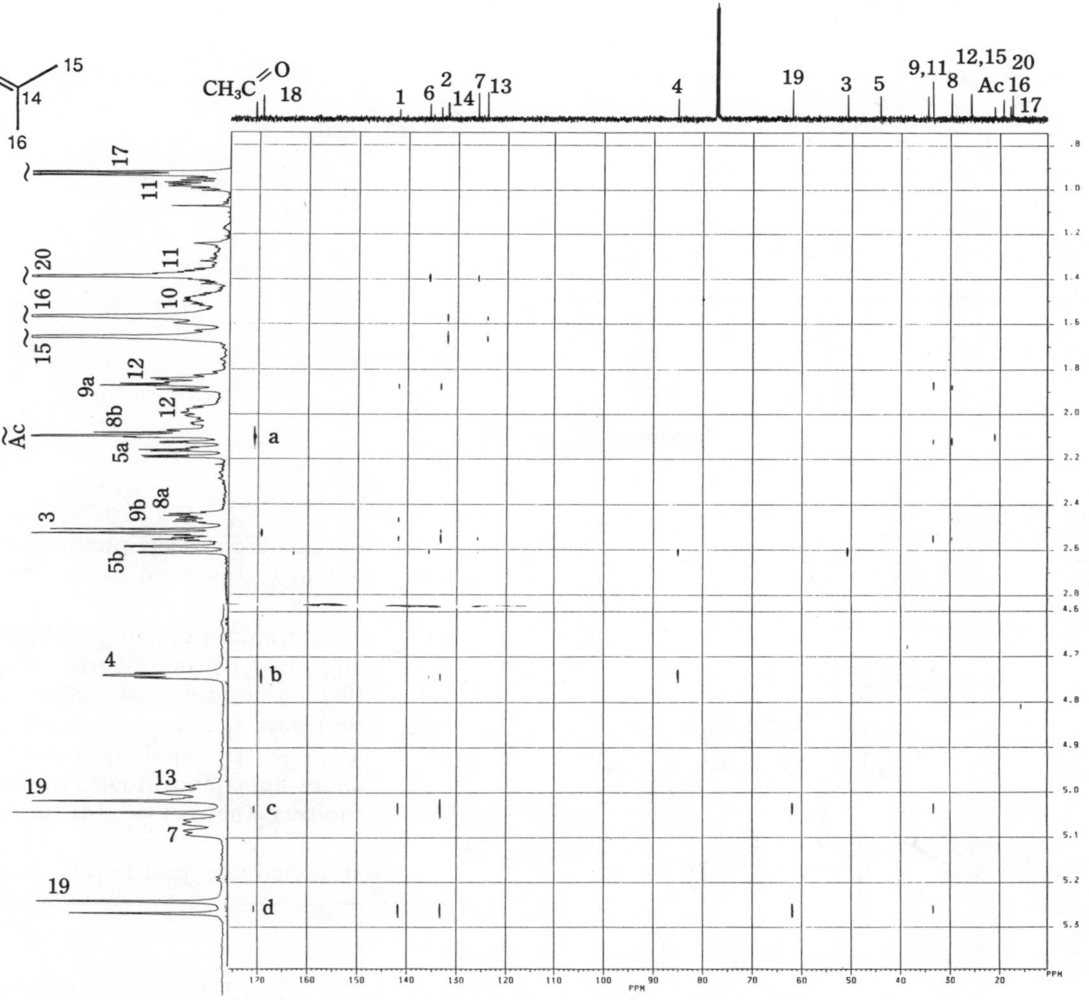

100 MHz ^{13}C-NMR (CDCl$_3$) [MW]

82 Long-range H,C-COSY Spectrum of Compactin

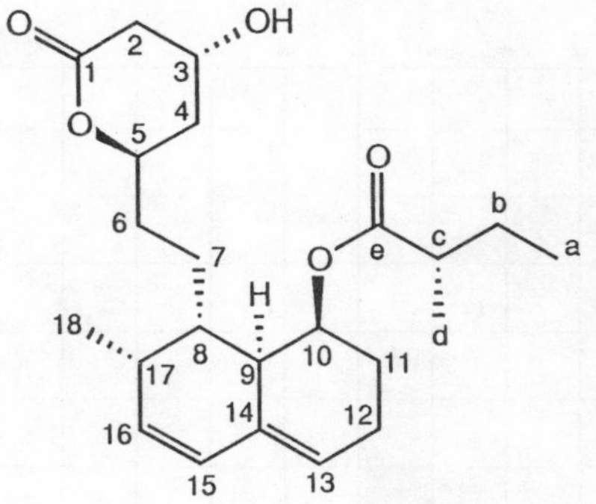

Assignment of the two carbonyl carbons of compactin is difficult by ordinary methods.

For the study of quaternary carbons such as carbonyl carbon, direct observation of the polarization transfer through long-range H-C coupling can also be done by using the pulse, H,C-COSY without decoupling (see Part III, section 2.4).[1)] It is seen that the higher field carbonyl carbon couples with proton 2 ($^2J_{CH}$) and the lower field carbonyl carbon with proton d ($^3J_{CH}$). The higher field carbonyl carbon is thereby assigned to 1 and the lower field carbonyl carbon to e.

The magnitude of $^2J_{CH}$ and $^3J_{CH}$ varies markedly with the kind of bonding, state of hybridization and electronegativity of substituent.

H,C-COSY without Decoupling

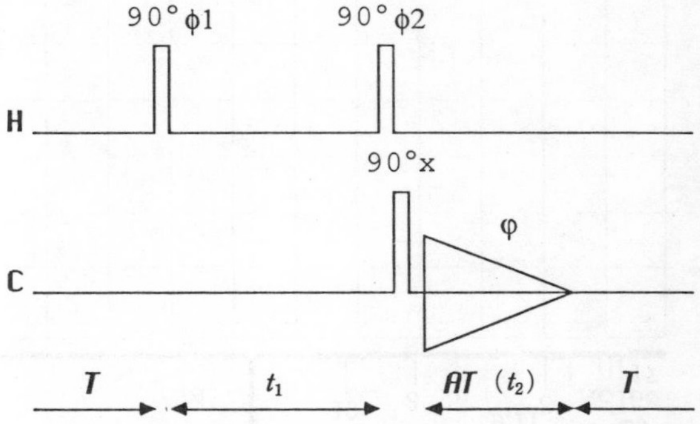

1) T. Iwashita and H. Naoki, *23rd Symposium on NMR*, Sendai, Japan (1984), *Abstracts*, p.33

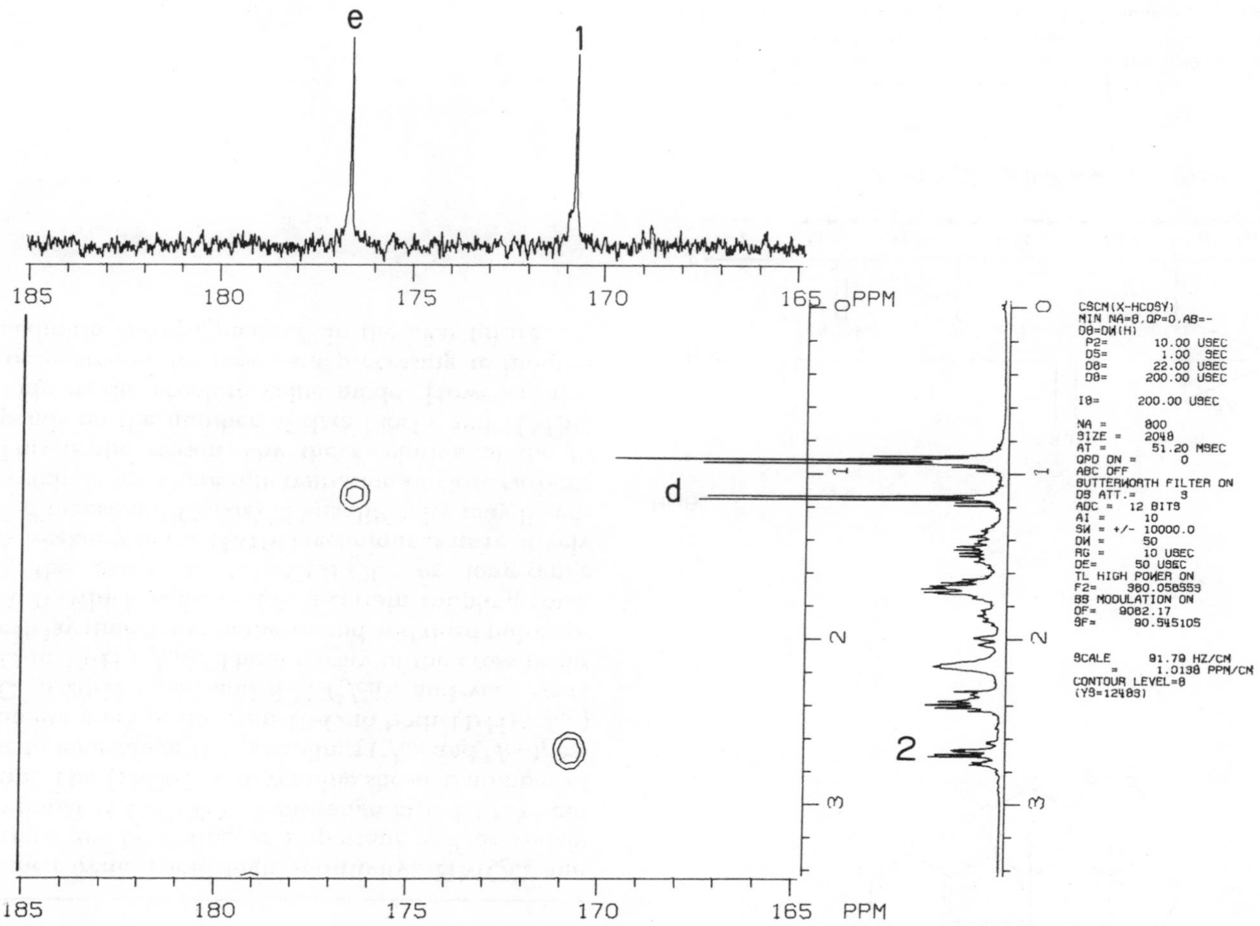

90MHz ^{13}C-NMR ($CDCl_3$) [TI]

83 HMBC Spectrum of Strychnine

Because of their remarkable high sensitivity, HMQC and HMBC[1, 2] spectra are becoming as important and as widely used as conventional H,C-COSY, long-range H,C-COSY and COLOC spectra. The HMBC of strychnine shows a number of cross peaks due to long-range H-C couplings ($^2J_{CH}$ and $^3J_{CH}$), especially, the intense cross peaks from 10-C to both 11-H_2 ($^2J_{CH}$) and from 14-C to 20-H ($^3J_{CH}$) and 8-H ($^3J_{CH}$), and very weak ones from 13-C to 15-H ($^3J_{CH}$). The intensity of the cross peaks depends on the delay time between the second and third pulse applied on ^{13}C (Δ 2), which is focused to a certain coupling constant ($1/2^nJ_{CH}$) the same as for COLOC or long-range H,C-COSY. A weakness of the HMBC technique is its relatively low resolved F_1 dimension (^{13}C-axis). Thus difficulty may be encountered in certain cases where differentiation of close carbons is required. This is the reason why the resolution of the F_1 dimension depends on the number of data blocks and HMBC can be done only in the absolute value mode. However, this problem will be overcome by new data processing techniques such as the maximum entropy method[3] in the near future.

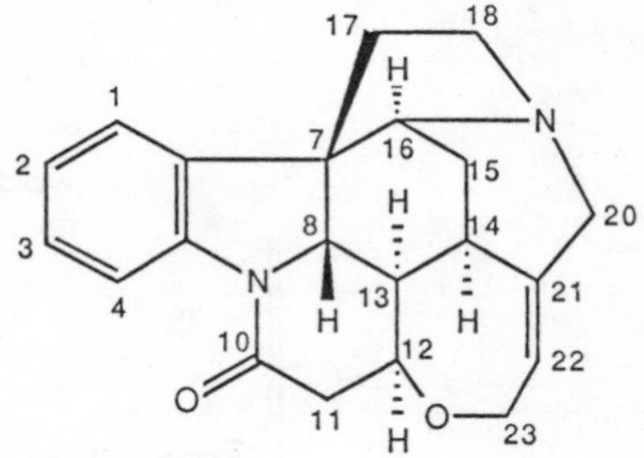

HMBC

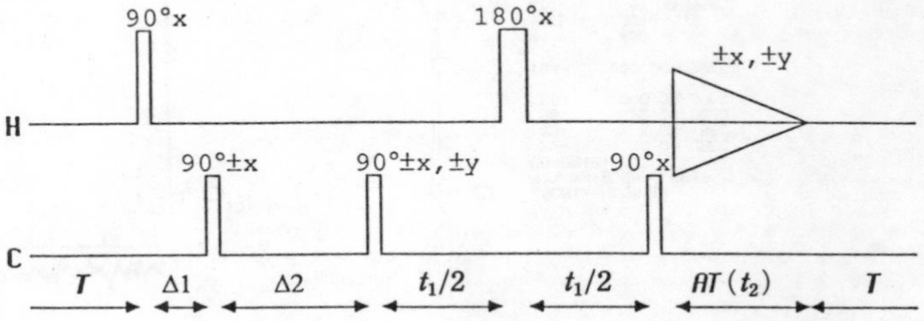

$\Delta_1 = 1/2^1J_{CH}$, In this case, Δ_2 is 50 ms.

1) A. Bax and M.F. Summers, *J. Am. Chem. Soc.*, **108**, 2093 (1986)
2) M.F. Summers, L.G. Marzilli and A. Bax, *J. Am. Chem. Soc.*, **108**, 4285 (1986)
3) E.D. Laue, M.R. Mayger, J. Skilling and J. Staunton, *J. Magn. Reson.*, **68**, 14 (1986)

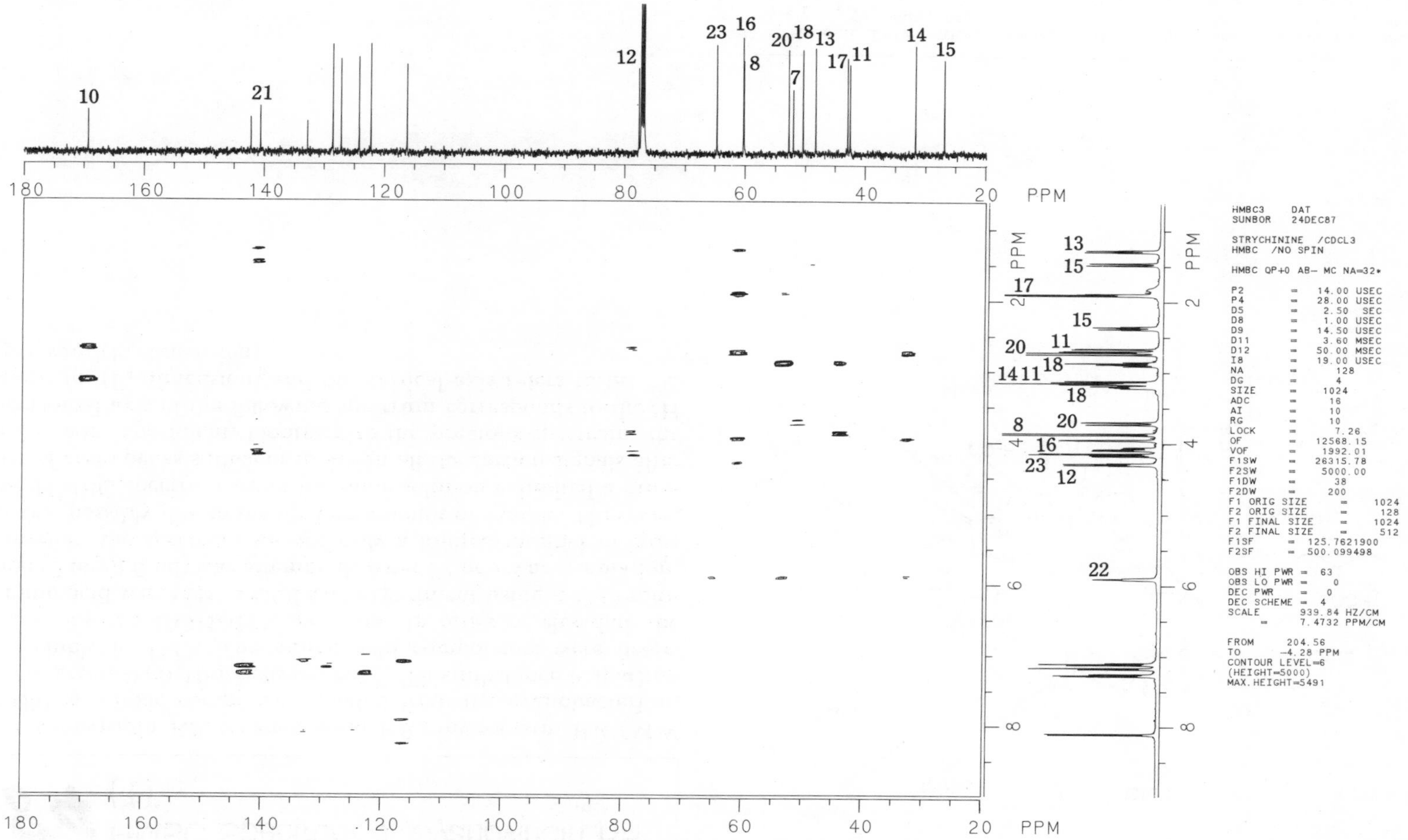

500 MHz ^{1}H-NMR ($CDCl_3$) [TI]

84 HMBC Spectrum of Cyanoviridin RR (I)

Cyanoviridin RR (cyanoginosin RR; microcystin RR)(MW 1038) is a toxic component isolated from the cyanobacterium (blue-green alga) *Microcystis viridis*.[1-3] This substance is sparingly soluble in D_2O. The amino acid components were determined by the HOHAHA spectrum. In order to elucidate the amino acid sequence, a COLOC experiment using a D_2O solution (3 mg/0.5 ml) was attempted. After 72 hours accumulation, however, the spectrum showed only a limited number of cross peaks, possibly due to insufficient amount of sample. However, the HMBC spectrum using the same solution exhibited a number of cross peaks sufficient to assign all the carbon signals after a 24-hour experiment. Contrary to the previous spectrum, the horizontal axis of the following spectrum corresponds to the ^1H spectrum (F_2 dimension) and the vertical axis refers to the ^{13}C spectrum (F_1 dimension).

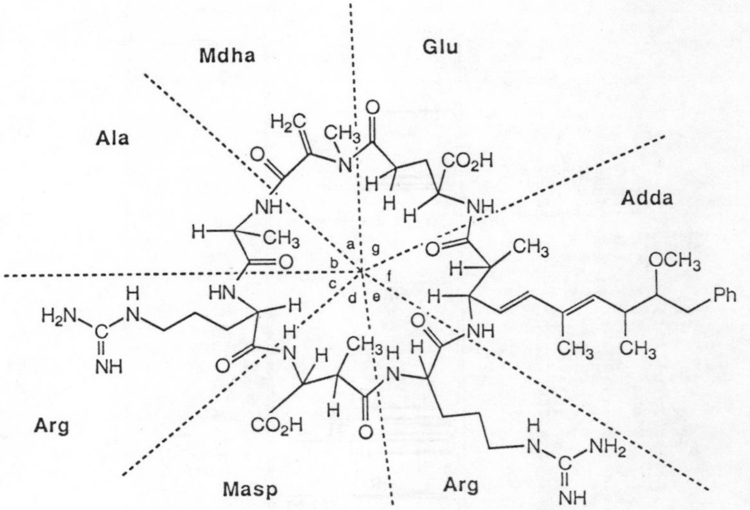

1) T. Kusumi, T. Ooi, M.M. Watanabe, H. Takahashi and H. Kakisawa, *Tetrahedron Lett.*, **28**, 4695 (1987)
2) T. Krishnamurthy, L. Szafraniac, E.W. Sarver, D.F. Hunt, J. Shabanowitz, W.W. Carmichael, S. Missler, O.M. Skulberg and G. Codd, *Proc. 34th Ann. Conference on Mass Spectrum and Allied Topics*, Cincinnati, Abstract, p. 93 (1986)
3) P. Painuly, R. Perez, T. Fukai and Y. Shimizu, *Tetrahedron Lett.*, **29**, 11 (1988)

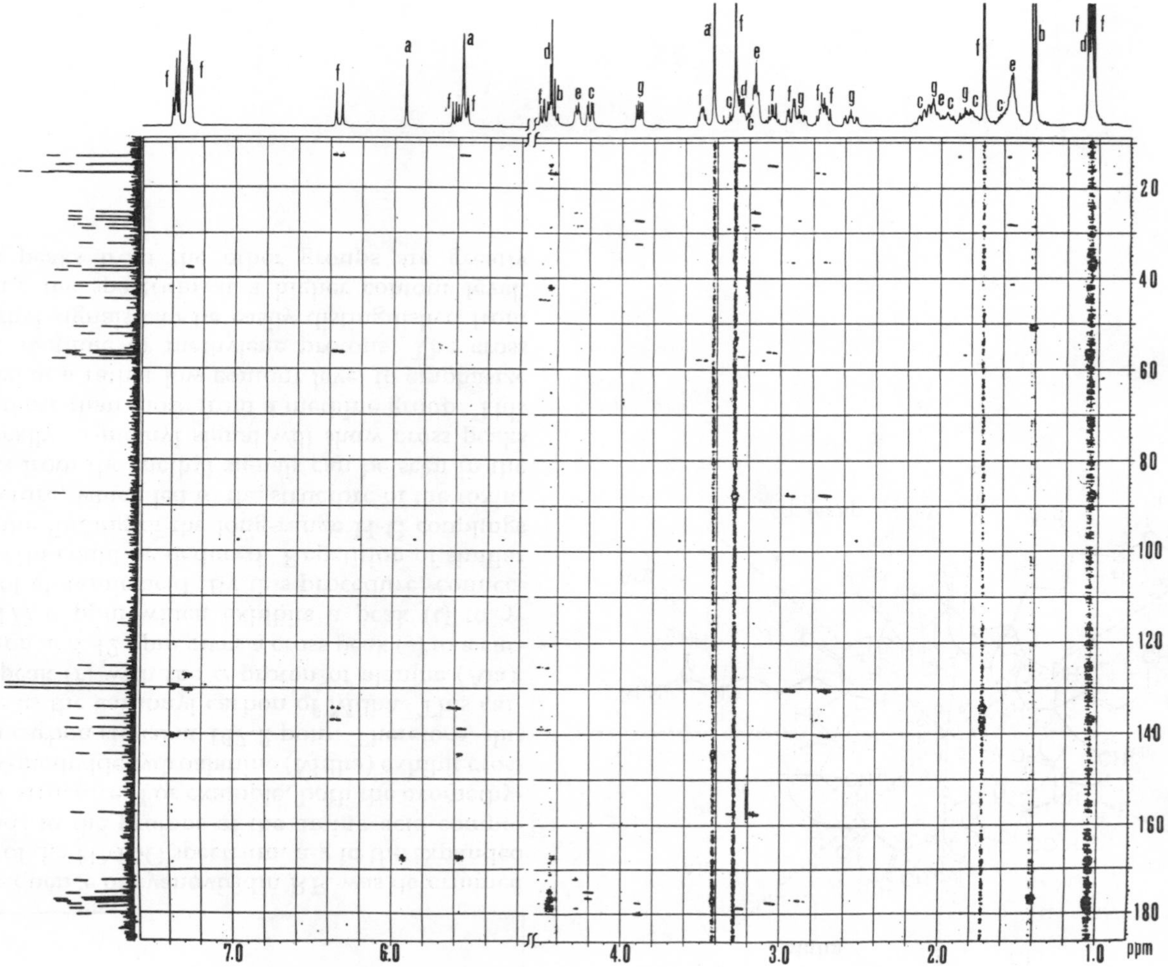

400 MHz ^{1}H-NMR (D_2O) [MW]

85 HMBC Spectrum of Cyanoviridin RR (II)

The amino acid sequence of cyanoviridin RR was determined by detailed analysis of the HMBC spectrum. a-g in the expanded spectrum corresponds to the protons of the amino acid components depicted in the structure. For example, both the exomethylene protons (a) of *N*-methyldehydroalanine (Mdha) exhibit cross peaks (p and q) to a carbon signal at 167.2 ppm. Therefore, this carbon is assignable as the carbonyl carbon of Mdha. This carbon shows a cross peak (r) with the α-proton of alanine (Ala). Also, *N*-methyl proton at 3.42 ppm gives a cross peak (s) to a carbon signal (g) at 177.0 ppm which exhibits a peak (t) to γ-methylene protons of glutamic acid. By this procedure, connectivity of Ala-Mdha-Glu could be deduced. Repetition of similar analyses permitted the finding of the long-range H-C couplings depicted in the structure, which led to the structure of the toxin.

Large noise peaks from the methyl signals can be seen in the spectrum. Theoretically, a methyl signal will show cross peaks three times more intense than those from a methine group. This spectrum was plotted at a rather low contour level to emphasize the peaks from the methine or methylene protons. The cross peaks from the methyl signals can be easily distinguished from the noise by plotting the spectrum at a higher contour level, although the cross peaks from the other groups are greatly weakened.

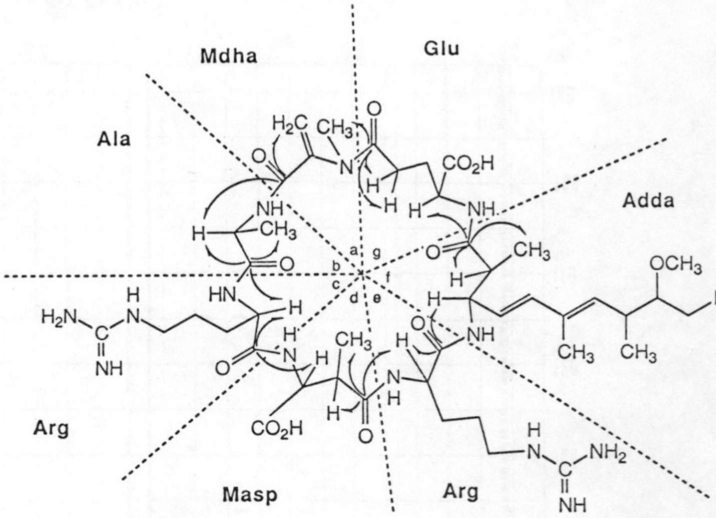

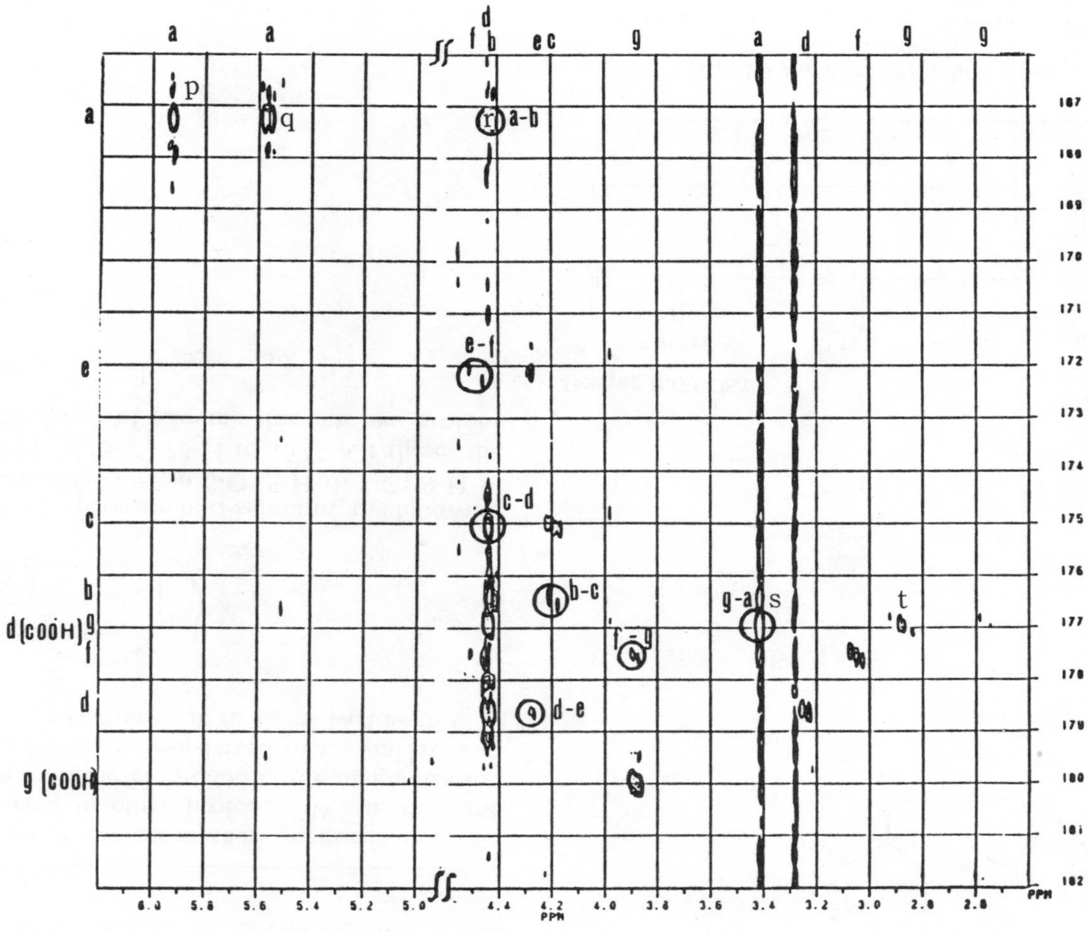

400 MHz ^{1}H-NMR (D_2O) [MW]

86 Relayed H,C-COSY of α-Santonin

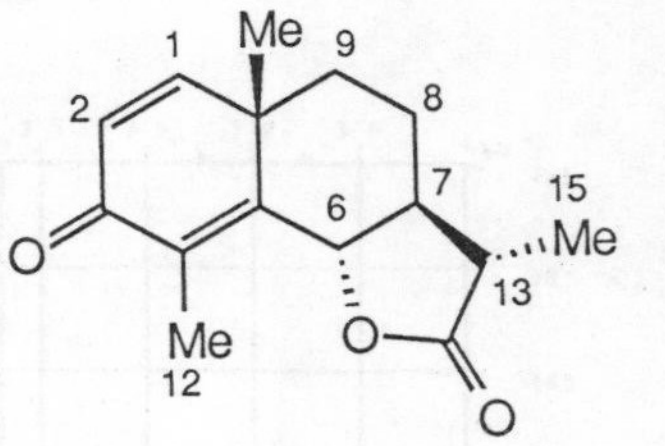

By the relayed coherence transfer method, magnetization of a proton can be transferred to other protons.[1] When the same technique is combined with the H,C-COSY, the magnetization of H_a of the following system is transferred to C_b via H_b, and therefore, H_a shows a correlation peak to C_b as well as to C_a.[2]

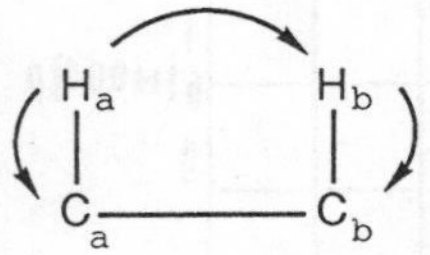

In the relayed H,C-COSY spectrum of α-santonin, the following relayed peaks are observable: 7-H to 6-C, 8-H to 7-C, 9-H to 7-C, 7-H to 13-C, 15-H to 13-C, 13-H to 15-C. Of these, the peak 9-H to 7-C is the result of relaying through two protons (8-H and 7-H).

Relayed H,C-COSY

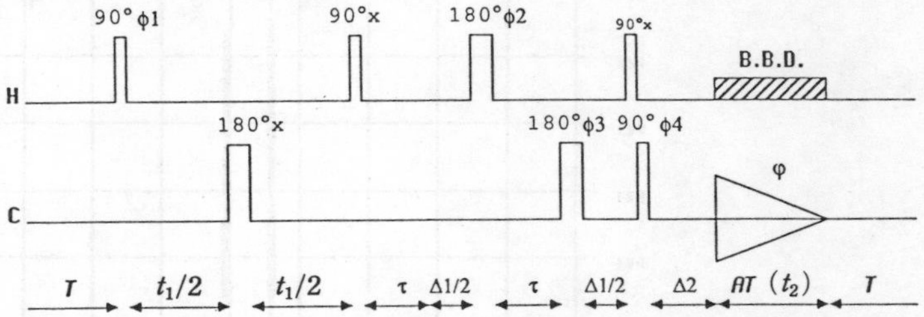

$\Delta_1 = 1/2{}^1J_{CH}$, $\Delta_2 = 1/3{}^1J_{CH}$, In this case, τ is 14.28 ms.

1) G. Wagner, *J. Magn. Reson.*, **55**, 151 (1983)
2) H. Kessler, M. Bernd, H. Kogler, J. Zarbock, O.W. Sørensen, G. Bodenhausen and R.R. Ernst, *J. Am. Chem. Soc.*, **105**, 6944 (1983)

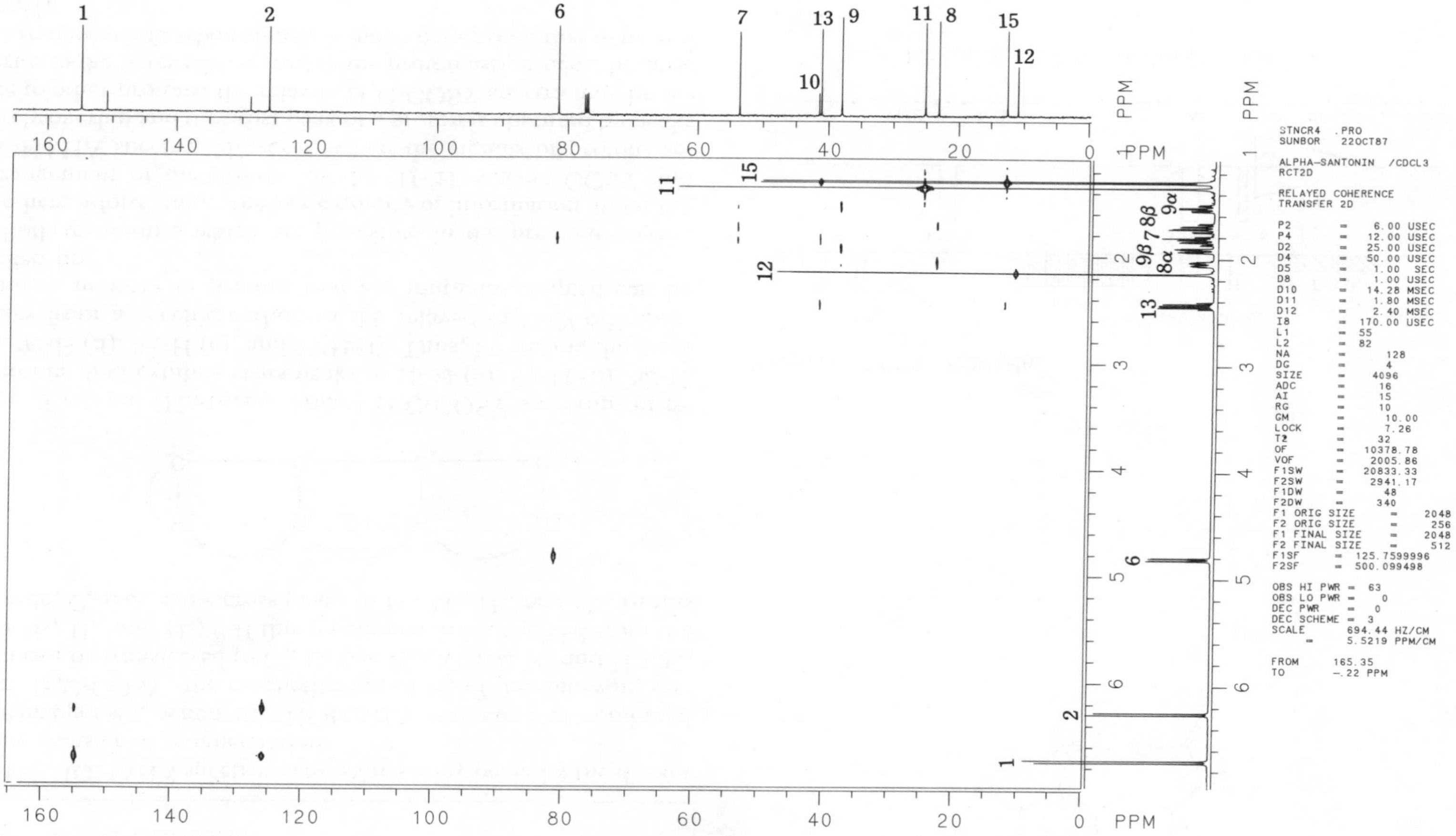

125 MHz ^{13}C-NMR (CDC$_6$) [TI]

87 Relayed H,C-COSY/HOHAHA of α-Santonin

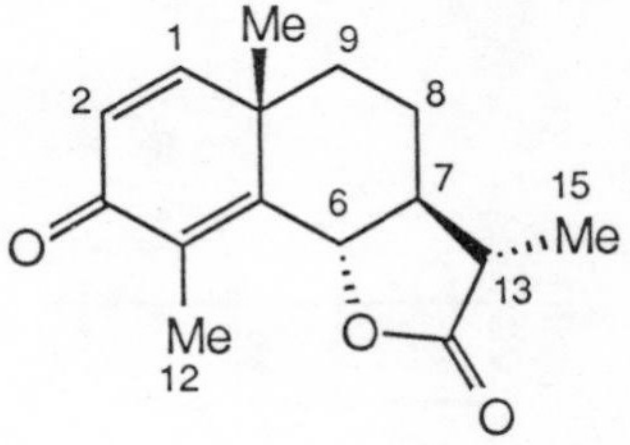

The HOHAHA spectrum also shows relay peaks by the distant relay transfer of magnetizations.[1]

Analogously, when the HOHAHA technique is combined with H,C-COSY, the magnetization of H_a of the following system can be transferred to C_a, C_b (*via* H_b), C_c (*via* H_b and H_c), C_d (*via* H_b, H_c, and H_d).[2] If this phenomenon is observed from the C_a side, C_a may show cross peaks to H_a, H_b, H_c, and H_d. In this

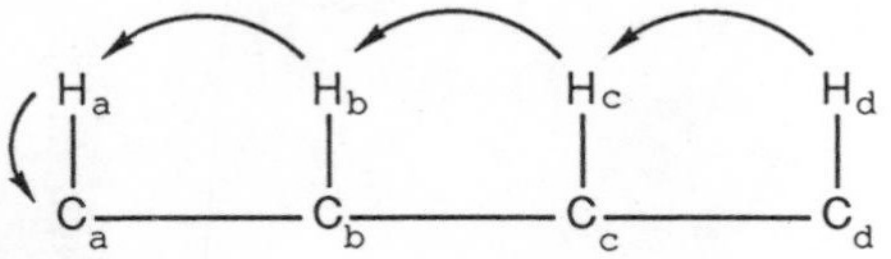

type of relayed (Hartmann-Hahn) H,C-COSY spectrum of α-santonin, 6-C exhibits cross peaks to 15-H (a), 9α-H (b), 8β-H (c), 9β-H (d), 8α-H (e), and 13-H (f). Thus, by tracing the cross peaks from a specific carbon in the relayed H,C-COSY spectrum, a network of protons that are mutually coupled can be picked up.

Both techniques which are presented in the previous section and here afford almost the same quality of information about the arrangement of the protons as the ^1H-^1H relayed COSY and HOHAHA spectra. However, when the signals of protons severely overlap and the cross peak of a proton is obscured by peaks due to other protons, the relayed H,C-COSY spectra may be superior to the latter with regard to the proton assignments because the resolution of carbon signals is much better than that of proton signals.

Relayed H, C-COSY/HOHAHA

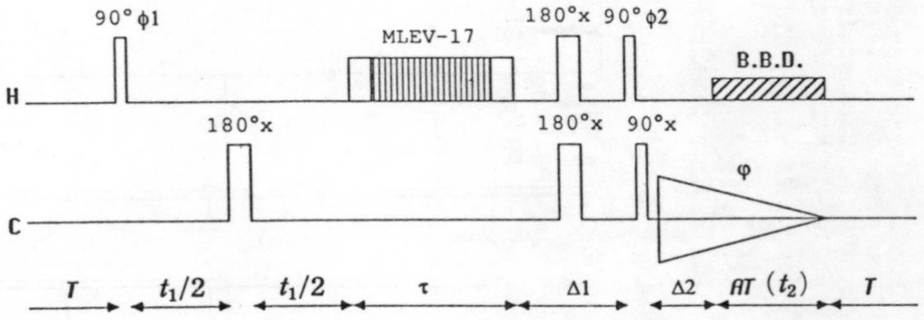

$\Delta_1 = 1/2^1J_{CH}$, $\Delta_2 = 1/3^1J_{CH}$, In this case, τ is 75 ms.

1) See section 55.
2) A. Bax, D.G. Davis and S.K. Sarkar, *J. Magn. Reson.*, **63**, 230 (1985)

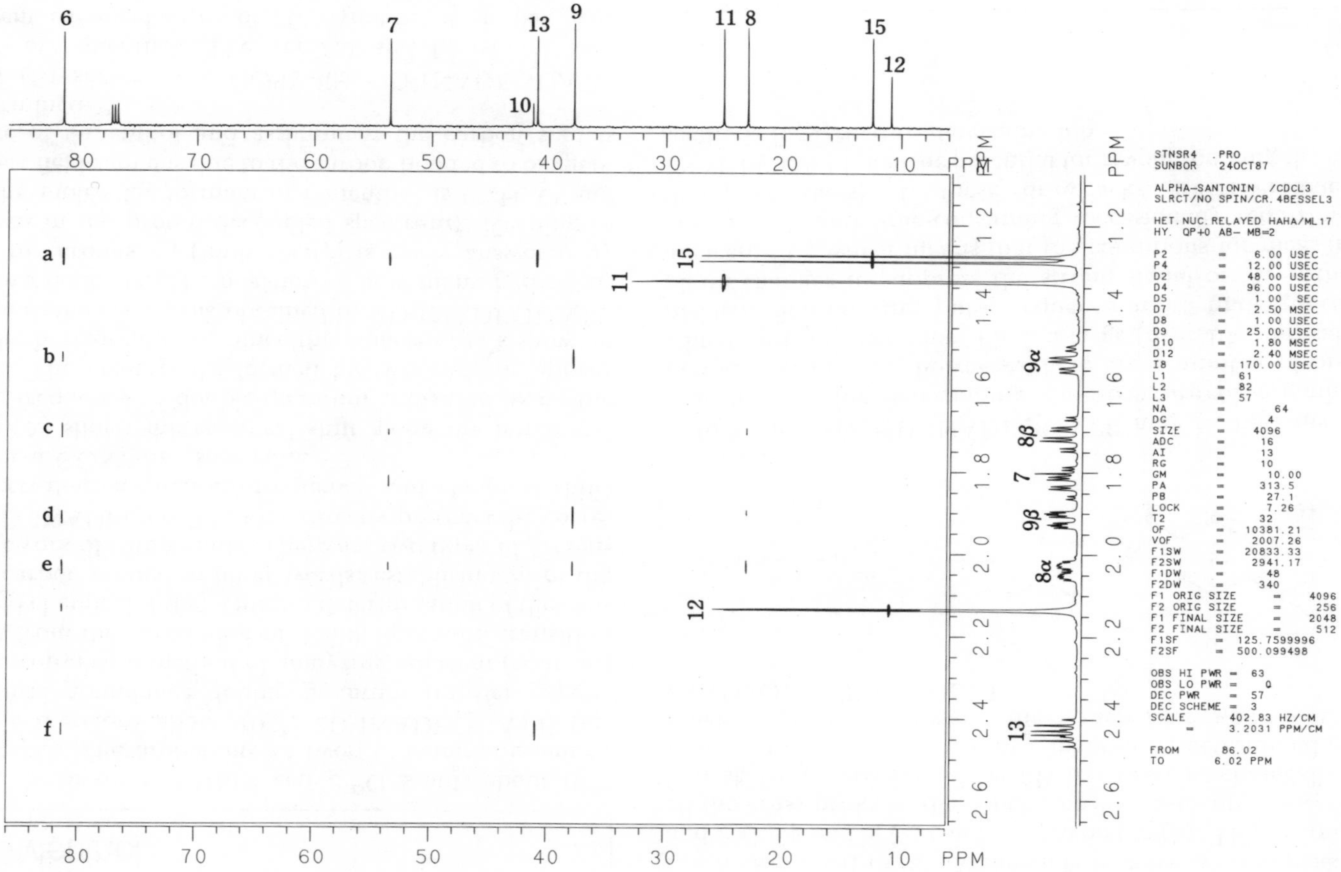

125 MHz ^{13}C-NMR ($CDCl_3$) [TI]

88 2D-INADEQUATE Spectrum of L-Menthol

In nature, the amount of NMR-active ^{13}C is only about 10^{-2} times that of ^{12}C. The probability of two ^{13}C being adjacent to each other is therefore about 10^{-4}. 2D-INADEQUATE (incredible natural abundance double quantum transfer experiment spectroscopy)[1] is a method of analyzing adjacent (coupled) ^{13}C-^{13}C pairs from the correlation of double quantum transition frequency and chemical shifts. This is a determination of the connectivity of carbon atoms, in other words, establishment of the carbon framework of compounds. There are two types of presentations for 2D-INADEQUATE. One directly presents the correlation between double quantum-frequency and chemical shift, and the other a COSY-like spectrum.

Spectrum (A) shows the chemical shift along the horizontal axis and the frequency of double-quantum transition along the vertical axis. The cross peaks formed by two carbons appear parallel to the horizontal axis. Since the adjacent ^{13}C's show an AX type signal, the cross peak obtained by 2D-INADEQUATE appears as two doublets. The question of how many hydrogens are bonded to carbons a-j (multiplicity) is easily answered by DEPT (shown in the proton decoupled spectrum). Keeping in mind that the molecular formula of L-menthol is $C_{10}H_{20}O$ and that the lowest field signal is due to the carbon bonded to oxygen, the structure of L-menthol and assignment the carbon signals can be determined.

Spectrum (B) shown is a COSY-like 2D-INADEQUATE spectrum[2–4] of L-menthol. The vertical and horizontal axes both represent chemical shifts of ^{13}C. Analysis is simple. The cross peaks corresponding to the adjacent 13C's appear symmetrically with respect to the diagonal line. The carbon framework of the molecule can be determined simply by taking the signal of one carbon as the origin and successively connecting adjacent carbons. Starting from j, a partial framework of (1) is obtained. Since e shows a cross peak with a and b and d with g, (1) can be extended to (2). Meanwhile h shows a cross peak with f and f with g and e, (2) can be expanded to (3). This encompasses all the cross peaks so the conclusion that L-menthol has the carbon skeleton represented by (3) has been mechanically determined. Thus the structure of L-menthol is determined without much difficulty from the molecular formula, 2D-INADEQUATE and DEPT.

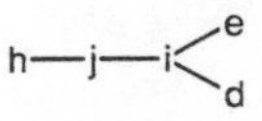

(1)

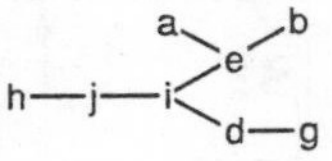

(2)

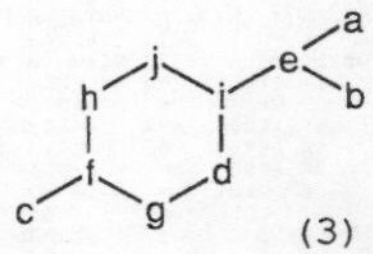

(3)

In some ways, 2D-INADEQUATE may be regarded as the best method for determining the structure of organic compounds. At present, however, there are a number of problems which must be overcome. First, a large (over 100 mg) sample is needed. Second, the pulse sequence must undergo complex phase changes to suppress the strong signal of ^{13}C without an adjacent ^{13}C, which means that the conditions for measurement are difficult and time-consuming (occasionally requiring more than two days). If these drawbacks are overcome, 2D-INADEQUATE has the potential for revolutionizing the existing methods for structure determination.

1) A. Bax, *Two-Dimensional Nuclear Magnetic Resonance in Liquids*, Delft University Press, Delft, Holland (1982) pp.155-174
2) D.L. Turner, *J. Magn. Reson.*, **53**, 259 (1983)
3) D.L. Turner, *J. Magn. Reson.*, **49**, 175 (1982)
4) A. Bax and T.H. Mareci, *J. Magn. Reson.*, **53**, 360 (1983)

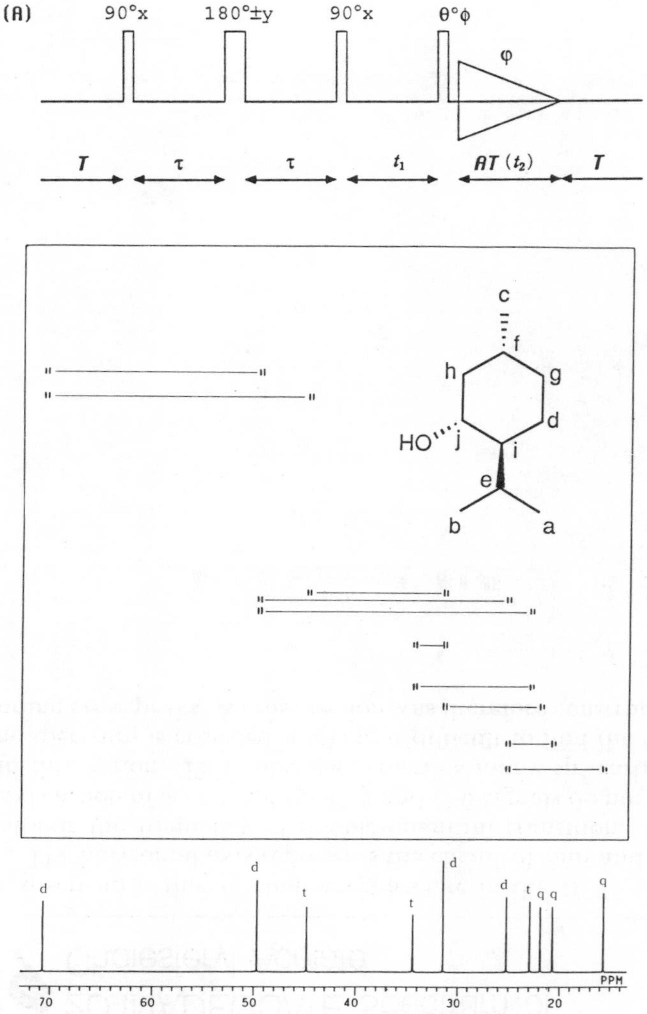

100 MHz ^{13}C-NMR (C_6D_6) [JEOL]

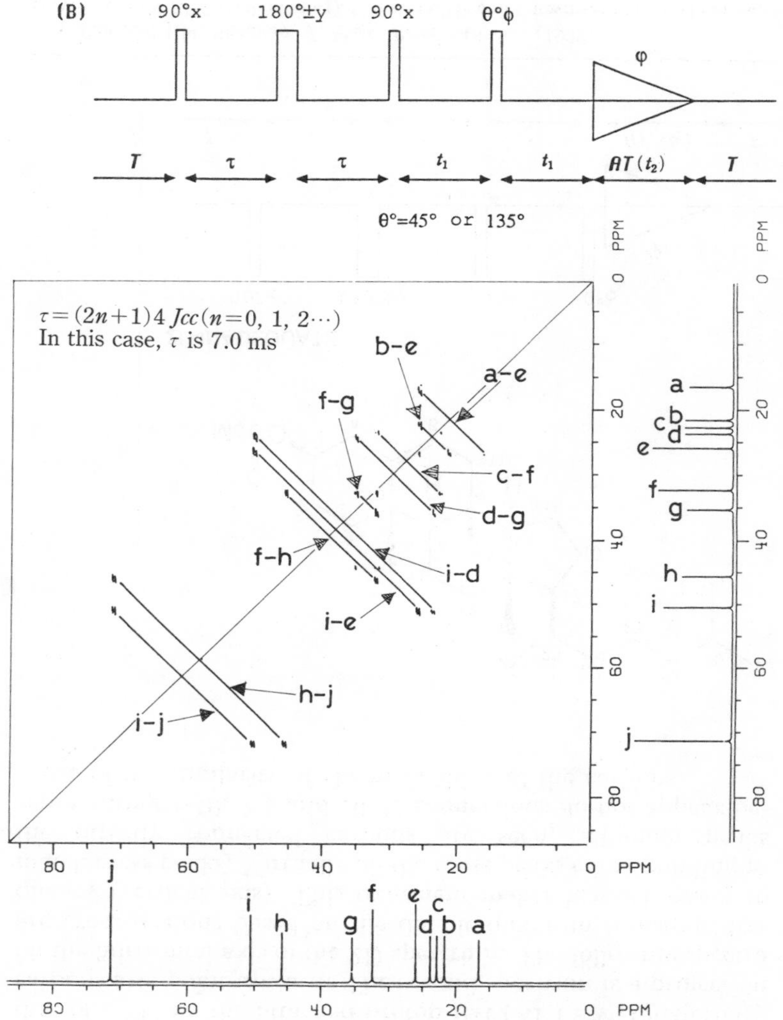

90 MHz ^{13}C-NMR ($CDCl_3$) [TI]

89 2D-INADEQUATE Spectrum of Cholesteryl Acetate

The spectrum is that of cholesteryl acetate in the 10-75 ppm region. The horizontal axis represents the chemical shift and the vertical axis the frequency of double-quantum transition. The carbonyl carbon of acetyl and the C-5 and C-6 signals do not appear in this region. This substance contains many sp^3 carbons and the spectrum is crowded, making it difficult to find the corresponding cross peaks. A cross section was therefore constructed to obtain the spectra shown at the bottom. The top spectrum is that of a ^{13}C by the attached proton test (APT).[1, 2] Quaternary carbons also point upward. The second spectrum is a projection on the horizontal axis of the 2D spectrum. The following spectra are cross sections based on the double-quantum transition frequency (vertical axis). This treatment makes it much easier to match cross peaks. Virtually all the cross peaks corresponding to the directly connected carbons are seen, although peaks representing 1-10, 7-8 and 20-22 connections do not appear because of the similarity of chemical shifts of the carbons.

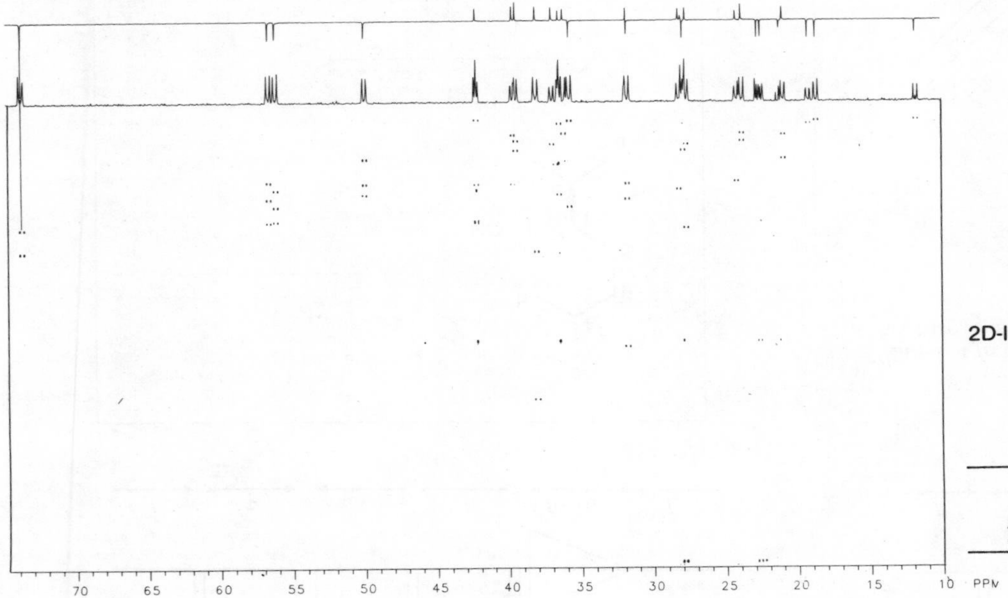

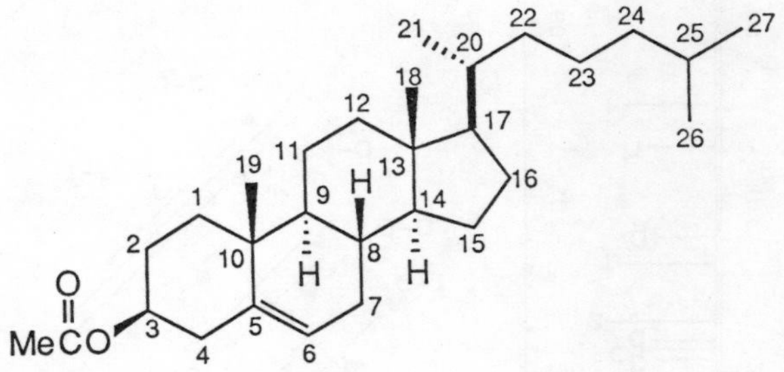

2D-INADEQUATE

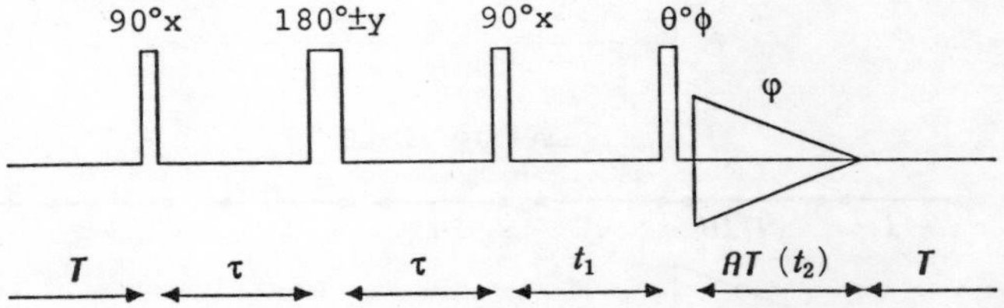

1) S. Patt and J.N. Shoolery, *J. Magn. Reson.*, **46**, 535 (1982)
2) As in the IN-EPT method (3/4J), CH and CH_3 point downward and CH_2 upward.

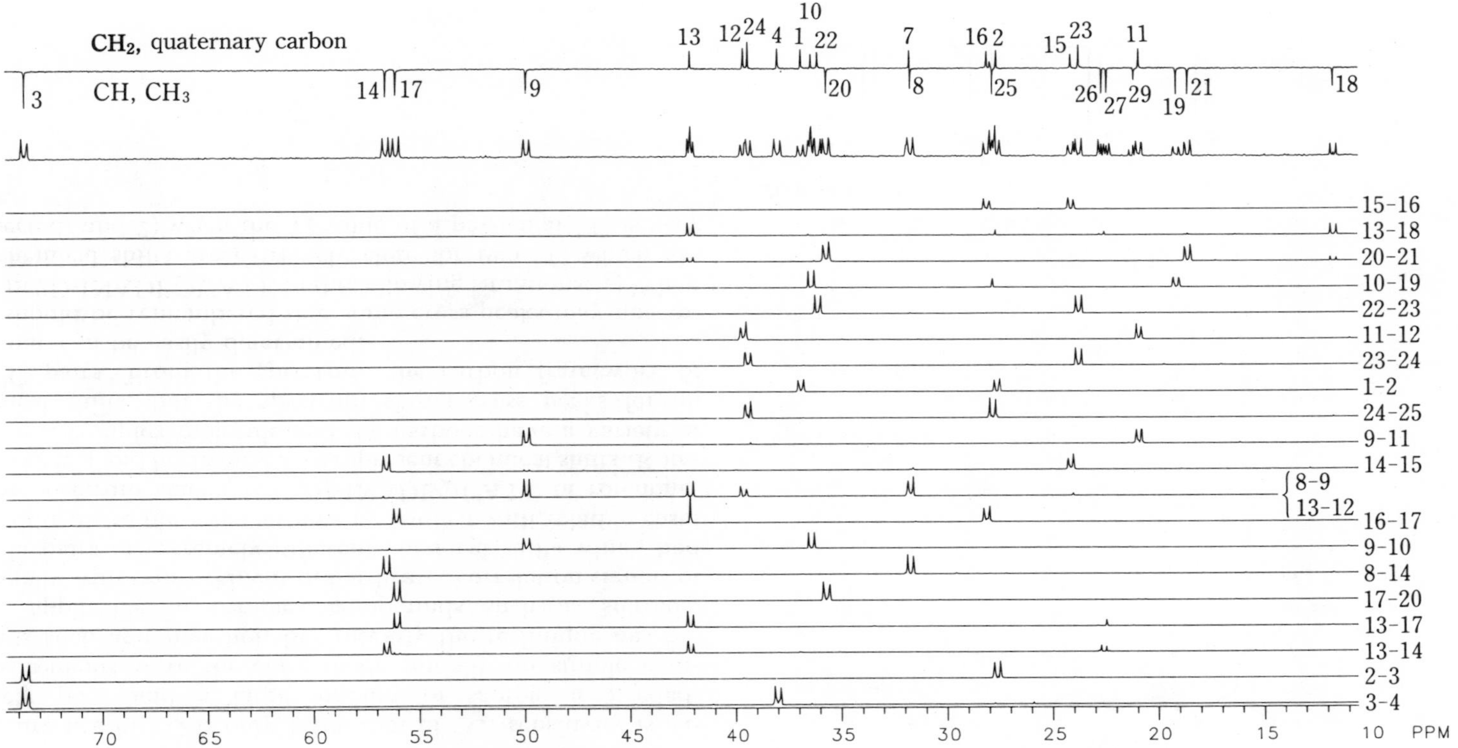

125MHz ^{13}C-NMR ($CDCl_3$) [LJ]

90 2D-INADEQUATE Spectrum of Rotenone

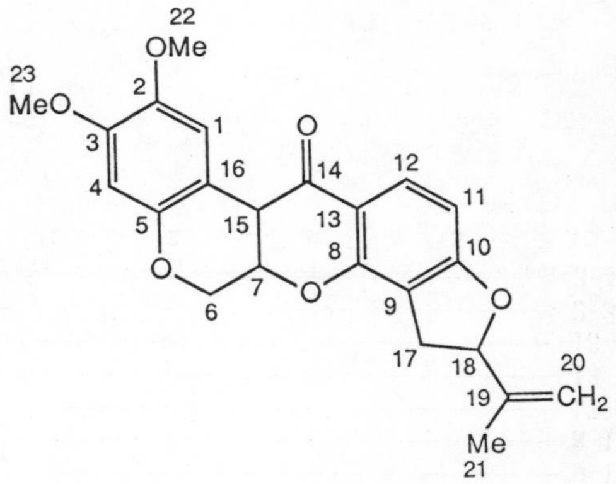

As stated earlier, 2D-INADEQUATE spectroscopy detects the double-quantum transition of adjacent ^{13}C-^{13}C. Since the natural occurrence of ^{13}C is only 1/100 that of ^{12}C, sensitivity is extremely poor and a large amount of sample is needed. Measurements were therefore made initially on simple compounds such as L-menthol, but recently the technique has become applicable to complex compounds such as steroids. Especially with improvement in hardware, correlation signals of ^{13}C-^{13}C pairs of extremely different chemical shifts which had been difficult to observe can now be studied with relative ease.

The spectrum shown is a 2D-INADEQUATE of rotenone. Both vertical and horizontal axes represent chemical shifts. Rotenone is a complex molecule and the carbons have a variety of chemical shifts, but the spectrum shows cross peaks for all ^{13}C-^{13}C pairs. From the spectrum, the carbon framework of rotenone can be easily determined.

It should be remembered that some cross peaks may not appear in 2D-INADEQUATE in the following situations: 1) When the chemical shifts are extremely close for two ^{13}C which are connected, and 2) when the T_1 value of a particular ^{13}C is very large.

2D-INADEQUATE

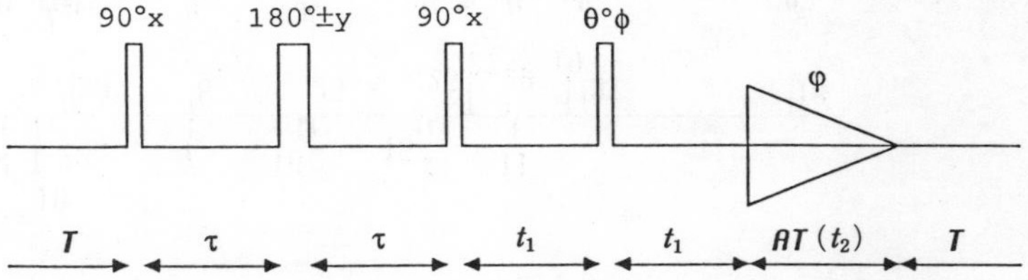

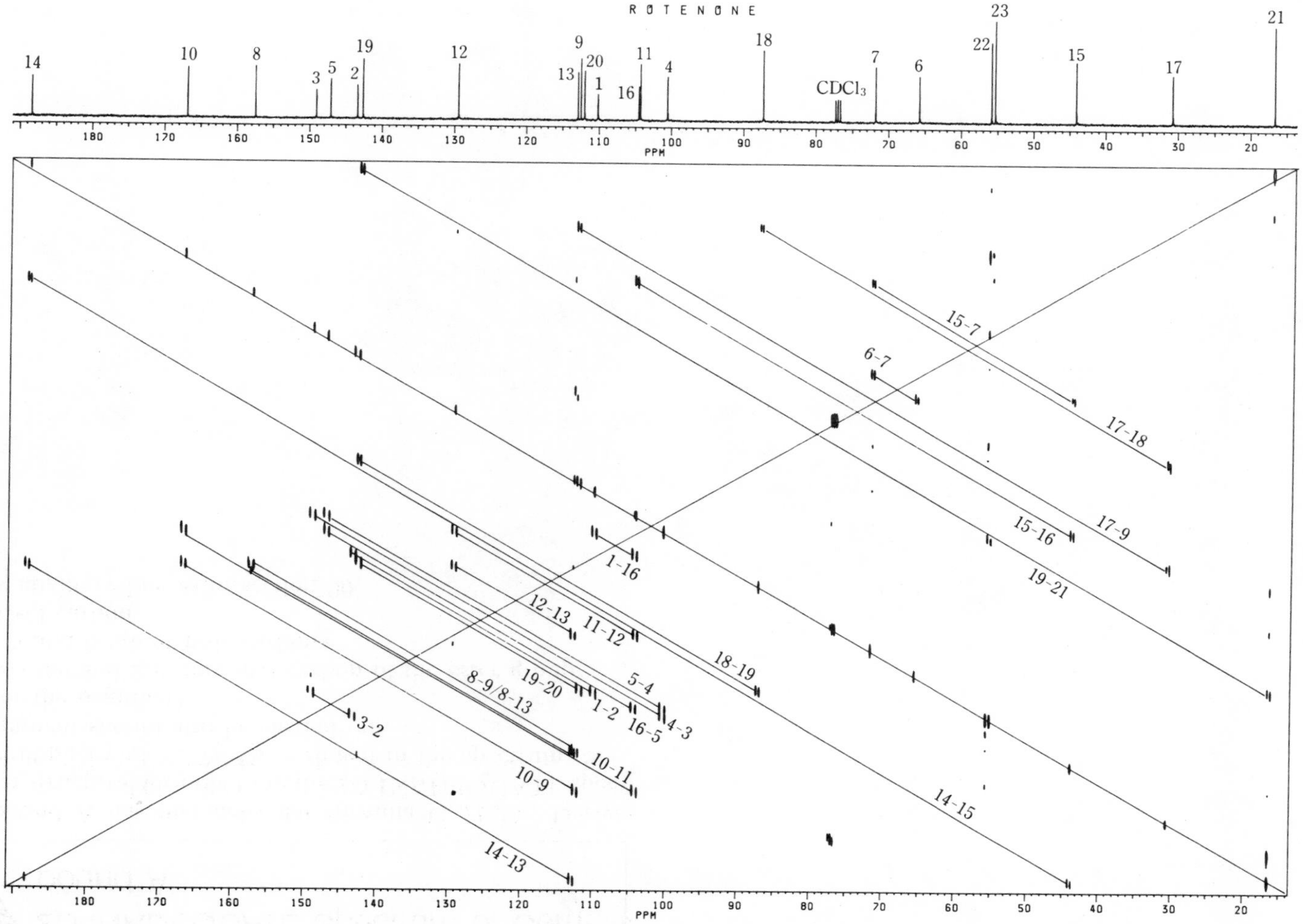

100MHz ^{13}C-NMR ($CDCl_3$) [MW]

91 2D-INADEQUATE Spectrum of Compound A

Compound A has the molecular formula $C_{15}H_{18}O_3$. Derive the planar structural formula from the 2D-INADEQUATE spectrum. Multiplicity of ^{13}C-NMR is shown in the spectrum. Signal assignment should also be carried out.

(Hints for the beginner)

1) 1 is ketone and 2 is carbonyl carbon of the ester group.
2) 3, 4, 5 and 6 are olefinic carbons.
3) 7 is C-O carbon.
4) Draw auxiliary line as in section 90.

Compound A ($C_{15}H_{18}O_3$)

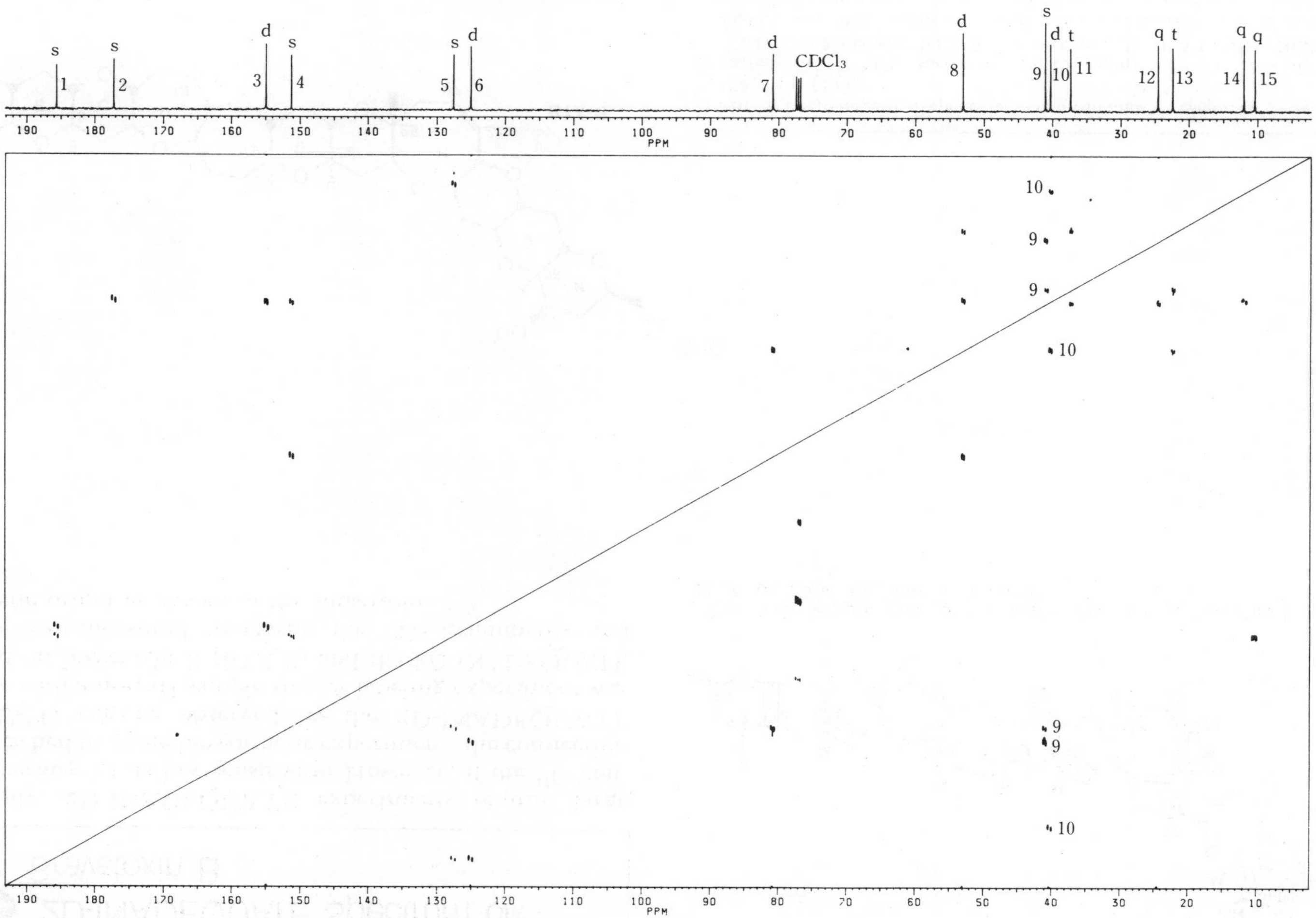

100MHz ^{13}C-NMR ($CDCl_3$) [MW]

92 2D-INADEQUATE Spectrum of Brevetoxin B

Generally, 2D-INADEQUATE experiments require large samples because of its low sensitivity. However, if the ^{13}C content is enriched by some biosynthetic experiment, the connectivity of ^{13}C-^{13}C can be observed by the 2D-INADEQUATE technique with a normal sample size. A labeling experiment was performed on brevetoxin B (BTX-B) and the 2D-INADEQUATE spectrum was measured to clarify the ^{13}C assignments and biosynthetic origin as shown in the illustration.[1, 2)]

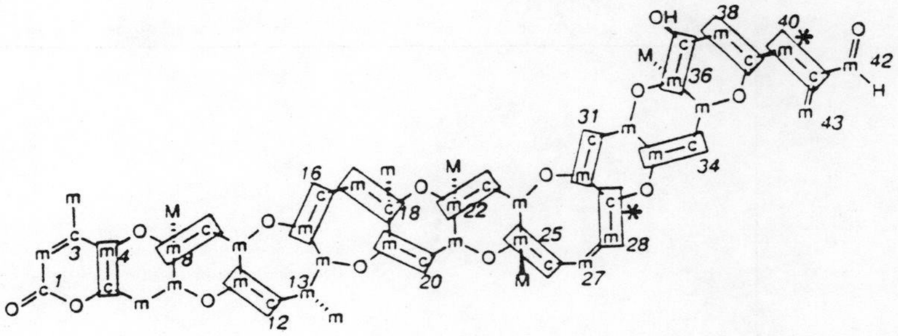

(Reproduced by permission from Lee, Min S., Repeta, D. J., Nakanishi, K and Zagorski, M., *J. Am. Chem. Soc.*, **108,** 7856 (1986))

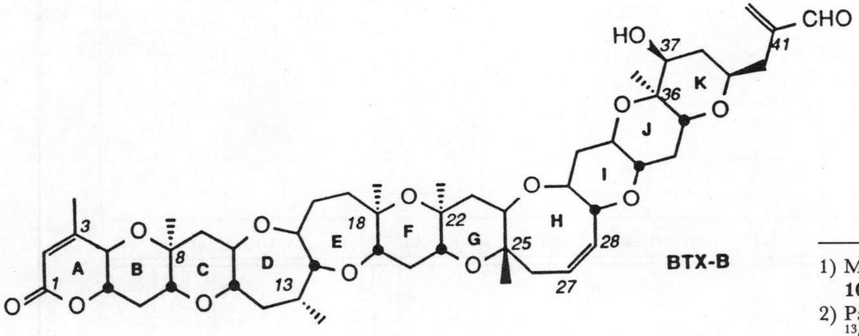

BTX-B

1) Min S. Lee, Daniel J. Repeta, K. Nakanishi and M. Zagorski, *J. Am. Chem. Soc.*, **108**, 7855 (1986)
2) Partial ^{13}C NMR spectrum (90-25 ppm) of 2D INADEQUATE of $^{13}CH_3{}^{13}CO_2Na$ labeled BTX-B (3.2 mM) in C_6D_6. 125.13 MHz Bruker AM-500. The C_2 units that originate from the same acetates are enclosed in squares. The asterisked carbon pairs 28/29 and 40/41 are not in the figure because they fall out of the range shown, but the connectivities were evident. m = CH_3CO_2Na. c = $CH_3{}^{13}CO_2Na$. and M = $^{13}CH_3SCH_2CH_2CH(NH_2)CO_2H$.

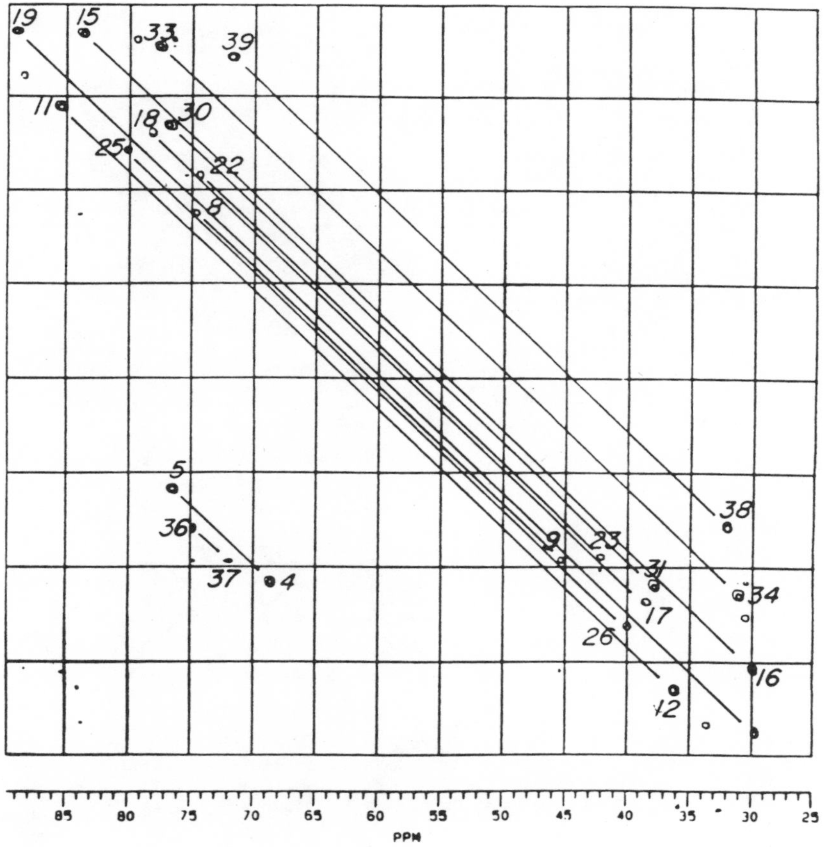

125 MHz ^{13}C-NMR (C_6D_6) [MD]

(Reproduced by permission from Lee, Min S., Repeta, D. J., Nakanishi, K. and Zagorski, M., *J. Am. Chem. Soc.*, **108,** 7856 (1986))

Part III

Principles of FT-NMR

1 FT-NMR

In an atomic nucleus containing a spin quantum number, I, there is a magnetic moment $\boldsymbol{\mu}$, and in a static magnetic field there are $(2I+1)$ energy levels. Nuclear magnetic resonance spectroscopy (NMR) is concerned with the transition between these levels. In ^1H and ^{13}C, $I=1/2$, which means that there are two energy levels, each atomic nucleus being in one or the other energy state (the stable state being $\alpha[-1/2\gamma\hbar B_o]$ and the unstable state being $\beta[+1/2\gamma\hbar B_o]$, where γ is the magnetogyric ratio, $\hbar$ is Planck's constant divided by 2π and B_o is the strength of the static magnetic field of a spectrometer).[1] The differences in energy level increase with the strength of the static magnetic field ($\boldsymbol{B}_o$), and the probability of the presence of atomic nuclei in the various states is determined by the Boltzmann distribution. The property of the atomic nucleus as a magnet (magnetic dipole moment $\boldsymbol{\mu}$) is in the realm of quantum dynamics, but much of the phenomenon may be explained using the spinning top as a model. If we think of the atomic nucleus as a spinnig magnet, the two energy states refer to whether the magnet precesses are parallel or anti-parallel to the static magnetic field $\boldsymbol{B}_o$ (Equation 1).

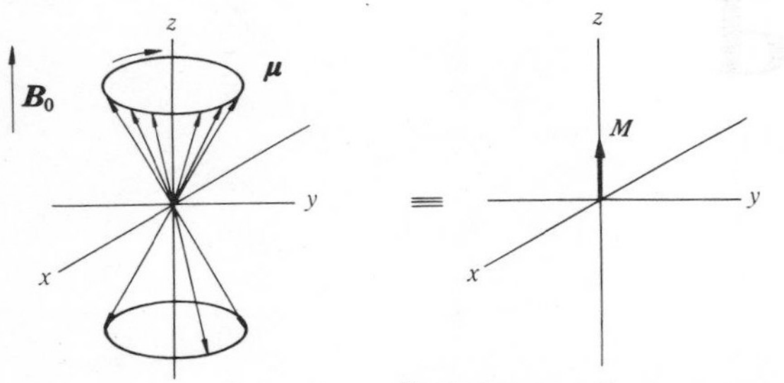

Fig. 1.1

$$d\boldsymbol{\mu}/dt = \gamma(\boldsymbol{\mu} \times \boldsymbol{B}_o) \tag{1}$$

In a Cartesian coordinate system (a laboratory frame) in which the direction of the static magnetic field is the z axis, the state is as shown in Fig. 1.1. Even if the nuclei have the same precession frequency, macroscopically the phases of the magnetic moments are dispersed and x and y components cancel out, leaving only the z component to indicate differences in energy levels. This is called the macroscopic magnetization vector $\boldsymbol{M}$. $\boldsymbol{M}$ is important in the NMR experiments to be described, and can be perturbed by magnetic field $\boldsymbol{B}$. Changes in $\boldsymbol{M}$ are represented in the equation:

$$d\boldsymbol{M}/dt = \gamma(\boldsymbol{M} \times \boldsymbol{B}) \tag{2}$$

The perturbed magnetization $\boldsymbol{M}$ is relaxed to the thermal equilibrium state by various interactions. Bloch presumed that the components of $\boldsymbol{M}$ decay to the thermal equilibrium magnetization $\boldsymbol{M}_o$ exponentially with longitudinal and transverse relaxation time constants T_1 and T_2 as follows.

$$\begin{aligned} \frac{d}{dt} M_z &= -(M_z - M_o)/T_1 \\ \frac{d}{dt} M_x &= -M_x/T_2 \\ \frac{d}{dt} M_y &= -M_y/T_2 \end{aligned} \tag{3}$$

Here the field $\boldsymbol{B}$ on spins is

$$\boldsymbol{B} = B_1 \cdot \cos\omega t \cdot \boldsymbol{i} - B_1 \cdot \sin\,\omega t \cdot \boldsymbol{j} + B_o \cdot \boldsymbol{k} \tag{4}$$

Where $\boldsymbol{i}$, $\boldsymbol{j}$, and $\boldsymbol{k}$ are the coordinates unit vectors, and B_1 is the strength of the oscillating magnetic field which has angular velocity ω.

Equation (2) should be combined with equation (3) to obtain the complete Bloch's equation (5).[2)]

$$\begin{aligned} \frac{d}{dt} M_z &= -\gamma[B_1 \cdot \cos \omega t \cdot M_y + B_1 \cdot \sin \omega t \cdot M_x] - (M_z - M_o)/T_1 \\ \frac{d}{dt} M_y &= -\gamma[B_o \cdot M_x - B_1 \cdot \cos \omega t \cdot M_z] - M_y/T_2 \\ \frac{d}{dt} M_x &= \gamma[B_1 \cdot \sin \omega t \cdot M_z + B_o M_y] - M_x/T_2 \end{aligned} \qquad (5)$$

If one observes $\boldsymbol{M}$ in a rotating frame at the same velocity as the precession frequency of $\boldsymbol{M}$, upon the application of a short pulse, $\boldsymbol{M}$ is subjected to the oscillating magnetic field $\boldsymbol{B}_1$ which has the same precession frequency as $\boldsymbol{M}$ (Fig. 1-2). If the oscillating magnetic field is applied from the direction of x′, precession

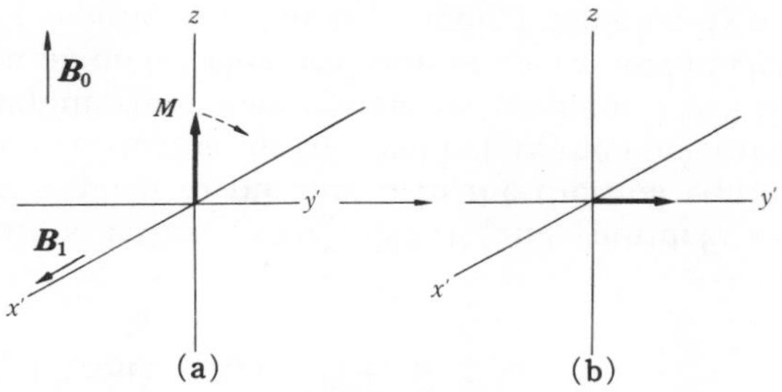

(a) (b)

Rotating frame

Fig. 1.2 (a) Thermal equilibrium state; Applying radiofrequency pulse. (b) Transverse magnetization.

begins with x′ as the center axis and proceeds toward the y′ axis. With the appearance of the y′ component (transverse magnetization vector), a signal is induced in the coil on the y′ axis. This is observed under the rotating frame (see QPD in section 1.1; Fig. 1.6) until thermal equilibrium is attained, and the response is an exponential decay line (free induction decay; FID) (Fig. 1.3

(A)

(B)

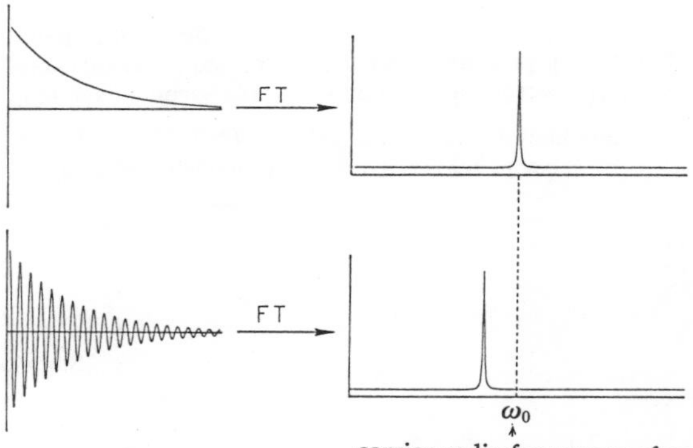

Fig. 1.3
(A) On-resonance magnetization $\boldsymbol{M}$.
(B) Off-resonance magnetization which has close percession frequency to that of $\boldsymbol{M}$.

(A)). The short pulse which is applied above has the broad frequency components of oscillating magnetic field around the carrier frequency.[3)] This can excite other magnetization which has close precession frequency to that of $\boldsymbol{M}$, and the FID has the same periodicity as the offset to the carrier radiofrequency of the pulse (Fig. 1.3 (B)). The FID is obtained as a function of time $s(t)$, and with a Fourier transform (FT) (see below) the frequency spectrum is obtained. (In practice, Fourier transform is carried on a finite number of data points. For details see appropriate textbooks on Fourier transform.[4–8)] This is the spectrum which presents chemical shift *vs.* signal intensity with which we are familiar.

$$S(\omega) = \int_{-\infty}^{+\infty} s(t)\exp(-i\omega t)dt$$

ω : Frequency $\qquad s(t)$: FID

t : Time $\qquad S(\omega)$: Spectrum

1) R.K. Harris, *Nuclear Magnetic Resonance Spectroscopy—A Physicochemical View*, Pitman

Books Ltd., London (1983), pp.8-11
2) *ibid.*, pp.66-70
3) *ibid.*, pp.74-75
4) R.N. Bracewell, *The Fourier Transform and Its Applications*, McGraw-Hill (1978)
5) R.J. Bell, *Introductory Fourier Transform Spectroscopy*, Academic Press (1972)
6) H.P. Hsu, *Fourier Analysis*, Simon & Schuster (1967)
7) N. Ahmed and K.R. Rao, *Orthogonal Transforms for Digital Signal Processing*, Springer Verlag (1975)
8) B. Gold and C.M. Rader, *Digital Processing of Signals*, McGraw-Hill (1969)

1.1 1 Pulse Experiment

This technique is basic to FT-NMR and routinely used. A short pulse is applied to the nuclei in the thermal equilibrium state, and the FID during the interval before the nuclei return to the thermal equilibrium state is collected as data. The FID contains information on the chemical shift of the excited nucleus and the spin-spin coupling constant as periodic functions. By a Fourier transform the information is converted into the NMR spectrum.[1)]

Pulse sequence
(1 Pulse Experiment)

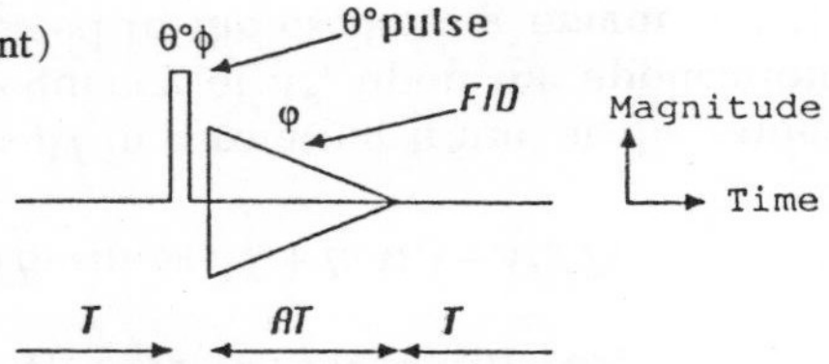

Pulse : Radio frequency pulse
θ° : Pulse rotation angle (flip angle)
FID : Free induction decay
AT : Acquisition time
T : Recovery interval for spin-lattice relaxation
ϕ, φ : Radiofrequency and receiver phase

Fig. 1.4

[Parameters Determined by the Spectroscopist]

1) Frequency and Observation Range

The procedure varies depending on whether a single phase detection (SPD) or quadrature phase detection (QPD)[2, 3)] is used (Fig. 1.5).

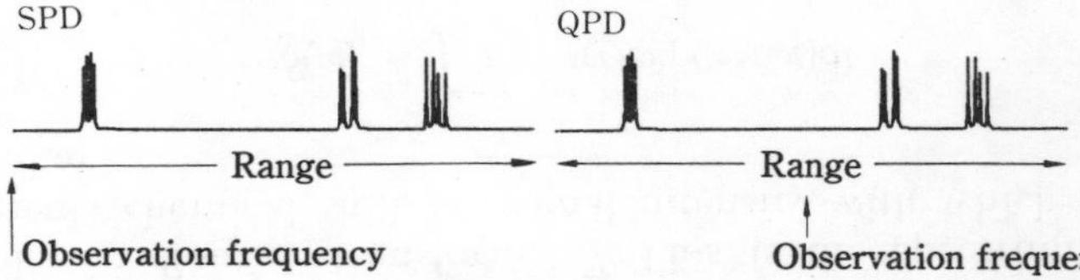

Fig. 1.5

In either instance, it should be remembered that the presence of a signal outside the range results in fold-back. At present, QPD is used in most devices. Hardware is outlined in Fig. 1.6.

(A) Quadrature Phase Detection (QPD)

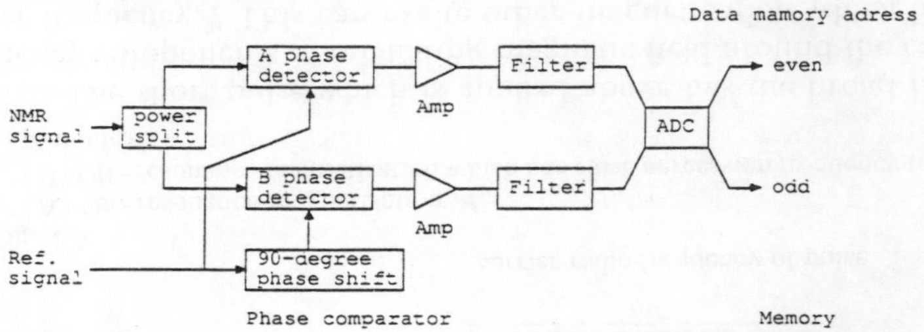

(B) Phase cycling

Acquisition Number →

	1	2	3	4	5	6	7	8
ϕ	x	x	-x	-x	y	y	-y	-y
φ	x	x	-x	-x	y	y	-y	-y

ϕ : Radiofrequency phase, φ : Receiver phase
x, −x, y, −y mean 0°, 180°, 90°, −90°, respectively.

The above phase cycling is an example of the 1 Pulse experiment. This eliminates DC offsets of FID and quadrature images.

Fig. 1.6

The QPD system has two channels of phase detection that detect the components of magnetization which are in-phase and 90° out-of-phase with the carrier radiofrequency. The signals from the two detector channels are stored in the CPU memory, which contains the information necessary to distinguish the signals having positive and negative offsets from the carrier radiofrequency by using the complex Fourier transform.

2) Pulse Width

When the magnetization is tipped 90°, the FID response becomes maximal. This pulse width is called a 90° pulse and is used as a standard. In actual experiments, a 30°-60° pulse width is used; for detecting quaternary carbons which are characterized by long relaxation time, a pulse width near 30° is preferred.

3) Delay Time for Recovery of Magnetization

A delay time should be about three times the longest T_1. [When AT in Fig. 1.4 is long, delay time should be $(AT + T)$. AT is determined automatically according to the sweep width and the size of memory required for the measurement.] It should be remembered that when the delay time is extremely short, integrated areas become less reliable.

4) Repetitions

The number of repetitions is determined by the sample concentration and dynamic range of the signal. It should be remembered that to double the S/N ratio, four times as many repetitions are required.

5) Number of Data Points and Resolution

When the sweep width is set, the resolution after FT is determined by the number of FID data points. Usually the fast FT (FFT) algorithm is used, and the number of data points must be 2^n. After FT, the spectrum is made up of real part only, and the number of data points is 2^{n-1}, and if the sweep width is A Hz, resolution is $A/2^{n-1}$ Hz.

6) Digital Filter

The digital filter (Fig. 1.7 (A)) improves the S/N ratio or increases resolution as shown in Fig. 1.7 (B). Usually, the NMR signal has a Lorentzian line shape, which corresponds to an exponentially decreasing FID.[4, 5] So exponential filtering ($\exp(-iK)$) improves the S/N ratio instead of somewhat broadening the lines (where K is the time constant and i is the number of the data points). If a Gaussian line shape is required, Gaussian multiplication ($\exp(-iK^2)$) can be used and increases resolution.

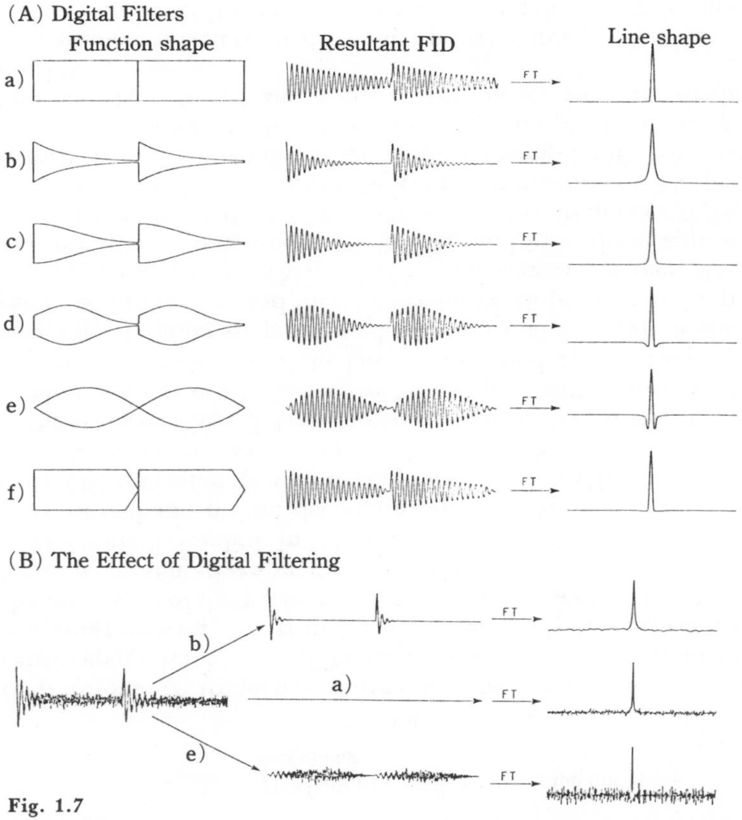

Fig. 1.7

a) Without digital filter b) Exponential multiplication c) Gaussian multiplication d) Double exponential multiplication (Lorentz-Gauss function) e) Sine-bell function f) Trapezoidal multiplication

Especially in the case of absolute mode 2D-FT NMR, which will be touched upon later, it is necessary to use sine bell filter or Lorentzian-Gaussian transformation to make a pseudo-echo-like FID, which has no "hump" or "skirt" on doing magnitude calculation because of its remarkably enhanced line shape resolution.[6, 7] For the phase-sensitive 2D-FT NMR, Gaussian or exponential multiplication can be used because of its good line shape.[6, 7] However, if there are only few data points in the F1 dimension, the sharp peak should have a "wiggle" on its foot. At this time, FID is apodized by trapezoidal multiplication to obtain a good line shape

1) T.C. Farrar and E.D. Becker, *Pulse and Fourier Transform NMR—Introduction to Theory and Methods*, Academic Press, New York and London (1971)
2) J.D. Ellétt, M.G. Gibby, U. Haeberlen, L.M. Huber, M. Mehring, A. Pines and J.S. Waugh, *Advan. Magn. Resonance*, **5**, 117 (1971)
3) E.O. Stejskal and J. Shaefer, *J. Magn. Reson.*, **14**, 160 (1974)
4) R.K. Harris, *Nuclear Magnetic Resonance Spectroscopy—A Physicochemical View*, Pitman Books Ltd., London (1983) pp.70-71
5) E.D. Becker, J.A. Ferreti and P.N. Gambhir, *Anal. Chem.*, **51**, 1413 (1979)
6) A. Bax, *Two-Dimensional Nuclear Magnetic Resonance in Liquids*, Delft University Press, Delft, Holland (1982) pp.34-46
7) W.R. Croasmun and R.M.K. Carlson (Ed.), *Two-Dimensional NMR Spectroscopy—Applications for Chemists and Biochemists*, VCH Publishers Inc., New York (1987) pp.109-121

1.2 Spin Decoupling

When two or more nuclei are interacted by spin-spin coupling, the individuals appear as two or more peaks and the spectrum becomes complex. Conversely, when the nuclei responsible for the multiplet structure are determined, the molecular structure can be analyzed on the basis of information provided by these multiplets. Spin decoupling is a method used for this purpose.[1, 2] In spin decoupling, specific nuclei (usually ^{1}H) are irradiated at

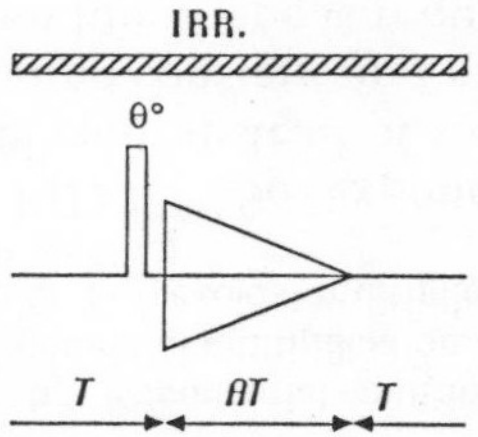

▨ : This indicates the irradiation time for decoupling.

Fig. 1.8

their resonance frequency during the data acquisition, which cause rapid interchanges of the spin states, thereby disrupting the couplings between nuclei under observation. By this procedure, the resonance of the partner is decoupled, *i.e.*, seen in simplified form. Representative spin decoupling methods are the following.

1) Homodecoupling in ^{1}H-NMR[3]
2) Broadband proton decoupling in ^{13}C-NMR[4—6]
3) Off-resonance proton decoupling in ^{13}C-NMR[7]
4) Selective proton decoupling in ^{13}C-NMR[8]

(Spin decouplings 2-4 are not restricted to ^{13}C-NMR)

Methods 1 and 4 link a ^{1}H or ^{13}C coupled with a specific ^{1}H by selective irradiation of the latter. In method 2, the irradiation frequency is noise or pulse-modulated so that it covers a wide range of frequencies and thereby decouples all protons in a typical spectrum. In ^{13}C-NMR, ^{13}C-^{13}C coupling is not seen since ^{13}C in nature is only about 1% of ^{12}C, and everything appears as a singlet. In method 3, a single irradiating frequency (offset by several hundred Hz from the central frequency of ^{1}H spectrum) is used to obtain a spectrum showing NOE but with reduced couplings. By this procedure, the multiplicity of the carbon signals is found and methyl, methylene, etc., are distinguished.

[Parameters Determined by the Spectroscopist]

In all instances it is necessary to determine the decoupler field strength and the site of irradiation. Decoupler output level should be selected experimentally by using some simple compound.

1) J.D. Baldeschwieler and E.W. Randall, *Chem. Rev.*, **63**, 81 (1963)
2) R.A. Hoffman and S. Forsen, *Prog. Nucl. Magn. Reson. Spectrosc.*, **1**, 15 (1966)
3) J.W. Akitt, *NMR and Chemistry—An Introduction to the Fourier Transform-Multinuclear Era*, Chapman and Hall, London and New York (1983), p.149, Fig. 7.7
4) *ibid.*, p.148
5) A.J. Shaka, J. Keeler and R. Freeman, *J. Magn. Reson.*, **53**, 313 (1983)
6) E. Breitmaier and W. Voelter, *^{13}C NMR Spectroscopy—Methods and Applications in Organic Chemistry*, Verlag Chemie, Weinheim and New York (1978), pp.37-38
7) *ibid.*, pp.42-43
8) *ibid.*, pp.36-37

1.3 Spin Decoupling Difference Spectrum (SDDS)

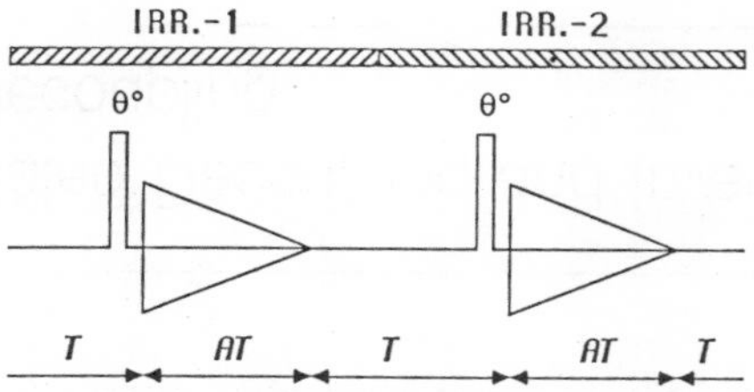

Fig. 1.9

In carrying out homodecoupling experiments (section 1.2, 1) on larger molecules, it often happens that individual protons cannot be selectively irradiated because of signal overlap. Under these conditions, if more than one proton is simultaneously irradiated, it is not easy to interpret the changes that occur. Analysis is facilitated when a control spectrum is subtracted from the decoupled spectrum to bring out the parts that have undergone changes.[1—3] This is called spin decoupling difference spectrum (SDDS). There are, however, instances in which clean subtraction cannot be made because of subtle differences in experimental conditions. To minimize the experimental differences as much as possible, a pulse sequence such as that shown in Fig. 1.9 is used, and after each pulse, irradiation of the ^{1}H in question (IRR-1) is switched with the control (IRR-2) in which the irradiation is applied elsewhere (no proton region). In addition, the experiment is carried out so that the FID in IRR-1 differs from the FID in IRR-2 by 180° out of phase. Then each FID can be combined from different memories or may be accumulated in the same memory to obtain only the difference.

1) R. Freeman, *J. Chem. Phys.*, **43**, 3087 (1965)
2) W.A. Gibbsons, C.F. Beyer, J. Dadok, R.F. Sprecher and H.R. Wyssbrod, *Biochemistry*, **14**, 420 (1975)
3) J.K.M. Sanders and J.D. Mersh, *Prog. Nucl. Magn. Reson.*, **15**, 353 (1982)

1.4 Long-range Selective Proton Decoupling (LSPD)

In ^{13}C-NMR, long-range coupling (coupling of ^{1}H-^{13}C separated by two or more bonds) is an important feature in determin-

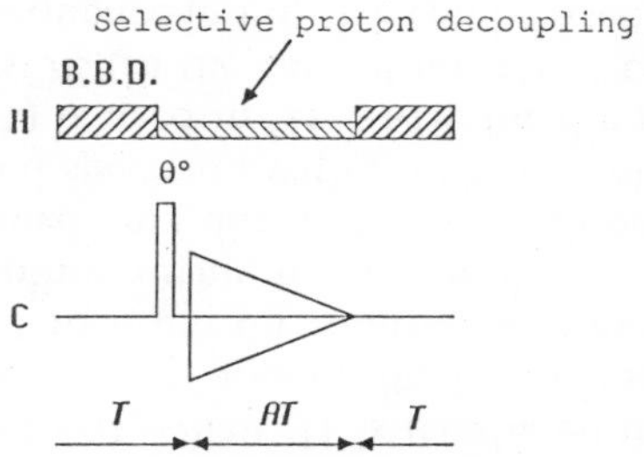

Fig. 1.10

ing the connectivities between molecules. Long-range selective proton decoupling (LSPD)[1, 2] is a method for detecting this interaction. The experimental procedure is essentially the same as what was described in section 1.2, 4, but whereas in selective decoupling the coupling constant is in the vicinity of 150 Hz, in LSPD it is under 10 Hz. In some instruments the decoupler output is decreased by means of a special attenuator.

1) S. Takeuchi, J. Uzawa, H. Seto and H. Yonehara, *Tetrahedron Lett.*, **1977**, 2943
2) J. Uzawa and S. Takeuchi, *Orgn. Magn. Reson.*, **11**, 502 (1978)

1.5 Gated Decoupling and Inverse Gated Decoupling

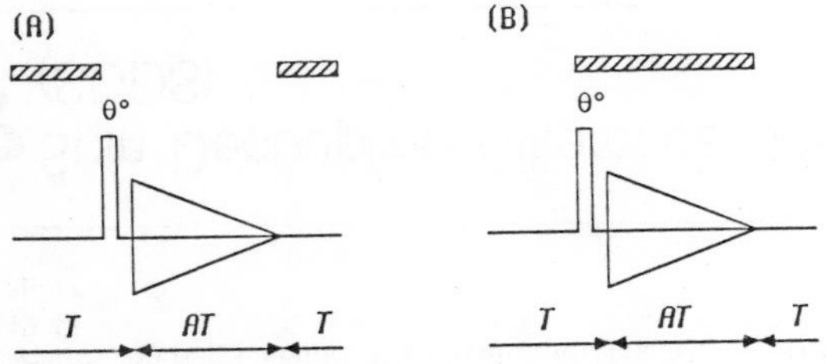

Fig. 1.11

When ^{1}H is irradiated both during the delay time (T) and data acquisition time (AT) as in the experiment described in section 1.2, not only does decoupling occur but the relaxation properties over the entire spin system changes. In most instances, cross-relaxation between nearby ^{1}H-^{1}H and between ^{1}H-^{13}C occurs, and the nuclear Overhauser effect (NOE) appears. When the molecular weight is small and the magnetogyric ratio is positive in the nucleus under observation with irradiation of ^{1}H, the NOE is observed as an increase in signal strength. As shown in Fig. 1.11 (A), when ^{1}H is irradiated during the delay time and no irradiation is applied during the acquisition time, NOE is largely retained but the multiplicities are observed since there is no decoupling during the accumulation of data.[1] When irradiation is applied only during the acquisition time as shown in (B), a decoupled spectrum with no NOE is obtained.[2−5] The pulse sequence (A) is used in ^{1}H-NMR for the measurement of the steady state NOE (in the case of no ^{1}H-^{1}H coupling system, the simple pulse sequence in Fig. 1.8 can be used for NOE measurement of these protons), and in ^{13}C-NMR it is used to increase sensitivity to facilitate the determination of spin-spin coupling constants (J_{CH}). Pulse sequence (B) is used when the purpose is to determine the number of carbons by integration in ^{13}C-NMR.

1) O.A. Gansow and W. Schittenhelm, *J. Am. Chem. Soc.*, **95**, 4294 (1973)
2) R. Freeman, H.D.W. Hill and R. Kaptein, *J. Magn. Reson.*, **7**, 327 (1972)
3) S.J. Opella, D.J. Nelson and O. Jardetgky, *J. Chem. Phys.*, **64**, 2533 (1976)
4) D. Canet, *J. Magn. Reson.*, **23**, 361 (1976)
5) R.K. Harris and R.H. Newman, *J. Magn. Reson.*, **24**, 449 (1976)

1.6 NOE Difference Spectrum (NOEDS), Selective Population Transfer (SPT), and Saturation Transfer

When, in measuring NOE, irradiation is not carried out during the acquisition time as discussed in section 1.5, the only change which occurs is in signal strength, and there is no decoupling. In a difference spectrum, therefore, only the signal with NOE is brought out which retains its multiplicity.[1] The pulse sequence shown in Fig. 1.12 (A) is used for this purpose. In other words, irradiated and control spectra are studied under approximately the same conditions as in spin decoupling difference spectroscopy. When it is arranged so that the phase of FID

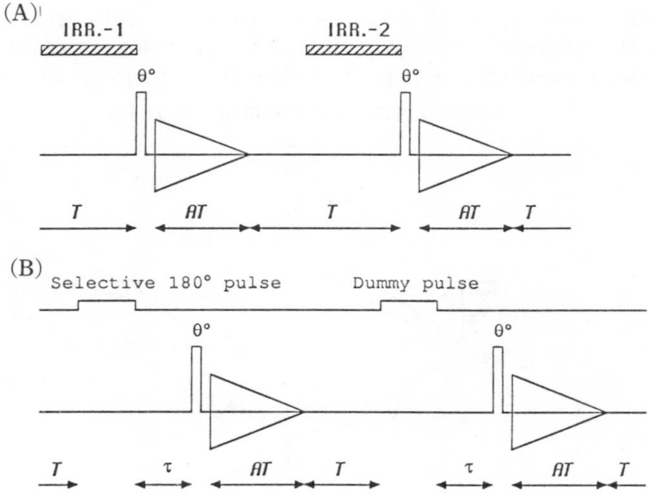

Fig. 1.12

accumulation changes by 180° per pulse, the spectrum of only the part containing the NOE can be measured by subtraction. When the change in signal strength needs to be expressed numerically, the FIDs of the first and the last half are handled as separate files and the integrated areas are compared. The decoupler output may be relatively weak, just enough for saturation of the irradiated nucleus.

There is another method of measuring NOE, based on the observation of perturbed spin after a delay time when a selective 180° pulse is applied as in (B). This procedure is used when a more physicochemical (*i.e.*, kinetic) measurement (transient NOE) is desired in a detailed study of cross relaxation[2] or chemical exchange.

The pulse sequence used here is employed, in addition to the NOE measurement, in experiments on selective population transfer (SPT)[3, 4] and saturation transfer.[5, 6]

Let us suppose that the pulse sequence of (A) is used to irradiate one peak in a spin-coupled nucleus. Saturation occurs in the energy level corresponding to the irradiated resonance and the populations are equalized, so that the strengths of resonance related these energy levels change. The change may be progressive (increased strength) or regressive (decreased strength) as indicated in Fig. 1.13 (also called generalized NOE or spin pumping). This method is comparable to INDOR (internuclear double resonance) in CW-NMR and is called pseudo-INDOR (SPT). It is used, for example, for determining the signs of spin-spin coupling constants. For example, AMX of SPT spectrum of aspartic acid (see Part I, section 15) becomes as follows.

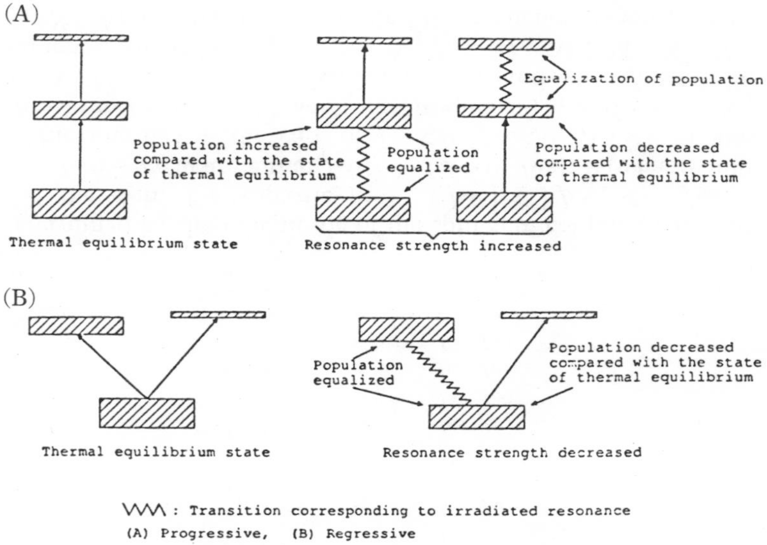

Fig. 1.13

Let us first examine carefully the 3-spin (AMX) energy levels and the positions of signals in the spectrum. The energy levels are as shown in Fig. 1.14 while the energy at each level is as indicated in Table 1.1. The allowable transfers in normal spectra are produced, for example, by the reversal of a single spin such as $\alpha\alpha\alpha$—$\alpha\alpha\beta$ in Fig. 1.14, and when the transfer energy corresponding to the allowable transfer lines is calculated, the

Table 1.1 Energy levels in AMX

AMX in spin		Energy
ϕ_1	$\alpha\alpha\alpha$	$-\frac{1}{2}(\nu_A+\nu_M+\nu_X)+\frac{1}{4}(J_{AM}+J_{AX}+J_{MX})$
ϕ_2	$\alpha\alpha\beta$	$-\frac{1}{2}(\nu_A+\nu_M-\nu_X)+\frac{1}{4}(J_{AM}-J_{AX}-J_{MX})$
ϕ_3	$\beta\alpha\alpha$	$-\frac{1}{2}(-\nu_A+\nu_M+\nu_X)+\frac{1}{4}(-J_{AM}-J_{AX}+J_{MX})$
ϕ_4	$\alpha\beta\alpha$	$-\frac{1}{2}(\nu_A-\nu_M+\nu_X)+\frac{1}{4}(-J_{AM}+J_{AX}-J_{MX})$
ϕ_5	$\beta\alpha\beta$	$-\frac{1}{2}(-\nu_A+\nu_M-\nu_X)+\frac{1}{4}(-J_{AM}+J_{AX}-J_{MX})$
ϕ_6	$\alpha\beta\beta$	$-\frac{1}{2}(\nu_A-\nu_M-\nu_X)+\frac{1}{4}(-J_{AM}-J_{AX}+J_{MX})$
ϕ_7	$\beta\beta\alpha$	$-\frac{1}{2}(-\nu_A-\nu_M+\nu_X)+\frac{1}{4}(J_{AM}-J_{AX}-J_{MX})$
ϕ_8	$\beta\beta\beta$	$-\frac{1}{2}(-\nu_A-\nu_M-\nu_X)+\frac{1}{4}(J_{AM}+J_{AX}+J_{MX})$

Energy is calculated from $h^{-1}U=-\sum_j \nu_j m_j+\sum_{j<k} J_{jk}m_j m_k$

ν_j : Larmor frequency of j nucleus.

m_j : $+1/2$ when spin is α, $-1/2$ when β.

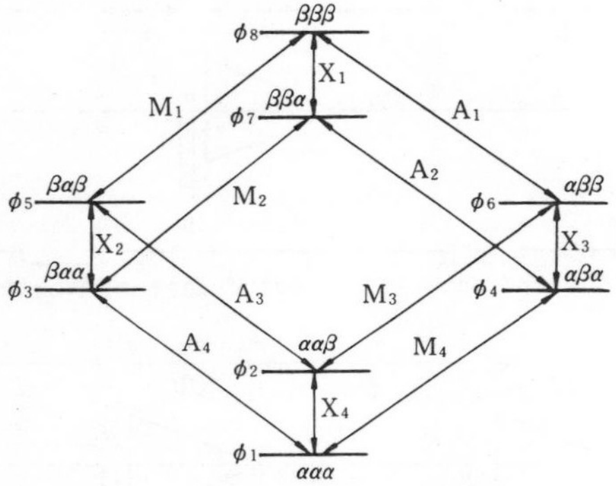

Fig. 1.14 Energy levels in AMX system

figures in Table 1.2 are obtained. Table 1.2 shows that each spectral line is formed by the items in the chemical shift term and those in the spin-spin coupling constant term. Thus transition is

Table 1.2 Spectral line and transfer energy

Transition		Transfer energy
A_1	$6\rightarrow 8$	$\nu_A+\frac{1}{2}(J_{AM}+J_{AX})$
A_2	$4\rightarrow 7$	$\nu_A+\frac{1}{2}(J_{AM}-J_{AX})$
A_3	$2\rightarrow 5$	$\nu_A+\frac{1}{2}(-J_{AM}+J_{AX})$
A_4	$1\rightarrow 3$	$\nu_A+\frac{1}{2}(-J_{AM}-J_{AX})$
M_1	$5\rightarrow 8$	$\nu_M+\frac{1}{2}(J_{AM}+J_{MX})$
M_2	$3\rightarrow 7$	$\nu_M+\frac{1}{2}(J_{AM}-J_{MX})$
M_3	$2\rightarrow 6$	$\nu_M+\frac{1}{2}(-J_{AM}+J_{MX})$
M_4	$1\rightarrow 4$	$\nu_M+\frac{1}{2}(-J_{AM}-J_{MX})$
X_1	$7\rightarrow 8$	$\nu_X+\frac{1}{2}(J_{AX}+J_{MX})$
X_2	$3\rightarrow 5$	$\nu_X+\frac{1}{2}(J_{AX}-J_{MX})$
X_3	$4\rightarrow 6$	$\nu_X+\frac{1}{2}(-J_{AX}+J_{MX})$
X_4	$1\rightarrow 2$	$\nu_X+\frac{1}{2}(-J_{AX}-J_{MX})$

determined by the magnitude of the spin-spin coupling constant and the sign. For aspartic acid, $|J_{AM}|<|J_{AX}|<|J_{MX}|$; therefore if we assume that $J_{AM}>0$, $J_{AX}>0$, $J_{MX}>0$ (what would be the findings in the case of $J_{AM}<0$, $J_{AX}<0$, $J_{MX}<0$?), the lines from the low field side (high frequency) would be A_1, A_3, A_2, A_4, M_1, M_3, M_2, M_4, X_1, X_3, X_2, X_4, and when $J_{AM}<0$ the sequence would be A_3, A_1, A_4, A_2, M_3, M_1, M_4, M_2, X_1, X_3, X_2, X_4. All one has to do is to find the sequences when $J_{AX}<0$, $J_{MX}<0$. Let us now consider the case of X spin when its lowest field lines is irradiated. When $J_{AM}>0$, $J_{AX}>0$, $J_{MX}>0$, this line would correspond to X_1, and according to Fig. 1.14, A_1, A_2, M_1, M_2 should be affected. Especially since, according to Fig. 1.14, M_1 and A_1 are in a regressive relation with X_1, line strength is weaker while M_2 and A_2 are stronger since they are in a progressive relation. Changes in strength can be predicted for other situations by the same method (Fig. 1.15 (B)).

Saturation transfer is a method used in a situation in which two signals of different chemical shifts become interchanged through chemical or conformational exchange. The method involves saturation of one signal by the pulse sequence shown in

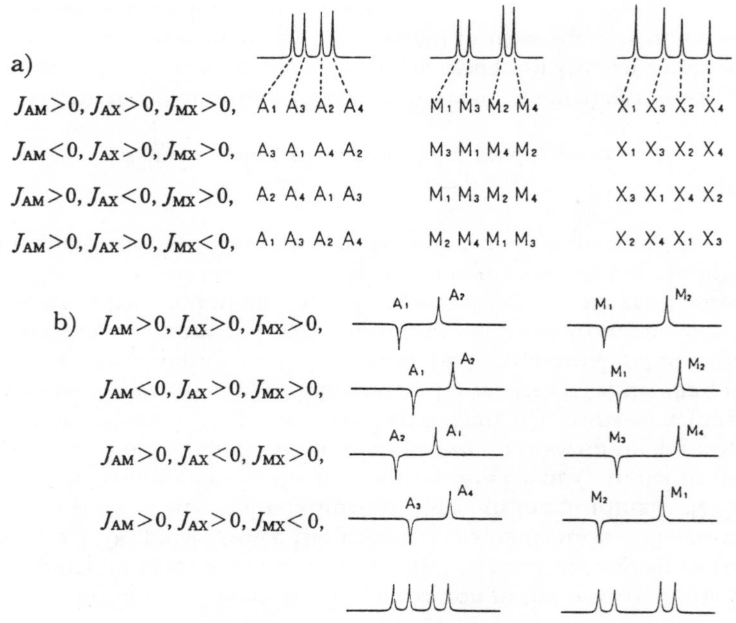

Assignment of lines in AMX system

Fig. 1.15

Fig. 1.12 (A) and the amount of transfer is observed as the amount of decrease in the signal strength of the opposite number. There are two methods to estimate the rate of exchange. One is the method in which the rate of exchange is obtained from the changing rate of recovery signal after the series of delay times from spin reversal by a procedure indicated in Fig. 1.12 (B).[7] The other method is that of calculation from the amount of saturation transfer and T_1 value.[8]

1) J.K.M. Sanders and J.D. Mersh, *Prog. Nucl. Magn. Reson.*, **15**, 353 (1982)
2) P. Dais and A.S. Perlin, *Adv. Carbohydr. Chem. Biochem.*, **45**, 125 (1987)
3) K.G.R. Pachler and P.I. Wessels, *J. Magn. Reson.*, **12**, 337 (1973)
4) S. Sørensen, R.S. Hansen and H.J. Jacobson, *J. Magn. Reson.*, **14**, 243 (1974)
5) S. Forsén and R.A. Hoffman, *J. Chem. Phys.*, **39**, 2892 (1963)
6) S. Forsén and R.A. Hoffman, *J. Chem. Phys.*, **40**, 1189 (1964)
7) I.D. Campbell, C.M. Dobson, R.G. Rateliffe and R.J.P. Williams *J. Magn. Reson.*, **29**, 397 (1978)
8) B.E. Mann, *J. Magn. Reson.*, **25**, 91 (1977)

1.7 Inversion Recovery (T_1 Measurement), PRFT, WEFT

Inversion recovery[1] is a well-known method for the determination of longitudinal relaxation time (T_1).[2] The pulse sequence

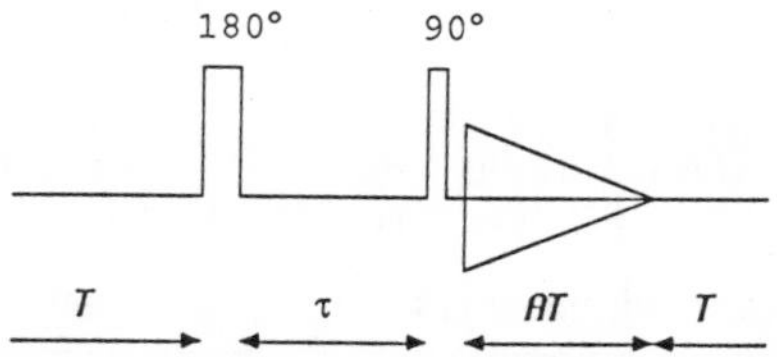

τ : Recovery time. This parameter is varied.

Fig. 1.16

(A)

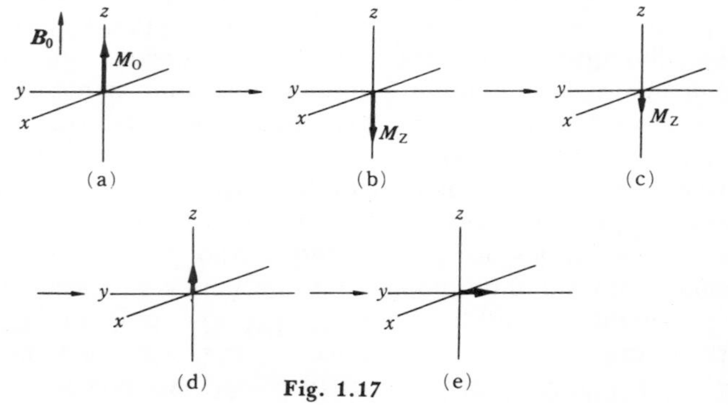

Fig. 1.17

(B)

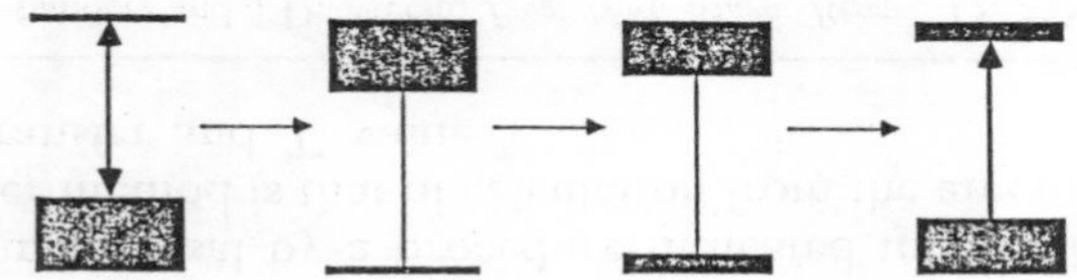

Fig. 1.17 Continued

(A) Spin vectors, (B) Energy level and spin population.

(a) Thermal equilibrium state; Applying 180° pulse, (b) Inverted magnetization, (c) Partially relaxed, (d) Further relaxed; Applying 90° pulse, (e) Detecting transverse magnetization

shown in Fig. 1.16 is used. Magnetization in the state of thermal equilibrium (Fig. 1.17, a) is converted to the state shown in (b) with a 180° pulse. Since the inverted magnetization will return to the state of thermal equilibrium through longitudinal relaxation, the component in the direciton of the *z* axis (parallel to the magnetic field) changes to the state shown in (c) with the passage of time. When a 90° pulse is applied to state (d), transverse magnetization shown in (e) is obtained and is measurable, the magnitude corresponding to the amount of longitudinal relaxation. This can be expressed by equation (6) derived from the Bloch equation (see equation (5)), $\boldsymbol{M}_0$ representing the magnetization at thermal equilibrium and $\boldsymbol{M}_z$ the magnitude of longitudinal magnetization at time τ which elapses between (b) and (d).

$$\boldsymbol{M}_z = \boldsymbol{M}_0\,(1 - 2\exp\,(-\tau/T_1)) \qquad (6)$$

T_1: Time constant, longitudinal relaxation time

When the magnitude of the transverse magnetization in (e) equals $\boldsymbol{M}_z$, T_1 can be obtained from equation (6). In practice, the calculation is made by the method of least squares based on a series of data for each τ value.

It may be seen from equation (6) that magnetization with a given T_1 would give $\boldsymbol{M}_z = 0$ when $\tau = T_1 \cdot \ln 2 = 0.69 T_1$ (Fig. 1.17, between c and d). This is utilized in the water eliminated FT (WEFT) method.[3] In WEFT, τ in the pulse sequence of Fig. 1.16 is selected for the water magnetization to be zero. Generally the T_1 of water is longer than that of the solute, so that magnetization of the solute returns to near thermal equilibrium state just before the 90° pulse is applied. The water signal is therefore suppressed and the solute signal is detected with a good S/N.

On the other hand, when τ is suitably adjusted, nuclei having a T_1 which is longer than $\tau/\ln 2$ exhibit signals which point down while those with short T_1 point up. By utilizing this feature in ^{13}C-NMR, carbons can be classified according to the length of T_1. This procedure is called partially relaxed FT (PRFT)[4] and is used to distinguish methine from methylene carbons.

1) R.L. Vold, J.S. Wangh, M.P. Klein and D.E. Phelps, *J. Chem. Phys.*, **48**, 3831 (1968)
2) R. Freeman and H.D. Hill, *J. Chem. Phys.*, **54**, 3367 (1971)
3) S.L. Patt and B.D. Sykes, *J. Chem. Phys.*, **56**, 3182 (1972)
4) D. Doddrell and A. Allerhand, *J. Am. Chem. Soc.*, **93**, 1558 (1971)

1.8 Spin Echo (T_2 Measurement)

For measuring the transverse relaxation time (T_2), either the Hahn's spin echo method[1] or its modification (CPMG method)[2, 3] is used. Fig. 1.18 (A) shows the pulse sequence of Hahn's spin echo method. In that procedure, magnetization at thermal equilibrium (Fig. 1.19, (a)) is converted to transverse magnetization by the application of a 90° pulse (b). This transverse magnetization evolves after time τ into a fast and a slow component as shown in Fig. 1.19 (c) through the effect of inhomogeneity of magnetic field, etc. When a 180° pulse is applied, the transverse magnetization begins to refocus as shown in

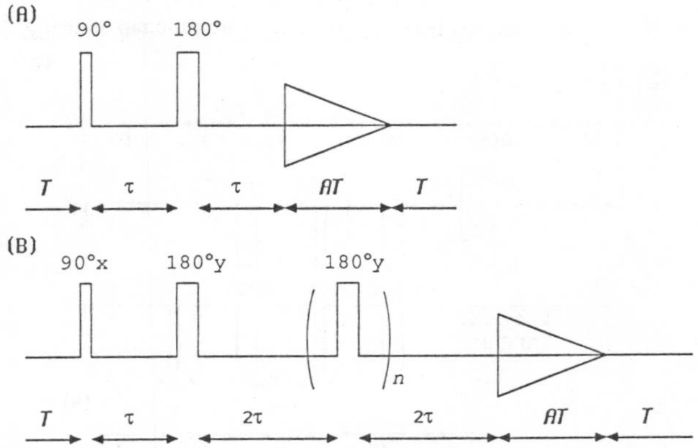

Fig. 1.18

(A) Hahn's spin-echo method, (B) Carr-Purecell-Meiboom-Gill method (CPMG)

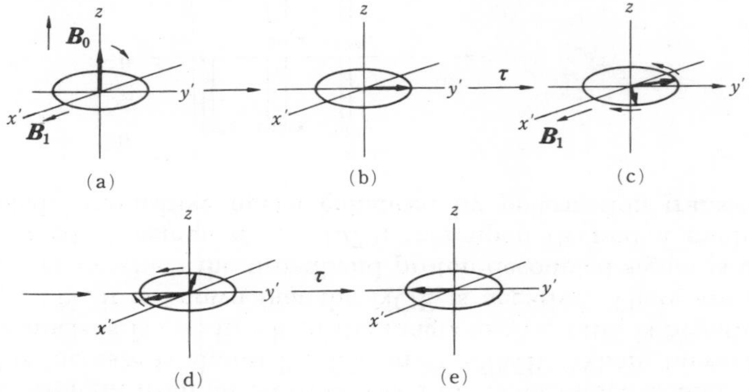

Fig. 1.19

Observation of nuclei in a rotating frame

(a) Thermal equilibrium state; Applying 90° pulse, (b) Transverse magnetization, (c) The transverse magnetization evolves after time τ into fast and slow components through the effect of inhomogeneity of magnetic field, etc., (d) Applying 180° pulse, (e) After time τ, detecting the refocused transverse magnetization.

Fig. 1.19 (d) and becomes completely refocused after time τ. The procedure, however, is subject to error caused by diffusion and imprecise pulse widths. Therefore the Carr-Purcell-Meiboom-Gill (CPMG) method, which is an improvement over Hahn's method, is used. In these procedures, a signal distortion called *J* modulation occurs due to homonuclear spin-spin coupling, but this has been developed to advantage in 2D *J*-resolved spectroscopy.

In ^{13}C-NMR in which there is heteronuclear spin-spin coupling, the procedure is also utilized in the gated spin echo method (Fig. 1.20).

The gated spin echo technique provides classification of the numbers of ^{1}H bonded to carbon.[4–6] Thus when τ is $1/J_{CH}$, the spectrum shows that quaternary carbons and methylene carbons point up and methine and methyl carbons point down.

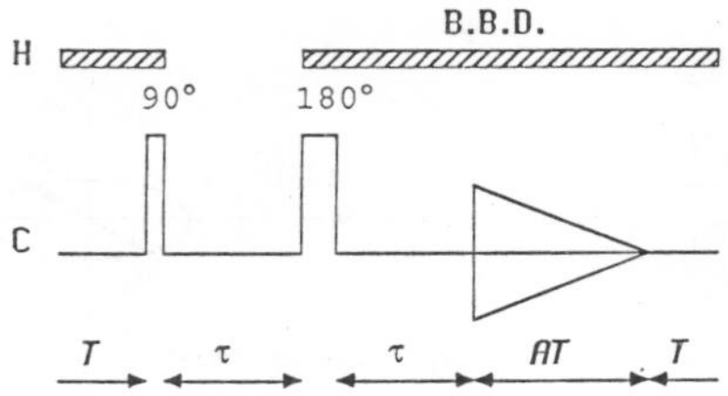

Fig. 1.20 The gated spin echo method

1) E.L. Hahn, *Phys. Rev.*, **80**, 580 (1950)
2) H.Y. Carr and E.M. Purcell, *Phys. Rev.*, **94**, 630 (1954)
3) S. Meiboom and D. Gill, *Rev. Sci. Instr.*, **29**, 688 (1958)
4) C. Le Cooq and J.-Y. Lallemand, *J. Chem. Soc. Chem. Commun.*, 150 (1981)
5) D.J. Cookson and B.E. Smith, *Org. Magn. Reson.*, **16**, 111 (1981)

1.9 INEPT

When we observe hetero-nuclei, we expect an increase in signal strength from an NOE between the hetero-nucleus and ^{1}H. This increase is almost 2.9-fold in ^{13}C-NMR. When, however, the nucleus is one in which the magnetogyric ratio is negative, *e.g.*, ^{15}N, it is known that the NOE is negative. There are instances in which the broadband proton-decoupled signal is obliterated. Freeman *et al.* (1979) thereupon devised a method called "insensitive nuclei enhanced by polarization transfer" (INEPT) in which the signal strength is increased by polarization transfer.[1] The method has a number of merits and has come to be used widely in ^{13}C-NMR.

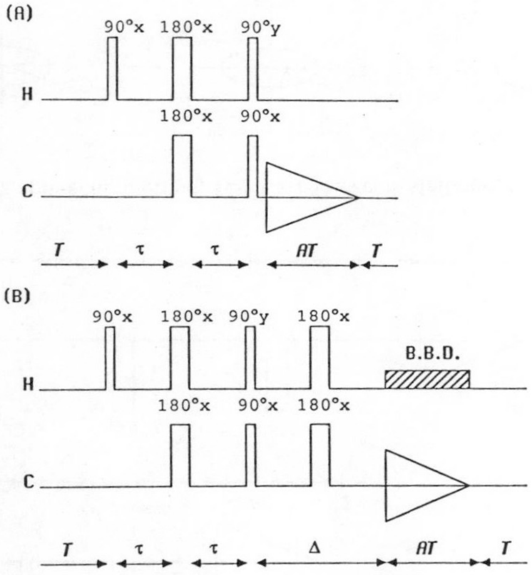

Fig. 1.21
(A) Without decoupling ; (B) INEPT with Refocusing
$\tau = 1/4\ {}^1J_{CH}$, $\Delta = 1/4\ {}^1J_{CH}$, $1/2\ {}^1J_{CH}$ or $1/3\ {}^1J_{CH}$

The pulse sequence is shown in Fig. 1.21 (A). For the sake of simplicity, let us take a 2-spin (^{13}C-^{1}H) system in a rotating frame. When transverse magnetization has been induced by the

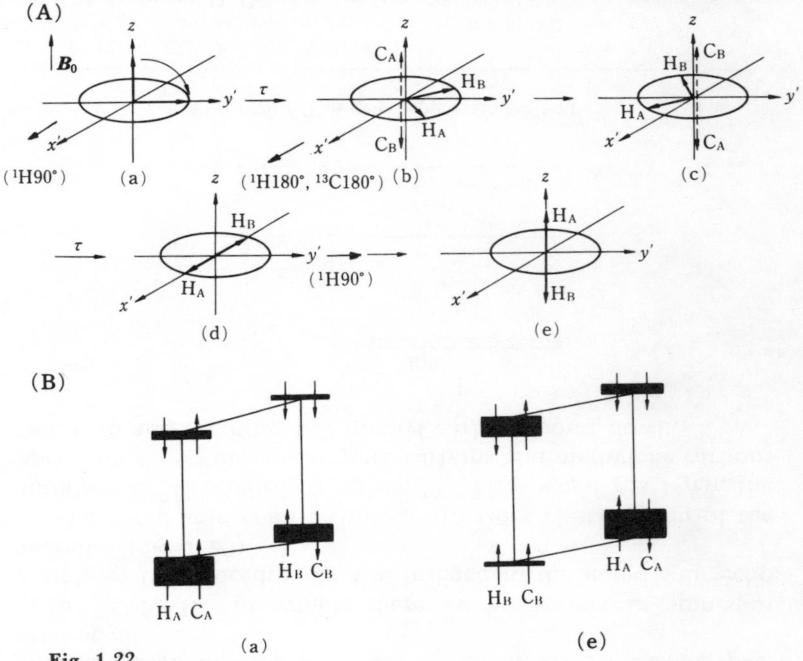

Fig. 1.22

(A) Spin vectors, (B) Energy level and spin population

(a) Thermal equilibrium state; Applying ^{1}H 90° pulse, (b) There is a spread into the fast and slow components through spin-spin coupling with ^{13}C. After time $\tau = 1/4\ {}^1J_{CH'}$, applying ^{1}H and ^{13}C 180° pulse, (c) The spread becomes more pronounced as the result of an interchange of the ^{13}C spins. (d) After time τ, applying ^{1}H 90° pulse differing by 90° out of phase from the initial pulse, (e) Retained and inverted ^{1}H magnetization.

application of a 90° pulse to ^{1}H in a state of thermal equilibrium (at which time the population at each energy level is as shown in Fig. 1.22 (B) (a) in accordance with the Boltzmann distribution), there is a spread into the fast and slow components through spin-spin coupling with ^{13}C. When, after $\tau = 1/4\ J_{CH}$, a 180° pulse is applied to both ^{13}C and ^{1}H, the spread becomes more pronounced as the result of an interchange of the spins, and after this τ the situation in Fig. 1.22 (A) (d) is obtained. When a 90° pulse differing by 90° out of phase from the initial pulse is now applied on ^{1}H, inversion of the ^{1}H spin occurs as in Fig. 1.22 (A) (e) depending on the spin state of ^{13}C. This signifies that the population of each energy level has become that shown in Fig. 1.22 (B) (e). Now when a 90° observation pulse for ^{13}C is applied, a signal with a fourfold increase in intensity is obtained. When the FID is sampled and FT is carried out in a pulse sequence shown in Fig. 1.22 (A), a spectrum retaining the spin-spin coupling is obtained. This pulse sequence is convenient when the focus is on spin-spin coupling constants, but when signals overlap it becomes difficult to distinguish methine, methylene, etc. The pulse sequence in Fig. 1.21 (B) was therefore developed to obtain decoupling without cancelling out signals through phase reversal.[2)] When this sequence is examined closely, it is seen that it comprises a part which causes polarization transfer, and a part similar to spin echo (below 90° pulse in ^{13}C). By setting the delay time Δ of the latter at $1/4\ J_{CH}$, J_{CH}, $1/2\ J_{CH}$ and $3/4\ J_{CH}$, carbons can be classified. Thus when $\Delta = 1/4\ J_{CH}$, methine, methylene and methyl all appear upright; when $\Delta = 1/2\ J_{CH}$, only methines are upright; and when $\Delta = 3/4\ J_{CH}$, methines and methyls point up and methylenes are inverted.

[Parameters Determined by the Spectroscopist]

1) Decoupler output

Residual spin-spin coupling constant is measured by off-resonance decoupling, and $\gamma B_2/2\pi$ is calculated from equation (7). Methine carbon is observed from the pulse sequence of Fig. 1.21 (A) to make certain that the appearance of the signals up and down is balanced.

$$\gamma B_2/2\pi = [\Delta\nu^2 \cdot (J/J_R)^2 - 1)]^{1/2} \qquad (7)$$

$(\gamma B_2/2\pi)^{-1}$ should be equal to 360° pulse width

$\Delta\nu$: Offset from ^{1}H resonance frequency

J: C-H coupling constant

J_R: Residual spin-spin coupling constant

2) Repetition time in pulse sequence

Repetition may be made at the ^{1}H relaxation time.

---

1) G.A. Morris and R. Freeman, *M. Am. Ghem. Soc.*, **101**, 761 (1979)
2) D.P. Burum and R.R. Ernst, *J. Magn. Reson.*, **39**, 163 (1980)

1.10 DEPT

Distortionless enhancement by polarization transfer (DEPT)[1–3)] is a modification of INEPT. Like INEPT, the increase of signal strength is achieved through polarization transfer. In INEPT, the delay time (Δ) was varied by $1/4J_{CH}$, $1/2J_{CH}$ and

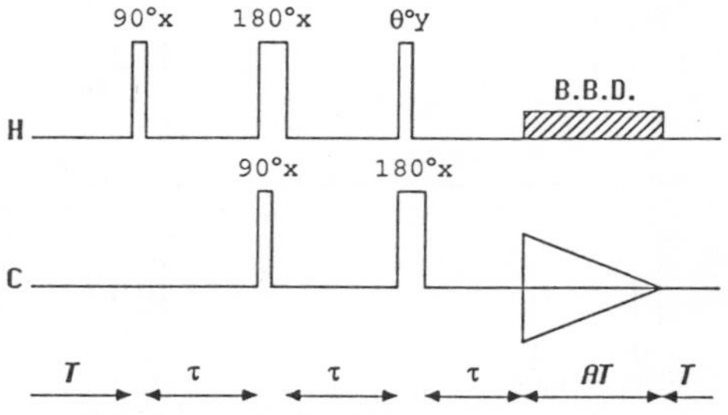

Fig. 1.23 $\tau = 1/2\ ^1J_{CH}$

$3/4J_{CH}$ before the FID accumulation to distinguish methyl, methylene and methine. In DEPT, the same spectra are obtained by varying the pulse width (θ) by 45°, 90° and 135° on the third ^{1}H pulse. The peaks obtained by DEPT have better line shapes and the procedure is more quantitative than INEPT. By appropriate data processing, spectra for methyl, methylene or methine alone can be displayed.

1) D.M. Doddrell, D.T. Pegg and M.R. Bendall, *J. Magn. Reson.*, **48**, 323 (1982)
2) D.T. Pegg, M.R. Bendall and D.M. Doddrell, *J. Magn. Reson.*, **44**, 238 (1981)
3) M.R. Bendall, D.T. Pegg, D.M. Doddrell and D.M. Thomas, *J. Magn. Reson.*, **46**, 43 (1982)

1.11 Selective Excitation: DANTE, Redfield 214, Jump and Return (JR), 1-3-3-1

The pulse widths used in experiments presented thus far (*e.g.*, single pulse) are very short and cover a wide frequency range. This may be inconvenient when, for example, a spin-spin coupling constant is sought for a single carbon signal in ^{13}C-NMR where many signals are close together, and when a proton spectrum is desired without inducing large and unnecessary signals such as solvents. In the former instance, a pulse sequence called DANTE is used (Fig. 1.24 (A)).[1] When a short $\theta°$ pulse is repeated n times at l second intervals, the frequency range that can be excited is $1/\theta n$ Hz, and the region of excitation is $\pm 1/l$ Hz removed from the center frequency. Thus when the pulse interval (l seconds) is properly selected during broadband proton decoupling, the desired carbon can be selectively excited. When the decoupler is turned off and the FID is sampled, the spin-spin coupling constant appears, so that the splitting pattern of a given carbon can be observed without interference by the signals of other carbons. (see Fig. 1.25)[2] For elimination of unwanted signals, the Redfield 214-pulse (see Fig. 1.24 (B))[3, 4] and other

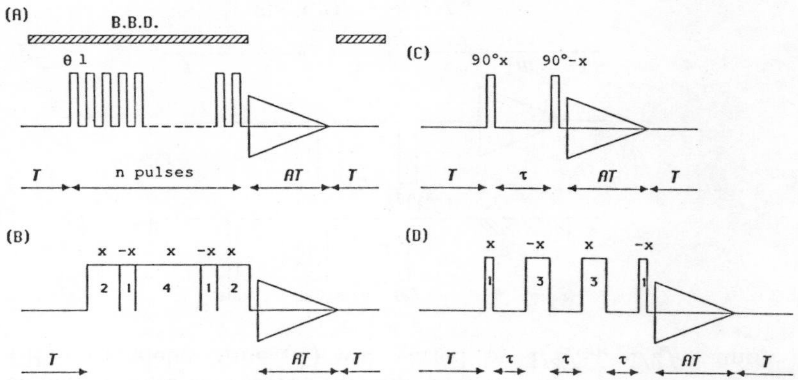

Fig. 1.24

(A) DANTE, (B) Redfield 214 (2,1,4: relative pulse lengths) (C) JR (1-1), (D) 1-3-3-1

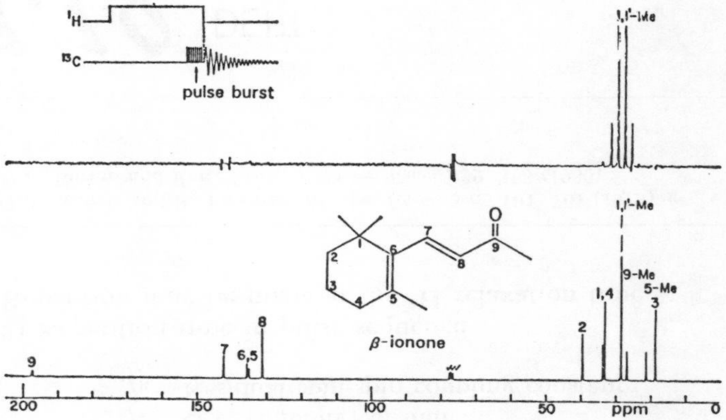

Fig. 1.25

methods have been described. The Redfield 214 pulse covers a narrow frequency range, as shown in Fig. 1.26. Measurements are made with the solvent signal coming at the zero point. For

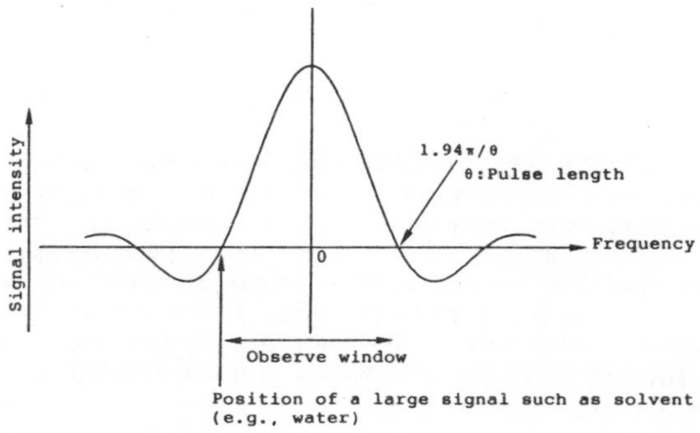

Fig. 1.26

the same purpose, Jump and Return (JR) and 1-3-3-1 pulse sequence [6, 7] were developed and are much easier to use since hard pulses are employed and the transmitter frequency can be set to the solvent signal in these pulse sequences. Both techniques have the null point region of the signal at the carrier frequency of the transmitter. These methods, however, are not very quantitative.

1) G.A. Morris and R. Freeman, *J. Magn. Reson.*, **29**, 433 (1978)
2) K. Nakanishi, T. Goto, S. Itô, S. Natori and S. Nozoe (Ed.), *Natural Product Chemistry*, vol. 3, Kodansha Ltd., Tokyo (co-published by) University Science Books, California (1983) p.5
3) A.G. Redfield, S.D. Kunz and E.K. Ralph, *J. Magn. Reson.*, **19**, 114 (1975)
4) K. Roth, B.J. Kimber and J. Freeney, *J. Magn. Reson.*, **41**, 302 (1980)
5) P. Plateau and M. Guéron, *J. Am. Chem. Soc.*, **104**, 7310 (1982)
6) D.L. Turner, *J. Magn. Reson.*, **54**, 146 (1983)
7) P.J. Hore, *J. Magn. Reson.*, **55**, 283 (1983)

1.12 INADEQUATE

Since the amount of ^{13}C in the natural environment is about 1% that of ^{12}C, the chance of two ^{13}C nueclei being adjacent or close by is only 0.01%. When, however, ^{13}C nuclei are separated by one to about four covalent bonds, spin-spin coupling will of course occur, and if the spin-spin coupling constants can be determined, information on the connectivity of carbons and molecular conformation can be obtained. In normal proton-decoupled ^{13}C-NMR, virtually all the signals are those of isolated ^{13}C, and the signals of coupled ^{13}C are obscured. A method for selectively observing coupled ^{13}C nuclei would be most useful. Freeman *et al.* (1980) achieved this objective using double quantum coherence.[1–3] The pulse sequence is shown in Fig. 1.27, in which τ is $(2n+1)/4J_{CC}$ (n = 0, 1, 2...) and Δ is about 10 μs (the time necessary for changing the pulse phase).

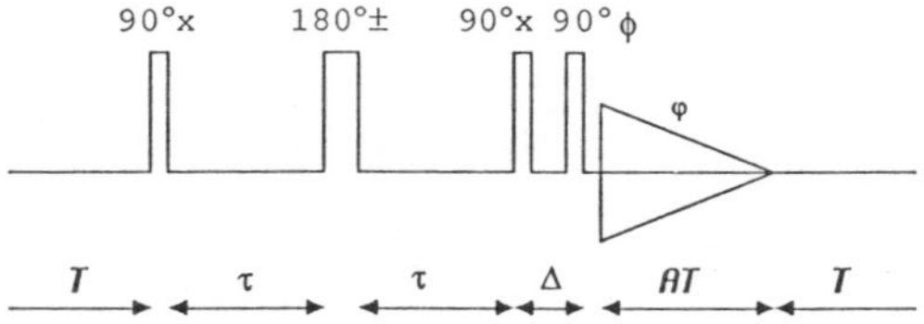

ϕ	x	y	−x	−y
φ	x	−y	−x	y

$\tau = (2n+1)/4J_{CC}$ (n=0, 1, 2···)
$\Delta = 10$ μsec.

Fig. 1.27

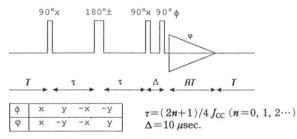

Double quantum coherence can occur in systems having two or more spins, and represents a state in which a certain relationship exists between the wave functions of energy levels differing by two spin quantum units as shown in transition $\alpha\alpha$—$\beta\beta$ in Fig. 1.28. In the pulse sequence shown in Fig. 1.27, double quantum coherence is evolved by the three pulses in the first half, and this is converted into a transverse magnetization by the final pulse. The events during this period cannot be precisely described without resorting to quantum mechanics based on density matrix. Details are presented in the texts by Bax (1982)[4] or Ernst (1987),[5] which should be consulted by those who are interested.

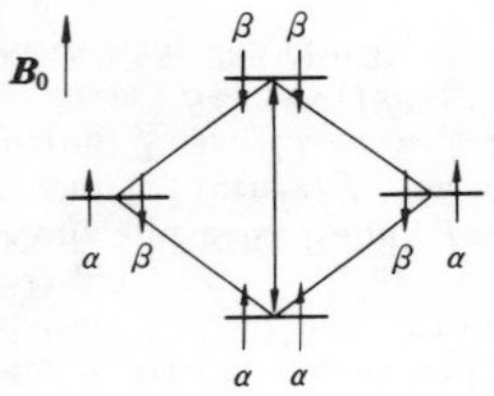

Fig. 1.28

The method is very attractive since it permits direct observation of C-C connectivities, but it has the drawback of poor sensitivity. To correct this deficiency, a combination of INEPT-INADEQUATE has been proposed.[6] In any event, since the sensitivity of ^{13}C probes is constantly increasing, this method may eventually become highly useful. (At present, this experiment is possible if the sample is of relatively low molecular weight and available in quantities of about 50 mg or more.) [TI]

1) A. Bax, R. Freeman and S.P. Kempsell, *J. Am. Chem. Soc.*, **102**, 4849 (1980)
2) A. Bax, R. Freeman and S.P. Kempsell, *J. Magn. Reson.*, **41**, 349 (1980)
3) A. Bax and R. Freeman, *J. Magn. Reson.*, **41**, 507 (1980)
4) A. Bax, *Two-Dimensional Nuclear Magnetic Resonance in Liquids*, Delft University Press, Delft, Holland (1982)
5) R.R. Ernst, G. Bodenhausen and A. Wokaun, *Principles of Nuclear Magnetic Resonance in One and Two Dimensions*, Oxford University Press, Oxford (1987)
6) H. Kessler, W. Bermel and C. Greisinger, *J. Magn. Reson.*, **62**, 573 (1985)

2

Two-dimensional FT-NMR

The nuclei (^1H, ^{13}C and so on) we study by NMR have various types of interactions such as spin-spin coupling and cross relaxation which provide the basis for inferences concerning molecular structure and motion. These interactions have been studied by perturbation of a certain nucleus using a weak oscillating magnetic field. Selective decoupling and NOE measurement are typical cases. However, there are some difficulties in the study of a complex molecule. Namely, such a molecule has many protons to be irradiated which overlap each other, and it is very difficult to distinguish these protons. Two-dimensional FT-NMR (2D-FT NMR) solves this problem.

In 2D-FT NMR, we assume four periods of preparation, evolution, mixing and detection as shown in Fig. 2.1. Frequency labeling on each magnetization is accomplished in the evolution period, the magnitude of magnetization is transferred to another magnetization by a certain interaction in the mixing period, and the modulated magnetization is detected in the detection period.

Take, for example, the NOESY experiment of a two spin-system which contains the on-resonance and off-resonance signals for the observation frequency as shown in Fig. 2-2. NOESY has two 90° pulses separated by t_1 in the evolution period. In this period, the first 90° pulse converts the longitudinal magneti-

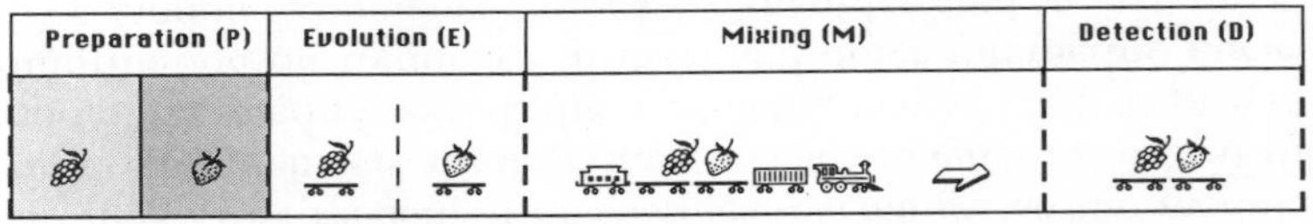

and … correspond to the information of each spin.

Fig. 2.1

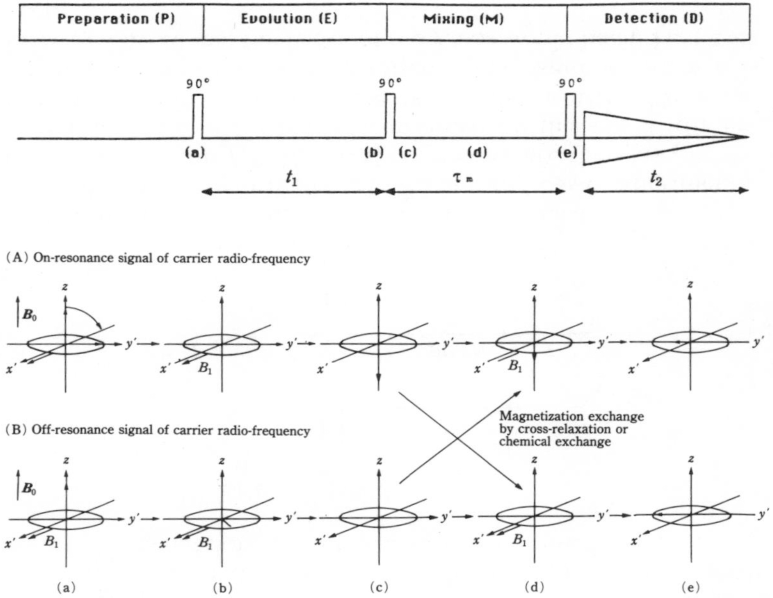

Fig. 2.2

t_1 : Length of evolution period
τ_m : Mixing time
t_2 : Elapsed time of recording FID

(a) Thermal equilibrium state; Applying first 90° pulse, (b) Transverse magnetization; After t_1, applying second 90° pulse, (c) Longitudinal and transverse magnetization components, (d) Magnetization exchange; Applying third 90° pulse, (e) Detecting the transverse magnetization

zations to transverse ones. The transverse magnetizations move on the rotating frame depending on the offset frequency from the observation (carrier) frequency. (The on-resonance magnetization does not move.) After t_1 (which is increased at fixed intervals), the second 90° pulse is applied to these magnetizations to form non-thermal equilibrium longitudinal components, which depend on the each offset frequency from the carrier. In the mixing period, these magnetizations are transferred to each other through cross relaxation or chemical exchange, so the modulation of frequency is accomplished on each magnetization. At the end of the mixing period, the modulated magnetizations provide information on frequency of itself and other interacting nuclei. The modulated magnetizations are then detected to give the set of FIDs formed in equation (1) in the detection period. When these FIDs are Fourier-transformed as shown in equation (2), two-dimensional spectra are obtained which show the values for each precession frequency.

$$s(t_1,t_2) = \mathrm{tr}[F\sigma(t_1,t_2)] \qquad (1)$$

s : Signal in time domain
t_1, t_2: Time variables
F: Spin operator
σ : Density matrix
tr means trace metric.

$$s(t_1,t_2) \xrightarrow{\text{FT over } t_2} S(t_1,\omega_2) \xrightarrow{\text{FT over } t_1} S(\omega_1,\omega_2)$$

or (2)

$$S(\omega_1,\omega_2) = \int_{-\infty}^{\infty} dt_1 \exp(-i\omega_1 t_1) \int_{-\infty}^{\infty} dt_2 \exp(-i\omega_2 t_2)\ S(t_1,t_2)$$

S: Signal in frequency domain
ω_1, ω_2 : Frequency variables

When the pulse sequence of 2D-FT NMR is designed for observing the apparent character of the interaction, many variations of this method, *e.g.* COSY, H,C-COSY, RELAY, 2D-INADEQUATE, are obtained.

2.1 J-Resolved Spectroscopy (^{1}H)

This method permits the separation of overlapping signals.[1,2] The pulse sequence shown in Fig. 2.3 is used. This sequence is very similar to Hahn's spin echo method and the spread of magnetization caused by inhomogeneity of the magnetic field is refocused. When observed in the rotating frame, however, the magnetization component which opens up as a result of spin-spin coupling opens still more since the second 180° pulse inverts the spin state of the partner nucleus (Fig. 2.4). Consequently when $t_1/2$ is increased at a fixed interval and the FID is collected at each value, a two-dimensional data matrix is obtained. When a 2D-FT is applied to two-dimensional data as in Fig. 2.5, a spectrum which shows the magnitude of the spin-spin coupling constant (J) in relation to chemical shift is obtained.

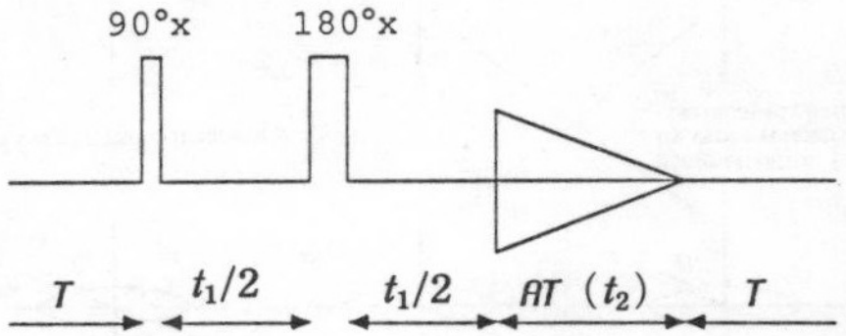

Fig. 2.3

[Parameters Determined by the Spectroscopist]

1 Pulse experiment is first carried out and basic experimental conditions for the *J*-Resolved spectroscopy are then set.

The increase in t_1 (Δt_1) is determined by the size of the expected spin-spin coupling constants. In other words, $1/\Delta t_1$ is the maximum width on the J coordinate. For example, when t_1 is increased by 20 ms, the width of the J axis would be 50 Hz ($\pm$25 Hz).

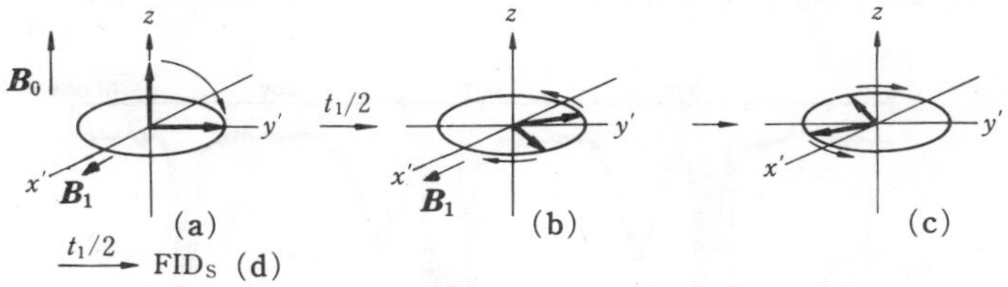

Fig. 2.4

(a) Thermal equilibrium state ; Applying 90° pulse, (b) There is a spread into the fast and slow components through spin–spin coupling. After time $t_1/2$, applying 180° pulse, (c) The magnetization components open stiil more since the 180° pulse inverts the spin of the partner nucleus. (d) After time $t_1/2$, detecting the transverse magnetization

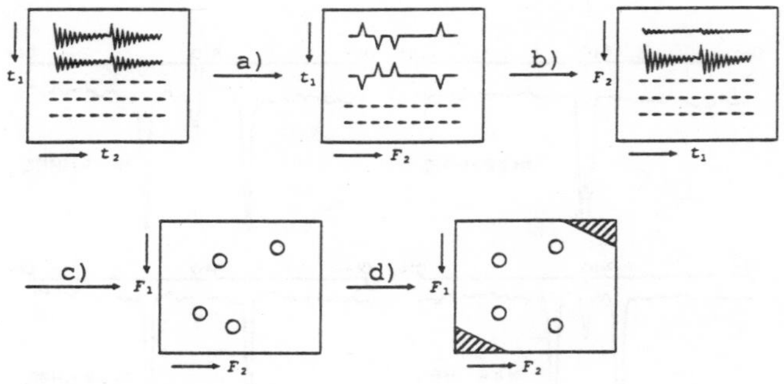

Fig. 2.5

a) FT for the time variable t_2
b) Coordinate change (some programs do not use this procedure and carry out the FT directly on t_1)
c) FT for time variable t_1
d) With 45° tilt

1) W.P. Aue, J. Karhan and R.R. Ernst, *J. Chem. Phys.*, **64**, 4226 (1976)
2) K. Nagayama, P. Bachmann, R.R. Ernst and K. Wüthrich, *Biochem. Biophys. Res. Commun.*, **86**, 218 (1979)

2.2 Heteronuclear *J*-Resolved Spectroscopy

This is a method of plotting the ^1H-^{13}C spin coupling constant against the chemical shift of ^{13}C. Basically it is the same as *J*-resolved spectroscopy for ^1H, but it can be divided into different methods (A)-(C) according to the manner of modulation of the spin-spin coupling constant ($^1J_{CH}$). So that modulation by spin-spin coupling is not cancelled out by refocusing, decoupling up to the application of 180° pulse (^{13}C) is carried out in (A) and (B).[1] In the pulse sequence in (A), the final tilt is necessary as in the *J*-resolved spectroscopy for ^1H. In (C) the 180° pulse for ^1H inverts the spin state of the spin-coupled partner nucleus at the same time as the application of 180° pulse to ^{13}C to prevent refocusing.[2]

Pulse sequences in (B) and (C) can be used for long-range coupling. In (B), the interval between the 90° and the 180° pulses is selectively decoupled, while broadband proton

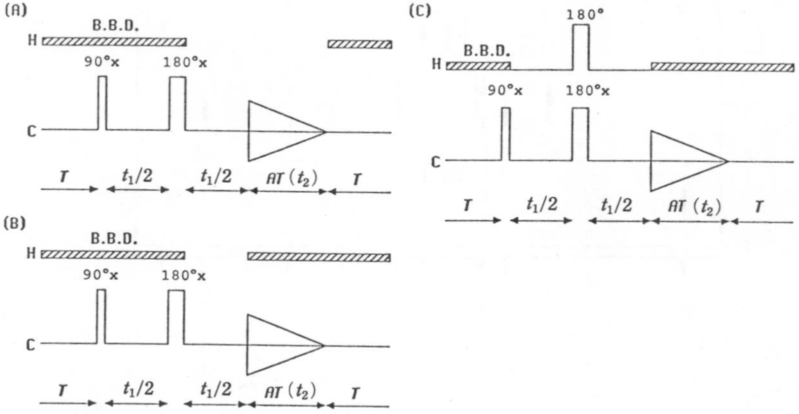

Fig. 2.6

decoupling is carried out in other areas. In (C) the selective 180° pulse is applied to ^{1}H. In all instances, the increase in t_1 should be large for small spin couplings.

The observation of long-range coupling by using of (C) was done by Ad Bax *et al.* (1982)[3, 4] for carvone (Fig. 2.7) and 2-acetonaphthone (Fig. 2.8), with application of a 180° pulse to the h-proton in carvone and to the proton on 1-carbon in 2-acetonaphthone, to identify the carbon in long-range coupling

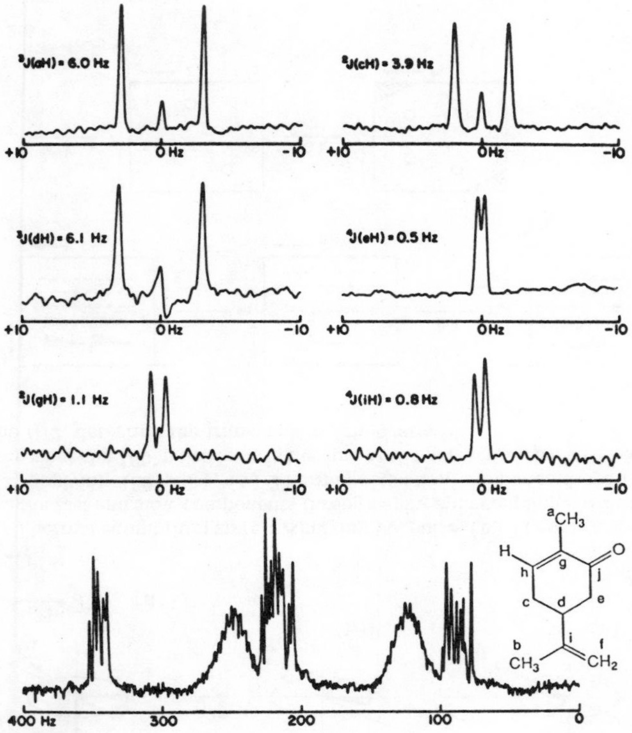

Fig. 2.7

(Reproduced by permission from Bax, Ad. and Freeman, R., *J. Am. Chem. Soc.*, **104**, 1099 (1982))

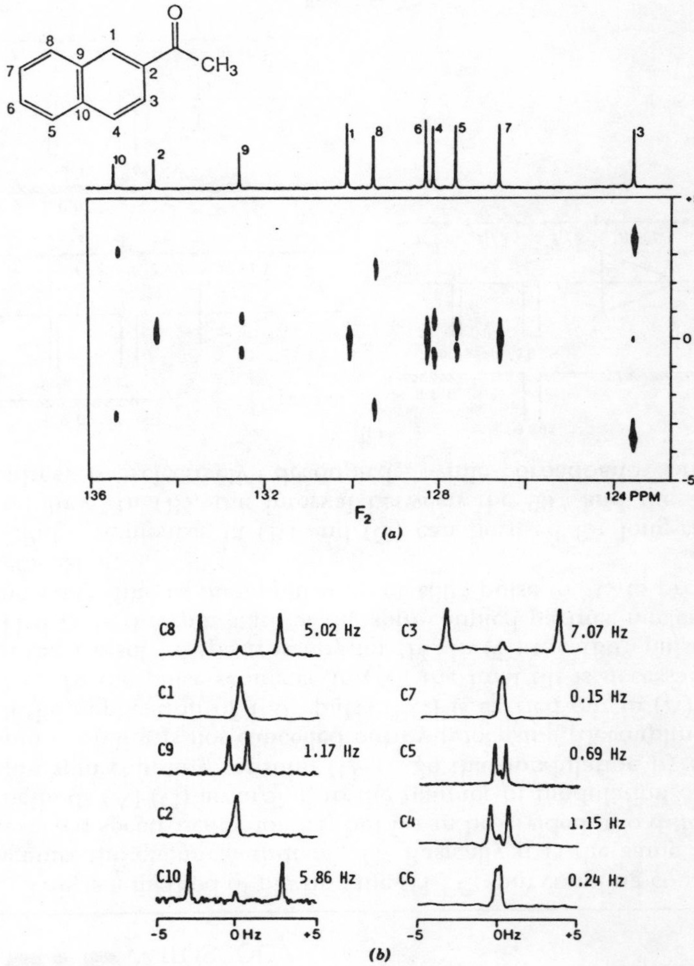

Fig. 2.8

(Reproduced by permission from Levy. G. C., *Topics in Carbon* 13 *NMR Spectroscopy*, vol. 4, p. 225, John Wiley & Sons (1984))

with that proton. The advantage of the procedure is that spin-spin coupling by protons other than that under irradiation becomes totally irrelevant, the J spectrum presenting nothing but simple doublets so that the J value can be easily determined.

1) G. Bodenhausen, R. Freeman and D.L. Turner, *J. Chem. Phys.*, **65**, 839 (1976)
2) G. Bodenhausen, R. Freeman, R. Niedermeyer and D.L. Turner, *J. Magn. Reson.*, **24**, 291 (1976)
3) A. Bax and R. Freeman, *J. Am. Chem. Soc.*, **104**, 1099 (1982)
4) G.C. Levy (Ed.), *Topics in Carbon-13 NMR Spectroscopy*, vol. 4, John Wiley & Sons., Inc., New York (1984) section 8, pp.222-226

2.3 COSY, COSY-45, PCOSY, DCOSY

When one thinks of "two-dimensional NMR," the first term that comes to mind may be COSY (correlation spectroscopy). It is most often used for ^{1}H-NMR, and provides information on complex spin-spin coupling system in the form of two-dimensional mappings. The pulse sequences are shown in Fig. 2.9. They include the sequence first proposed by Jeener (1971)[1] (A)[2] and its variants.[3, 4]

(A) COSY

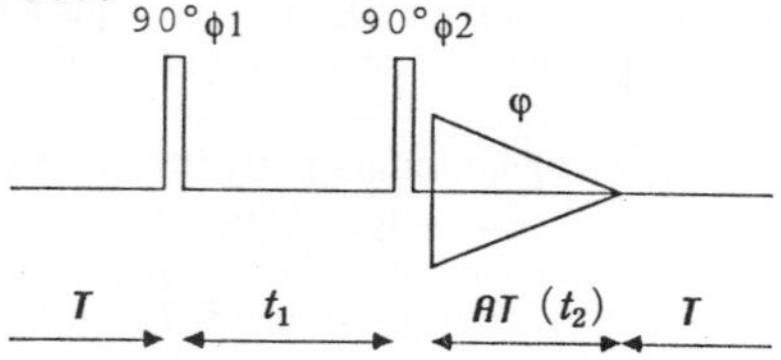

(1) Absolute value mode

φ1	x	x	x	x
φ2	x	y	-x	-y
φ	x	-x	x	-x

(2) Phase-sensitive mode

Real t_1

φ1	x	x
φ2	x	-x
φ	x	x

Imaginary t_1

φ1	x	x
φ2	y	-y
φ	y	y

(B) COSY45

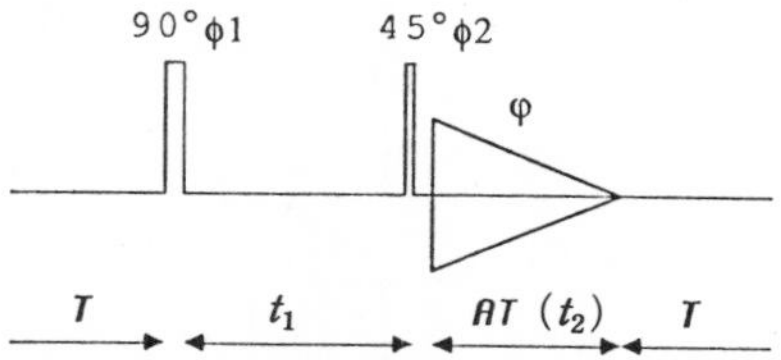

Fig. 2.9

Basically the procedure is to apply the first 90° pulse to a spin-coupled system in a state of thermal equilibrium to obtain transverse magnetization, and after t_1 during the non-equilibrium state, another 90° pulse is applied. In the case of amplitude modulation such as phase-sensitive COSY[5] by State's method or TPPI, which will be touched upon later, these two 90° pulses make longitudinal and transverse magnetization of various magnitudes depending on the offset frequency from the carrier and the spin-spin coupling constant as shown in Fig. 2.10. This means that the size of the longitudinal and transverse magnetization for another transition is altered, because there is a scalar coupling connectivity in the spin system. Since this transverse magnetization generates FID response in a detection coil, the resultant mutual modulation of frequency can be observed. A more precise expression can be given by the product operator approach, which is closely related to the density matrix calculation.

In a pulse sequence such as that shown in Fig. 2.9 (B) with the second pulse being 45° (COSY 45),[4] the pattern of magnetization transfer differs slightly from a 90° pulse, and the cross peaks

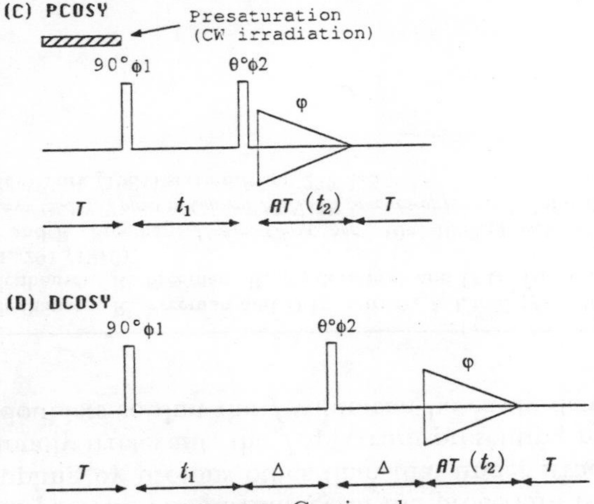

Fig. 2.9 Continued

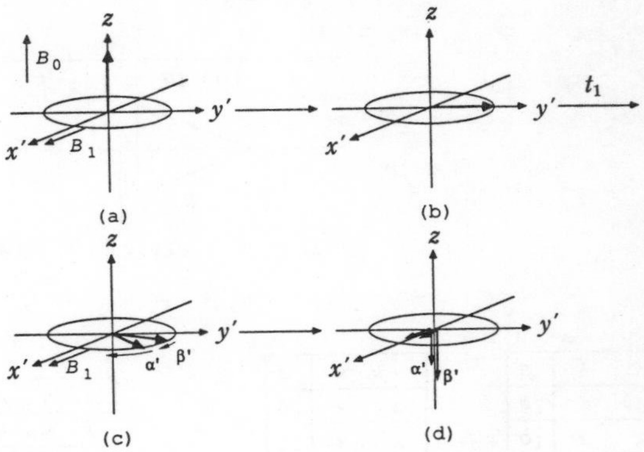

Fig. 2.10

(a) Thermal equilibrium state ; Applying first 90° pulse, (b) Transverse magnetization, (c) After t_1 applying second 90° pulse (The vectors marked α' and β' mean that their coupled partner is α or β state. The average velocity of α' and β' is precision frequency in a rotating frame, and the difference in velocity between α' and β' depends on the spin–spin coupling constant J.), (d) Longitudinal and transverse magnetization components

This shows the behavior of an off–resonance magnetization component in a spin–coupled system observed in a rotating frame. The appearance of the chage in the longitudinal magnetization component indicates the change in population at each energy level (see section on SPT and H, X–COSY).

may be tilted. By noting the direction of the tilt, the sign of the spin-spin coupling constant can be determined. Fig. 2.9 (C) shows a COSY obtained with selective saturation of large unwanted signals such as solvent, while in (D) the contribution from large spin-spin couplings are suppressed by inserting a delay time (Δ) after each pulse, to emphasize long-range coupling.[4] The data processing method is shown in Fig. 2.11.

Recently, some modifications as shown in Fig. 2.12[7–9] have been presented. The details of these methods are not touched upon here. Please see the references for further information.

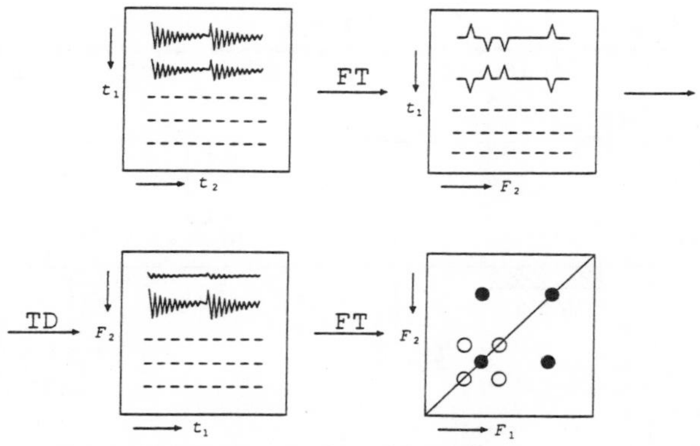

Fig. 2.11

Digital filtering is done before FT. The final F_1F_2 can be symmetrized.
TD=Coordinate change
FT=Fourier transform

1) J. Jeener, *Ampère International Summer School*, Basko Polje, Yugoslavia (1971)
2) W.P. Aue, E. Bartholdi and R.R. Ernst, *J. Chem. Phys.*, **64**, 2229 (1976)
3) A. Bax, R. Freeman and G.A. Morris, *J. Magn. Reson.*, **42**, 164 (1981)
4) A. Bax and R. Freeman, *J. Magn. Reson.*, **44**, 542 (1981)
5) H. Kessler, A. Müller and H. Oschkinat, *Magn. Reson. Chem.*, **23**, 844 (1985)
6) R.R. Ernst, G. Bodenhausen and A. Wokaun, *Principles of Nuclear Magnetic Resonance in One and Two Dimensions*, Oxford University Press, Oxford (1987)
7) A. Bax and R. Freeman, *J. Magn. Reson.*, **44**, 542 (1981)
8) H. Oschkinat, A. Pastore, P. Pfändler and G. Bodenhausen, *J. Magn. Reson.*, **69**, 559 (1986)
9) U. Piantini, O.W. Sørensen and R.R. Ernst, *J. Am. Chem. Soc.*, **104**, 6800 (1982)
C. Griesinger, O.W. Sørensen and R.R. Ernst, *J. Magn. Reson.*, **75**, 474 (1987)

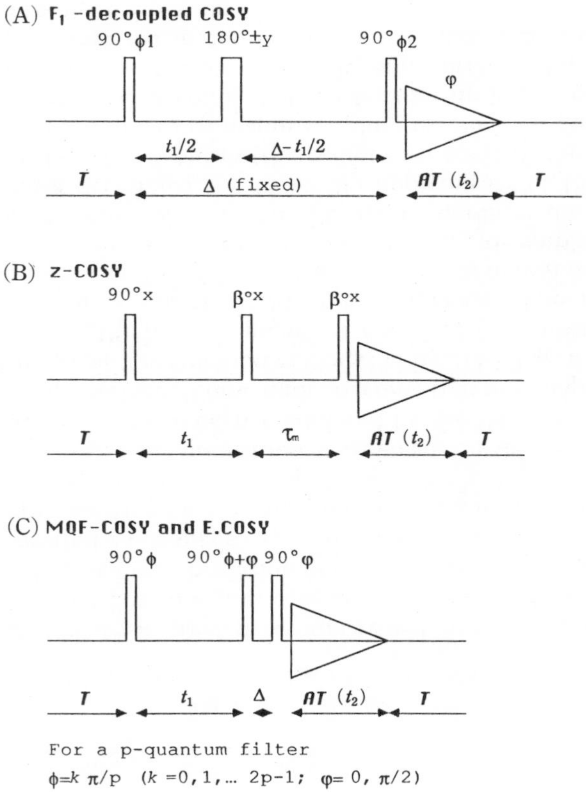

Fig. 2.12

[Parameters Determined by the Spectroscopist]

Ordinarily in COSY, the F_1 and F_2 axes have the same sweep width and therefore the increase in t_2 (dwell time: reciprocal of the sweep width) is set equal to t_1. The delay time (Δ) inserted in the pulse sequence of DCOSY (Fig. 2.9 (D)) is set at 1/4 J when the spin-spin coupling constant that one wants to emphasize is less than J. It should be noted that when the decay (FID) is rapid, S/N is poor.

2.4 H,X-COSY

Chemical shift correlation through spin-spin coupling such as COSY can be applied between heteroatoms such as ^{13}C and ^{1}H. The pulse sequence shown in Fig. 2.13 (A) is the first C-H correlation spectroscopy proposed by Ernst *et al.*[1] Basically it is the same as COSY; ^{1}H in a state of thermal equilibrium is converted to transverse magnetization by a 90° pulse, and by another 90° pulse after time t_1, longitudinal magnetization which includes the frequency information of proton (*i.e.*, the magnitude

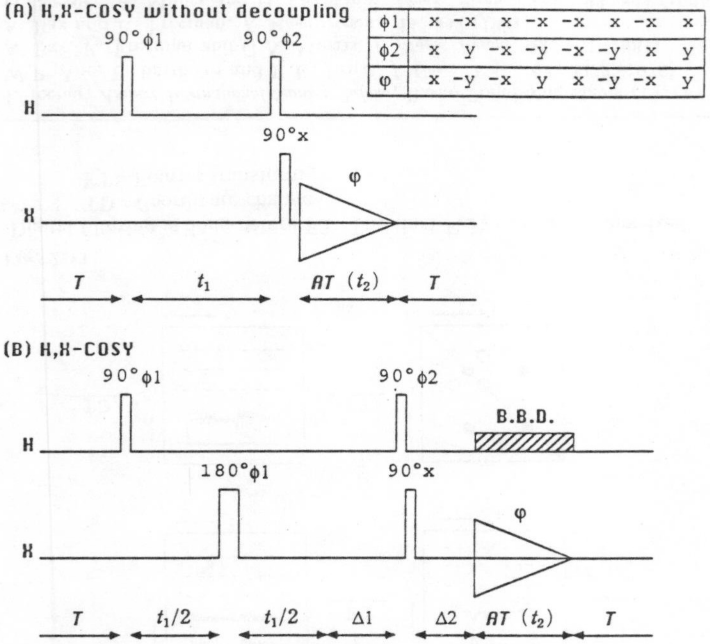

φ1	x	-x	x	-x	-x	x	-x	x
φ2	x	y	-x	-y	-x	-y	x	y
φ	x	-y	-x	y	x	-y	-x	y

Fig. 2.13

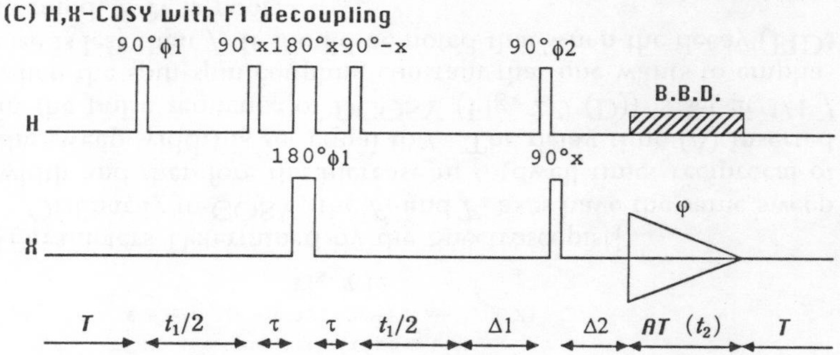

Fig. 2.13 Continued

$\Delta_1=1/2\,{}^1J_{XH}$, $\Delta_2=1/3\,{}^1J_{XH}$, $\tau=1/2\,{}^1J_{XH}$ $({}^1J_{XH} \gg J_{HH})$

varying with t_1) is created. In the two-spin system under consideration there is spin-spin coupling between C-H as shown in Fig. 2.14; thus the information of the longitudinal magnetization is transmitted as changes in the population of ^{13}C at the different energy levels, which means that there is a modulation by the ^{1}H precession frequency. By applying a 90° pulse to ^{13}C in this state, transverse magnetization is induced. In this way the ^{13}C FID data which have been modulated by the ^{1}H precession frequency are obtained. These data are converted to a map showing the chemical shift correlation of C and H by 2D-FT as in the case of COSY. In the pulse sequence of Fig. 2.13 (A), however, the ^{1}H-^{13}C coupling (about 120-250 Hz) appears. The pulse sequence shown in Fig. 2.13 (B) was designed to obtain decoupled spectra.[2-4] In (B), broadband proton decoupling was preformed while collecting the ^{13}C data, therefore the ^{13}C side results in a decoupled spectrum. By applying a ^{13}C 180° pulse between the ^{1}H 90° pulses (the center of t_1 period), the splitting magnetization due to spin-spin coupling is refocused and the ^{1}H side also becomes decoupled. In order that these procedures do not end up in cancellation of signals, appropriate delay times Δ_1 and Δ_2 are inserted to control the signal phases. Δ_1 and Δ_2 are set according to equation (3):

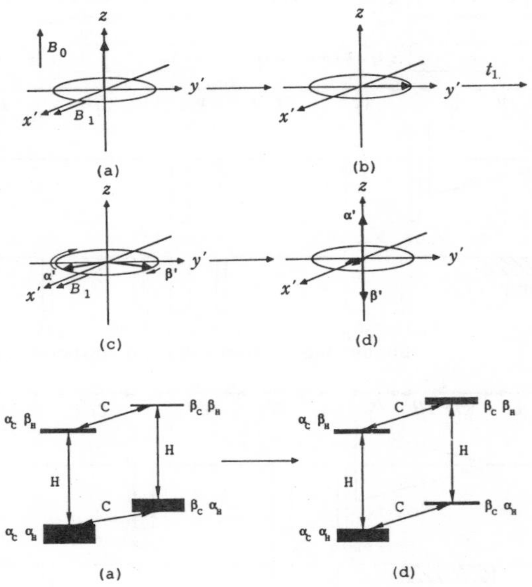

Fig. 2.14

(A) Spin vectors, (B) Energy level and spin population

(a) Thermal equilibrium state ; Applying first ^{1}H 90° pulse, (b) Transverse magnetization, (c) After t_1, applying second ^{1}H 90° pulse (The vectors maked α' and β' mean that thier coupled partner ^{13}C is α or β state. The average velocity of α' and β' is precession frequency of a proton in a rotating frame, and the difference of velocity between α' and β' is proportional to $1/{}^1J_{CH}$.), (d) Longitudinal and transverse magnetization components; The proton which is coupled with ^{13}C in β state is almost inverted in this situation.

$$\Delta_1 = 1/2J_{CH},\ \Delta_2 = 1/3J_{CH} \sim 1/4J_{CH} \tag{3}$$

assuming $J_{CH} = 150$ Hz, Δ_1 and Δ_2 are 3.3 and 2.2 ms, respectively.

By extending the delay time Δ_1 and Δ_2 to about 50 ms and 34 ms respectively and adding a J-selective filter as shown in Fig.

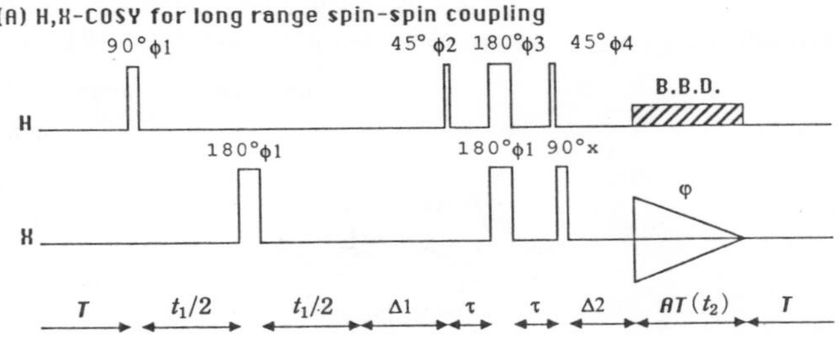

φ1	x	-x	x	-x	-x	x	-x	x
φ2	x	y	-x	-y	-x	-y	x	y
φ3	y	-x	-y	x	-y	x	y	-x
φ4	-x	-y	x	y	x	y	-x	-y
φ	x	-y	-x	y	x	-y	-x	y

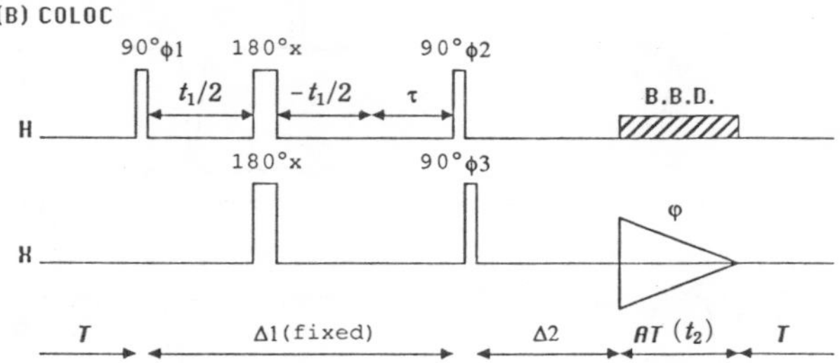

φ1	x	x	x	x	x	x	x	x	x	x	x	x	x	x	x	x
φ2	x	y	x	y	x	y	x	y	-x	-y	-x	-y	-x	-y	-x	-y
φ3	x	y	-x	-y	y	-x	-y	x	-x	-y	x	y	-y	x	y	-x
φ	x	x	-x	-x	y	y	-y	-y	x	x	-x	-x	y	y	-y	-y

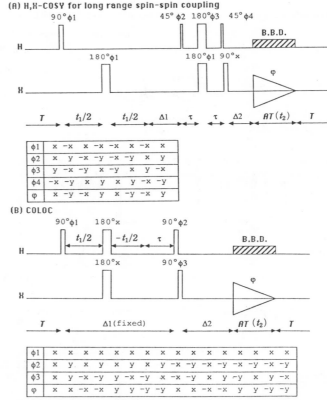

Fig. 2.15 (A) $\Delta_1 = 1/2\ {}^{1r}J_{XH}$, $\Delta_2 = 1/2\ {}^{1r}J_{XH} \sim 1/3\ {}^{1r}J_{XH}$, $\tau = 1/2\ {}^1J_{XH}$ (${}^{1r}J_{XH}$: long range spin-spin coupling between X and H)
(B) $\Delta_1 = 1/2\ {}^{1r}J_{XH}$, $\Delta_2 = 1/3 \sim 1/4\ {}^{1r}J_{XH}$

2.15 (A), a long-range H,C-COSY spectrum is obtained.[5] Additionally, the various pulse sequences for the detection of long-range scalar couplings such as COLOC (Fig. 2.15 (B))[6] are proposed and widely used. However, these sequences, when applied to small couplings (1-5 Hz), require lengthening of delay times even more, which results in appreciable relaxation so that the effect of polarization transfer is diminished. To overcome this problem, we have used the pulse sequence shown in Fig. 2.13 (A) only for carbonyl carbon to obtain a partial spectrum.[7] The protonated carbon such as methine or methylene is exclude, because it produces a very intense peak by C-H direct coupling to obscure small couplings. (In this sequense, the ^{13}C observation pulse can be replaced with a soft pulse of a narrow frequency band to detect only quaternary carbons even in the region of methylene carbons.)

With the pulse sequence of Fig. 2.13 (B), ^1H-^{13}C decoupling occurs but ^1H-^1H coupling remain. This often results in smaller cross peaks. The pulse sequence shown in Fig. 2.13 (C) was designed to overcome this problem and enable decoupling of

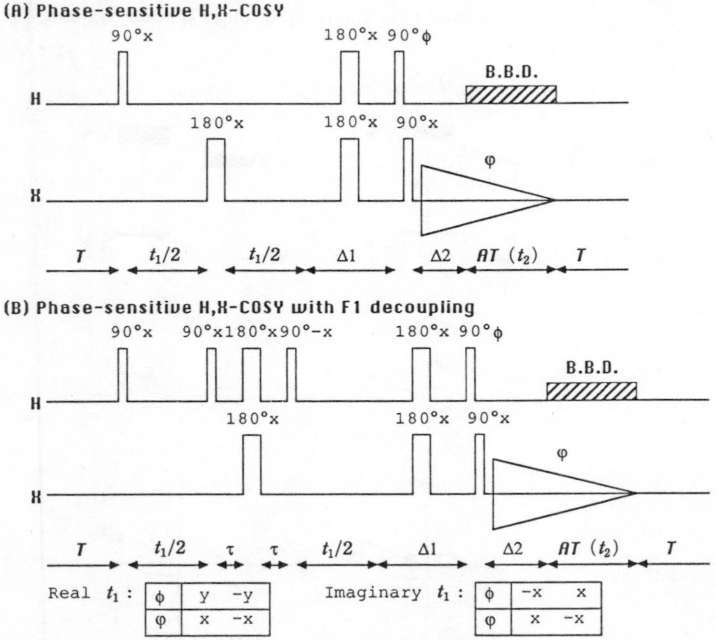

Fig. 2.16 $\Delta 1 = 1/2\,^1J_{XH}$, $\Delta 2 = 1/3\,^1J_{XH}$, $\tau = 1/2\,^1J_{XH}$ $(^1J_{XH} \gg J_{HH})$

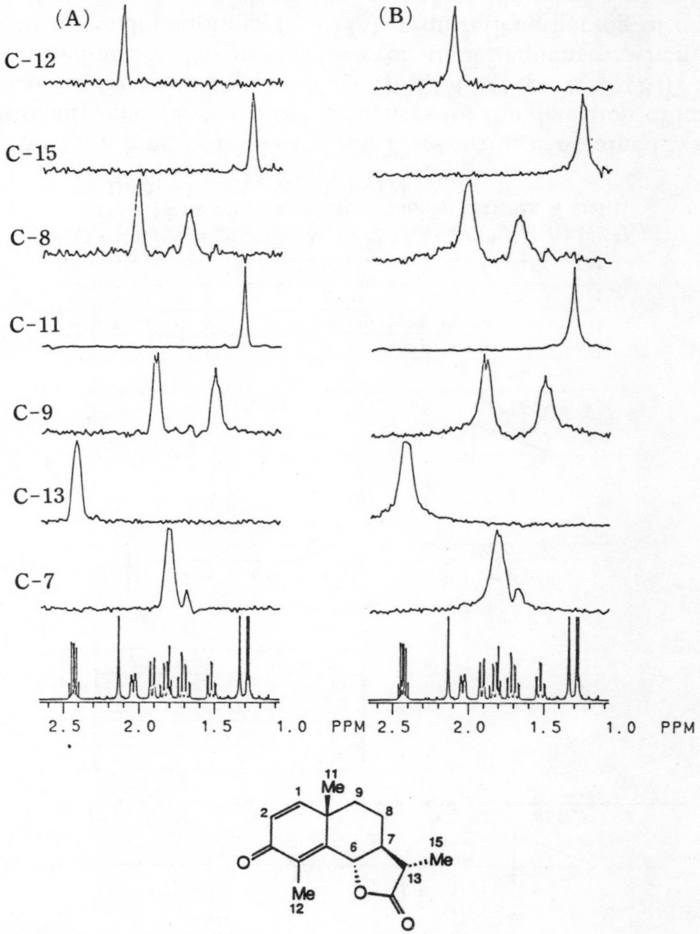

Fig. 2.17

The sample is α-santonin.

(A) : The cross sections of the phase-sensitive H, C-COSY spectrum

(B) : The absolute value mode description of (A)

^1H-^1H.[8] With this method, the S/N is improved and experimentation time is shortened.

In addition, phase-sensitive methods have recently been developed and widely used because of their good line-shape and high resolution. H,C-COSY and H,C-COSY with decoupling of ^1H-^1H can also be used in the phase-sensitive mode with the sequences as shown in Fig. 2.16.[9] As can be seen in Fig. 2.17, the line shape of these method is much better than that of absolute value mode.

[Parameters Determined the Spectroscopist]

1) Δ_1, Δ_2

Calculate from equation (3) using predicted J_{CH}.

2) Site of irradiation of decoupler

Let it coincide with the observation frequency of the control ^1H spectrum.

3) Decoupler output

Determine as for INEPT. Bax has suggested using the pulse sequence shown in Fig. 2.18 to determine the 90° pulse of the decoupler.[10]

4) t_1

As in COSY, this is determined by the sweep width on the ^1H side (set dwell time at the same length as that of the control ^1H spectrum).

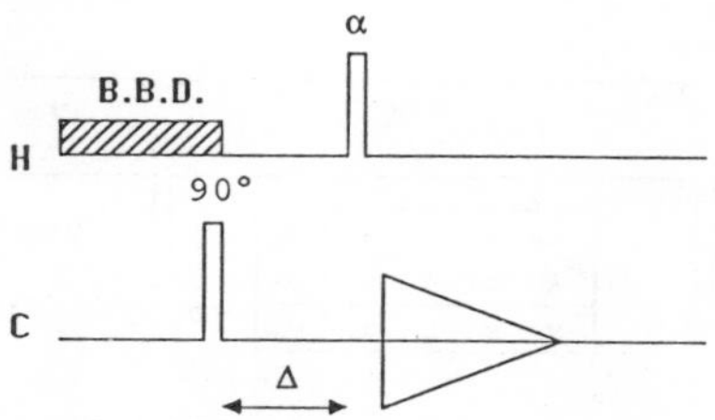

Fig. 2.18 $\Delta = 1/2\,J_{CH}$; signal is 0 when $\alpha = 90°$.

1) A.A. Maudsley and R.R. Ernst, *Chem. Phys. Lett.*, **50**, 368 (1977)
2) A.A. Maudsley, L. Müller and R.R. Ernst, *J. Magn. Reson.*, **28**, 463 (1977)
3) R. Freeman and G.A. Morris, *J. Chem. Soc. Chem. Commun.*, 684 (1978)
4) A. Bax and G.A. Morris, *J. Magn. Reson.*, **42**, 501 (1981)
5) M.J. Quast, A.S. Zektzer, G.E. Martin and R.N. Castle, *J. Magn. Reson.*, **71**, 554 (1987)
6) H. Kessler, C. Griesinger, J. Zarbock and H.R. Loosli, *J. Magn. Reson.*, **57**, 331 (1984)
7) T. Iwashita and H. Naoki, *23th Symposium on NMR*, Sendai, Japan (1984), *abstract* p.33
8) A. Bax, *J. Magn. Reson.*, **53**, 517 (1983)
9) A. Bax and S. Sarkar, *J. Magn. Reson.*, **60**, 170 (1984)
10) A. Bax, *J. Magn. Reson.*, **52**, 76 (1983)

2.5 NOESY

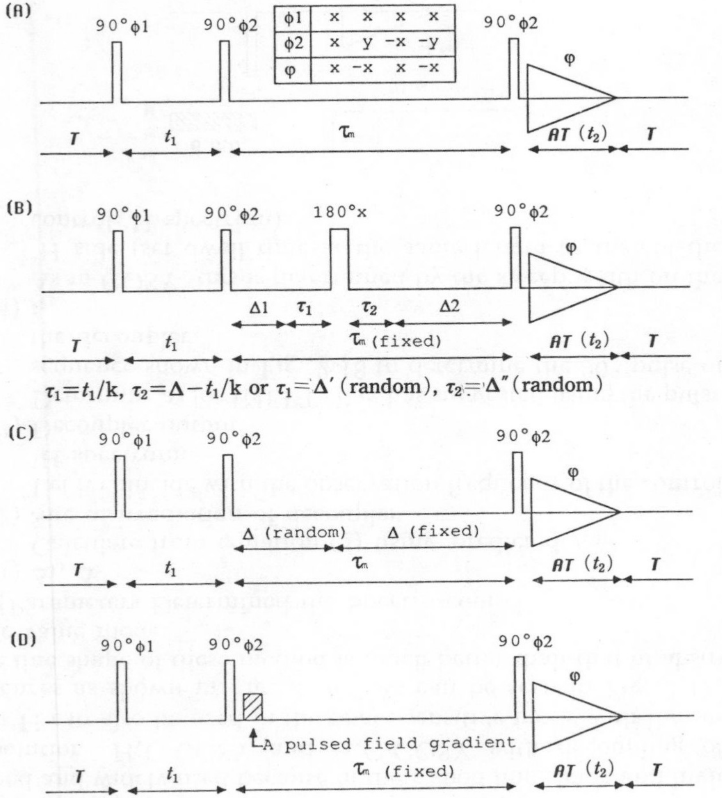

Fig. 2.19

This is a 2D NMR experiment used for the detection of NOE or of chemical exchange. The most basic sequence is that formed by the three pulses shown in Fig. 2.19 (A).[1–3]

Let us suppose that there is a cross relaxation (or chemical or conformational interchange) between A and X. Both A and X develop transverse magnetization from the state of thermal equilibrium by the application of the first 90° pulse. Examination of A shows that by applying another 90° pulse after t_1, the longitudinal magnetization component containing the information on precession frequency of A is obtained (refer to COSY). At the next mixing period (τ_m), the magnetization exchange due to cross relaxation (or chemical exchange) occurs between A and X, with the result that the longitudinal magnetization of X is modulated by the precession frequency of A. When this is converted to transverse magnetization by a 90° observation pulse and the experiment is allowed to evolve with increment time t_1, a two-dimensional FID data matrix is obtained. Thus the cross peak between A and X is obtained by the same process as for COSY.

Figure 2.19 (B)[4, 5] (C) and (D)[6, 7] are pulse sequences in which the cross peak due to spin-spin coupling is eliminated and correlation spectrum based only on NOE (or exchange) is produced. In (B), the residual transverse and longitudinal magnetization phase is coherence due to spin-spin coupling destroyed by the addition of a 180° pulse shifted systematically during the mixing period. In (C) and (D) the same process is used to introduce a random delay time or a field gradient pulse during the mixing period.

It has now become possible to measure the pure absorption (phase-sensitive) mode which distinguishes positive and negative NOE. (The pulse sequence is the same, but phase cycling and data processing are different, and there is no loss of information on the phase in F_1 dimensions corresponding to t_1.)

[Parameters Determined by the Spectroscopist]

Mixing time (τ_m)

The optimum level is related to the correlation time of the

molecule. There is therefore considerable range depending on molecular weight and solvent. For example, when the molecular weight is under 500 and the solution is in deuterochloroform, τ_m is in the order of the T_1 value of the nuclei involved. This must be shortened to 50 to hundreds ms when the molecular weight is much higher.

1) J. Jeener, B.H. Meier, P. Bachmann and R.R. Ernst, *J. Chem. Phys.*, **71**, 4546 (1979)
2) B.H. Meier and R.R. Ernst, *J. Am. Chem. Soc.*, **101**, 6441 (1979)
3) A. Kumar, R.R. Ernst and K. Wüthrich, *Biochem. Biophys. Res. Commun.*, **95**, 1 (1980)
4) S. Macura, K. Wüthrich and R.R. Ernst, *J. Magn. Reson.*, **47**, 351 (1982)
5) M. Rance, G. Bodenhausen, G. Wagner, K. Wüthrich and R.R. Ernst, *J. Magn. Reson.*, **62**, 497 (1985)
6) S. Macura, Y. Huang, D. Suter and R.R. Ernst, *J. Magn. Reson.*, **43**, 259 (1981)
7) S. Macura, K. Wüthrich and R.R. Ernst, *J. Magn. Reson.*, **46**, 269 (1982)

2.6 Phase-Sensitive NOESY

In COSY and NOESY, the dimension which corresponds to t_1 (F_1 axis) is expressed in absolute value (power spectrum) based only on the frequency of a given signal. Information on phase is therefore absent and the positive or negative sign of the NOEs, for example, cannot be determined. State *et al.* obtained signals of the entire absorption mode by devising the phase cycling of each pulse.[1] The pulse sequence of phase-sensitive

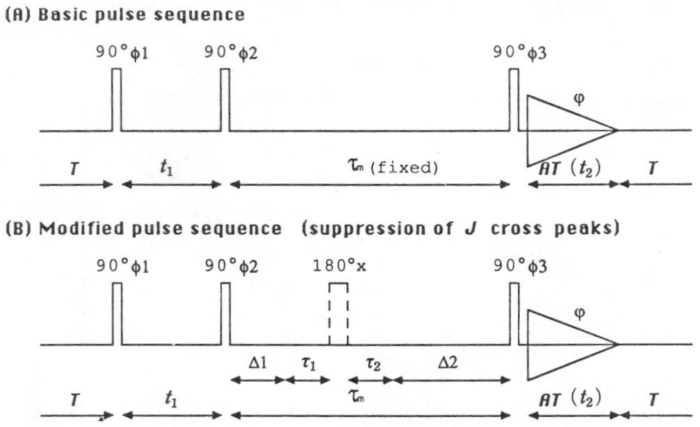

$\tau_1 = t_1/k$, $\tau_2 = \Delta - t_1/k$ or $\tau_1 = \Delta'$ (random), $\tau_2 = \Delta''$ (random)

The phase relationships of the pulses and receiver (Phase cycling)

ϕ1	ϕ2	ϕ3	φ	
0°(x)	0°(x)	0°(x)	+(x)	Real t_1
0°(x)	180°(−x)	0°(x)	−(−x)	
0°(x)	90°(y)	0°(x)	+(x)	Imaginary t_1
0°(x)	−90°(−y)	0°(x)	−(−x)	

Fig. 2.20

NOESY is basically the same as that of NOESY (Fig. 2.20). As shown in Fig. 2.21, two kinds of data corresponding to real t_1 and imaginary t_1 are acquired; FT for each t_2 is carried out, the real F_2 of both are combined to form a data set, and FT for t_1 is carried out to obtain a 2D data matrix. The 2D data obtained in this way consist of Lorentzian curves formed by the signals and there are no tails, so that there is less signal overlap and more phase information. There are advantages such as identifying the signal signs, but the drawback is that the procedure is time consuming.

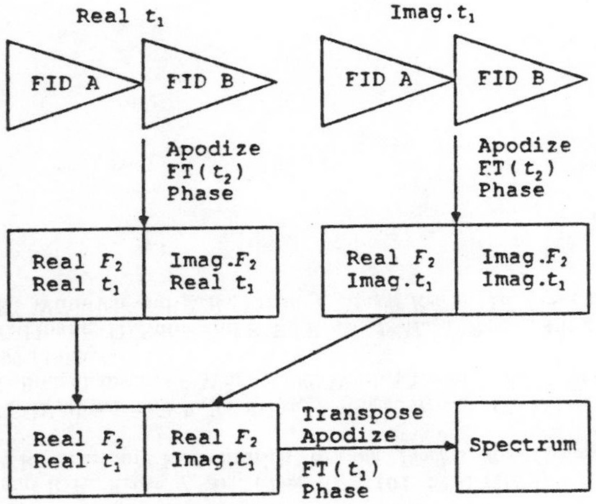

Fig. 2.21

Let us study the difference between phase-sensitive NOESY as developed by State *et al.* and the average NOESY. Figs. 2.22 and 2.23 are both NOESY spectra of bovine pancreatic trypsin inhibitor (BPTI), but Fig. 2.22 is a contour map in the absorption mode obtained by phase-sensitive NOESY, while the same data are presented in absolute values in Fig. 2.23. The latter has tails and the positions of cross peaks are difficult to determine, whereas they are easily located in Fig. 2.22.

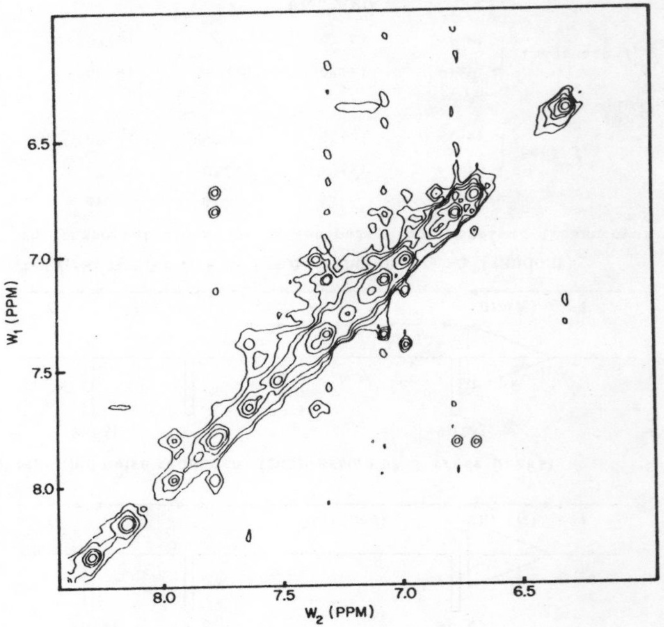

Fig. 2.22 (Reproduced by permission from State, D. J, *et al.*, *J. Magn. Reson.*, **48**, 288 (1982))

In addition to State's method, another method for pure absorption, TPPI (time proportional phase increment), was developed by Wüthrich and Ernst.[2] This method also causes amplitude modulation on each magnetization as in State's method.[3]

The NOE (nuclear Overhauser effect) depends on molecular motion (τ_c) and observation frequency (ω_0) as seen in section 62 and 64. For the two-spin system (I, S) which has dipole-dipole interaction, there are six relaxation routes as shown in Fig. 2.23. *W*s are transition probability. The NOE value ($f_I(S)$) can be estimated from these transition probabilities in accordance with equation (4), and the sign of the NOE depends on the cross-relaxation term $W_2 - W_0$.[4]

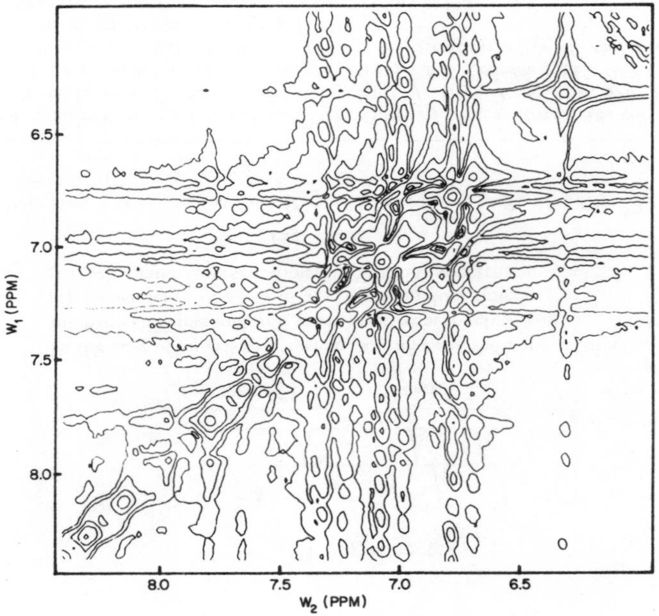

Fig. 2.23 (Reproduced by permission from State, D. J, *et al.*, *J. Magn. Reson.*, **48,** 290 (1982))

$$f_I(S) = \frac{W_2 - W_0}{2W_1^I + W_0 + W_2} \quad (4)$$

The dipole-dipole interaction contributes each transition probability as shown in equations (5) to (8),

$$W_1^I = \frac{3}{20} \cdot \frac{\gamma_I^2 \gamma_S^2 \hbar^2}{r} \cdot \frac{\tau_c}{1 + \omega_I^2 \tau_c^2} \quad (5)$$

$$W_1^S = \frac{3}{20} \cdot \frac{\gamma_I^2 \gamma_S^2 \hbar^2}{r^6} \cdot \frac{\tau_c}{1 + \omega_S^2 \tau_c^2} \quad (6)$$

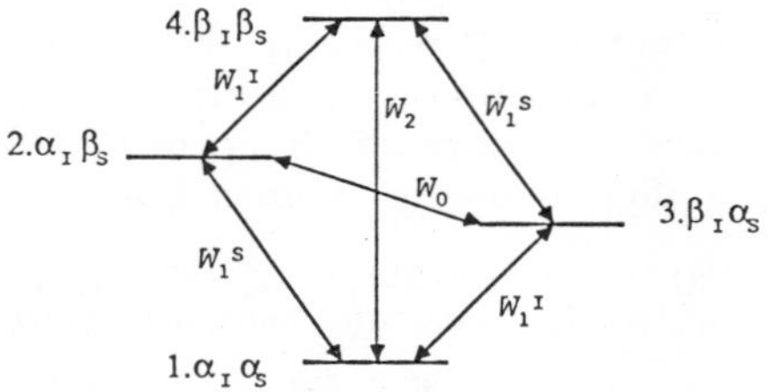

Fig. 2.24

Energy level diagram for the two-spin system (I, S) which has dipole-dipole interaction. $\alpha_I\beta_S$ means that spin I is α and spin S is β

$$W_0 = \frac{1}{10} \cdot \frac{\gamma_I^2 \gamma_S^2 \hbar^2}{r^6} \cdot \frac{\tau_c}{1 + (\omega_I - \omega_S)^2 \tau_c^2} \quad (7)$$

$$W_2 = \frac{3}{5} \cdot \frac{\gamma_I^2 \gamma_S^2 \hbar^2}{r^6} \cdot \frac{\tau_c}{1 + (\omega_I + \omega_S)^2 \tau_c^2} \quad (8)$$

where γ is the magnetogyric ratio, r the distance between I and S and $\hbar$ is Planck's constant divided by 2π. When I and S are protons, ω_I and ω_S are almost equal to the observation frequency ω_0, and $\gamma_I = \gamma_S = \gamma$. The cross-relaxation term $W_2 - W_0$ can be simplified as follows.

$$W_2 - W_0 = \frac{1}{10} \cdot \frac{\gamma^3 \hbar^2}{r^6} \cdot \left(\frac{6\tau_c}{1 + 4\omega_0^2 \tau_c^2} - \tau_c \right) \quad (9)$$

From equation (9), the relationship between the sign of NOE and $\omega_0\tau_c$ is derived.

$\omega_0\tau_c < 1.12$ $W_2 - W_0 > 0$: NOE is positive.
$\omega_0\tau_c = 1.12$ $W_2 - W_0 = 0$: NOE cannot be observed.
$\omega_0\tau_c > 1.12$ $W_2 - W_0 < 0$: NOE is negative.

Furthermore, Macura and Ernst showed the relationship between the mixing time of NOESY and $\omega_0\tau_c$ as shown in Fig. 2.25.[5)]

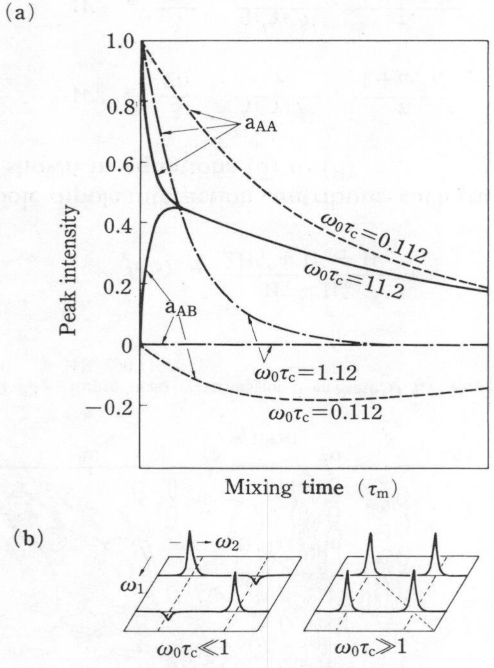

In the case of an AB spin system, a_{AA} and a_{AB} show the relative intensities of diagonal and cross peak respectively.
(a) Dependence of a_{AA} and a_{AB} on the mixing time τ_m.
(b) Schematic representations of NOESY **spectra.**

Fig. 2.25

(Reproduced by permission from Macura, S. and Ernst, R. R., *Mol. Phys.*, **41**, 103 (1980))

---

1) D.J. States, R.A. Harberkorn and D.J. Ruben, *J. Magn. Reson.*, **48**, 286 (1982)
2) G. Bodenhausen, H. Kogler and R.R. Ernst, *J. Magn. Reson.*, **58**, 370 (1984)
3) D. Marion and K. Wüthrich, *Biochem. Biophys. Res. Commun.*, **113**, 967 (1983)
4) J.H. Noggle and R.E. Schirmer, *The Nuclear Overhauser Effect—Chemical Applications*, Academic Press, New York (1971)
5) S. Macura and R.R. Ernst, *Mol. Phys.*, **41**, 95 (1980)

2.7 ROESY (or CAMELSPIN) and HOHAHA

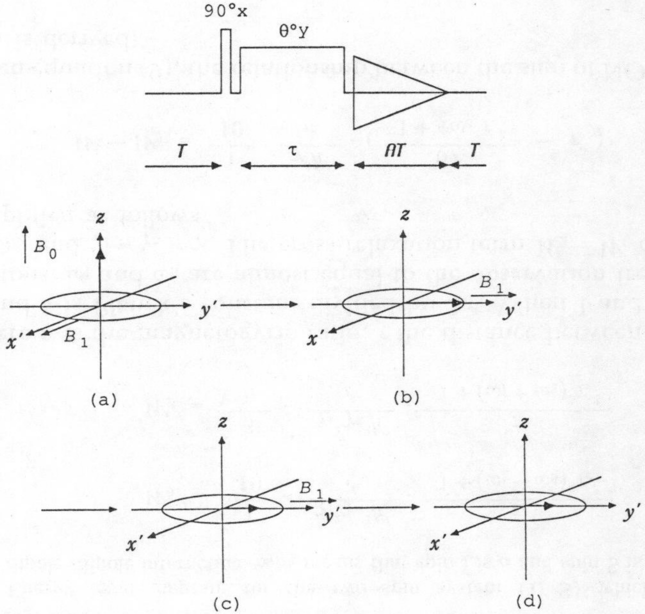

Fig. 2.26

(a) **Thermal equilibrium state ; Applying 90° pulse,** (b) **At the start of spin-locking,** (c) **During spin-locking time** (τ), (d) **Detection**

In the spin-lock experiment shown in Fig. 2.26,[1] the usual 90° pulse is applied to the longitudinal magnetization from the x' axis on the rotating frame to produce the transverse magnetization in the y' direction. If the oscillating magnetic field can be applied to the magnetization from the y' axis, it is fixed to the

y' direction and relaxed by the time constant $T_{1\rho}$. The HOHAHA (homonuclear hartmann-hahn spectroscopy) and ROESY are similar to the spin-lock experiment. The first 90° pulse and subsequent t_1 duration label the frequency of magnetization. During spin-locking time (mixing period), the magnetizations can be exchanged with each other through various interactions such as spin-spin coupling and cross relaxation, so the two-dimensional spectrum can be obtained.

ROESY (or CAMELSPIN) is a very simple experiment with a long weak pulse for spin-locking as shown in Fig. 2.27 (A).[2,3] The transverse NOE is measured by this experiment, which is always positive and becomes stronger with slower molecular motion.

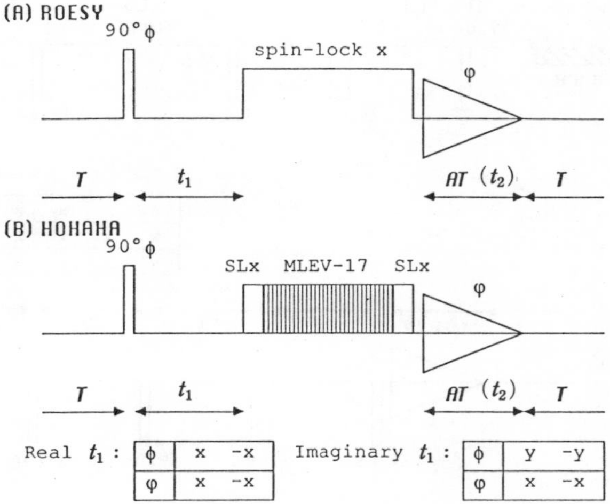

Fig. 2.27

In another experiment HOHAHA is used for determining the scalar coupling network.[4,5] For this purpose, the spin-lock duration is constructed with a short (2.5 ms) spin-lock field and MLEV-17 pulse sequences (mixing time) to attenuate the transverse NOE as shown in Fig. 2.27 (B). The advantage of HOHAHA is that net magnetization transfer occurs. In the case of the COSY experiment, differential magnetization transfer occurs (meaning that the components of the cross peaks are anti-phase to each other), and partial cancellation of the cross peaks is found, lowering sensitivity. Another feature of HOHAHA is to obtain relay peaks, which depend on the mixing time. The relay peaks are often used to assign overlapped protons.

---

1) M. Ohuchi, T. Fujito and M. Imanari, *J. Magn. Reson.*, **35**, 415 (1979)
2) A. Bax and D.G. Davis, *J. Magn. Reson.*, **63**, 207 (1985a)
3) A.A. Bothner-By, R.L.Stephens, J.-m. Lee, C.D. Warren and R.W. Jeanloz, *J. Am. Chem. Soc.*, **106**, 811 (1984)
4) A. Bax and D.G. Davis, *J. Magn. Reson*, **65**, 335 (1985b)
5) M.F. Summers, L.G. Marzilli and A. Bax, *J. Am. Chem. Soc.*, **108**, 4285 (1986)

2.8 RELAY

(A)

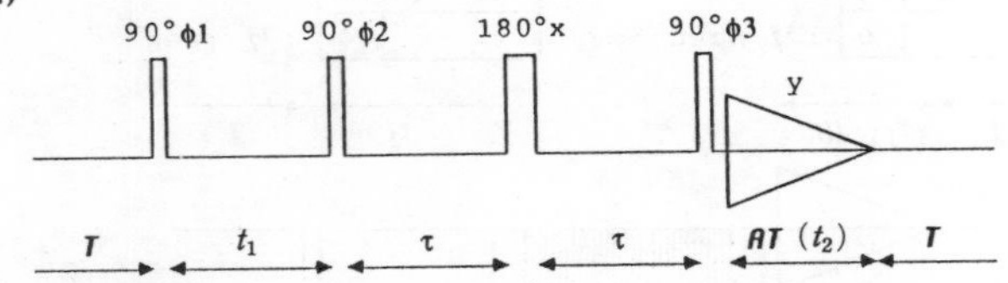

φ1	x	x	x	x
φ2	x	x	−x	−x
φ3	x	−x	x	−x

(B)

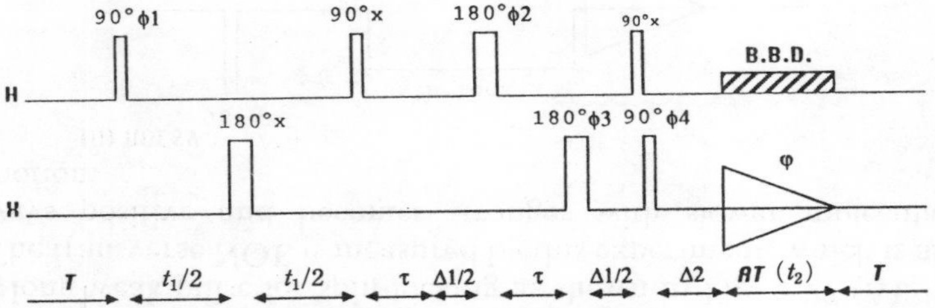

φ1	x	−x	y	−y	x	−x	y	−y
φ2	x	x	x	x	y	y	y	y
φ3	x	x	−x	−x	x	x	−x	−x
φ4	x	x	y	y	x	x	y	y
φ	x	−x	x	−x	x	−x	x	−x

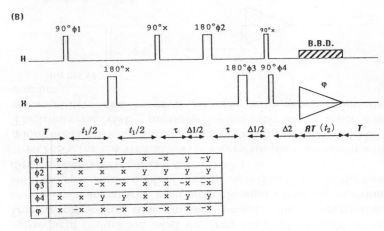

Fig. 2.28 (A) $\tau=1/J_{HH}$, (B) $\tau=1/10\,J_{HH}$, $\Delta_1=1/2\,{}^1J_{XH}$, $\Delta_2=1/3\,{}^1J_{XH}$

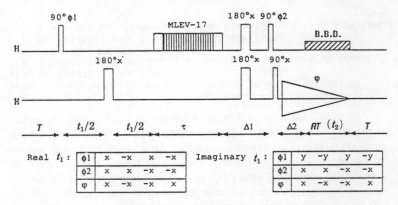

RELAY is a method of obtaining cross peaks in a spin system H_A-H_M-X through $J_{H_AH_M}$ and J_{H_MX} even when there is no spin coupling of H_A and X. When X is ^1H, the pulse sequence shown in Fig. 2.28 (A) is used.[1, 2] This is obtained by replacing the second 90° pulse in COSY with a sequence of 90°-τ-180°-τ-90°, during which time the information of H_A-H_M is transmitted through the H_M-X(^1H) bond. When X is ^{13}C the pulse sequence shown in Fig. 2.28 (B) is used.[3] This may also be regarded as a system in which the pulse sequence for transmitting information between H_A and H_M has been added to H,C-COSY pulse sequence. The second method is considerably more sensitive than 2D-INADEQUATE (see section 2.10) and clearly shows the ^{13}C connectivities.

Kessler *et al.* (1983) used an analog of somatostatin (see Fig. 2.29) to demonstrate RELAY (^1H_A-^1H_M-^{13}C). In Fig. 2.29 cross peaks are formed by the β carbon of Thr and the γ proton, β carbon of Pro and γ proton, and γ carbon and β proton through the transmission of spin information (coherence transfer).

Recently, the phase-sensitive H,C-COSY experiment was combined with the HOHAHA sequence by Bax, as shown in Fig. 2.30.[4] Many relay peaks can be observed by this sequence,

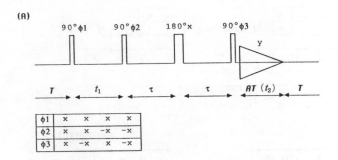

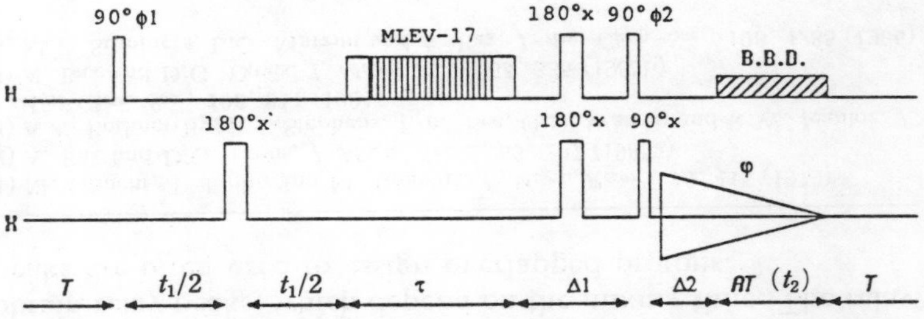

Real t_1:

φ1	x	−x	x	−x
φ2	x	x	−x	−x
φ	x	−x	−x	x

Imaginary t_1:

φ1	y	−y	y	−y
φ2	x	x	−x	−x
φ	x	−x	−x	x

Fig. 2.30 τ is mixing time. $\Delta_1=1/2\,{}^1J_{CH}$, $\Delta_2=1/3\,{}^1J_{CH}$

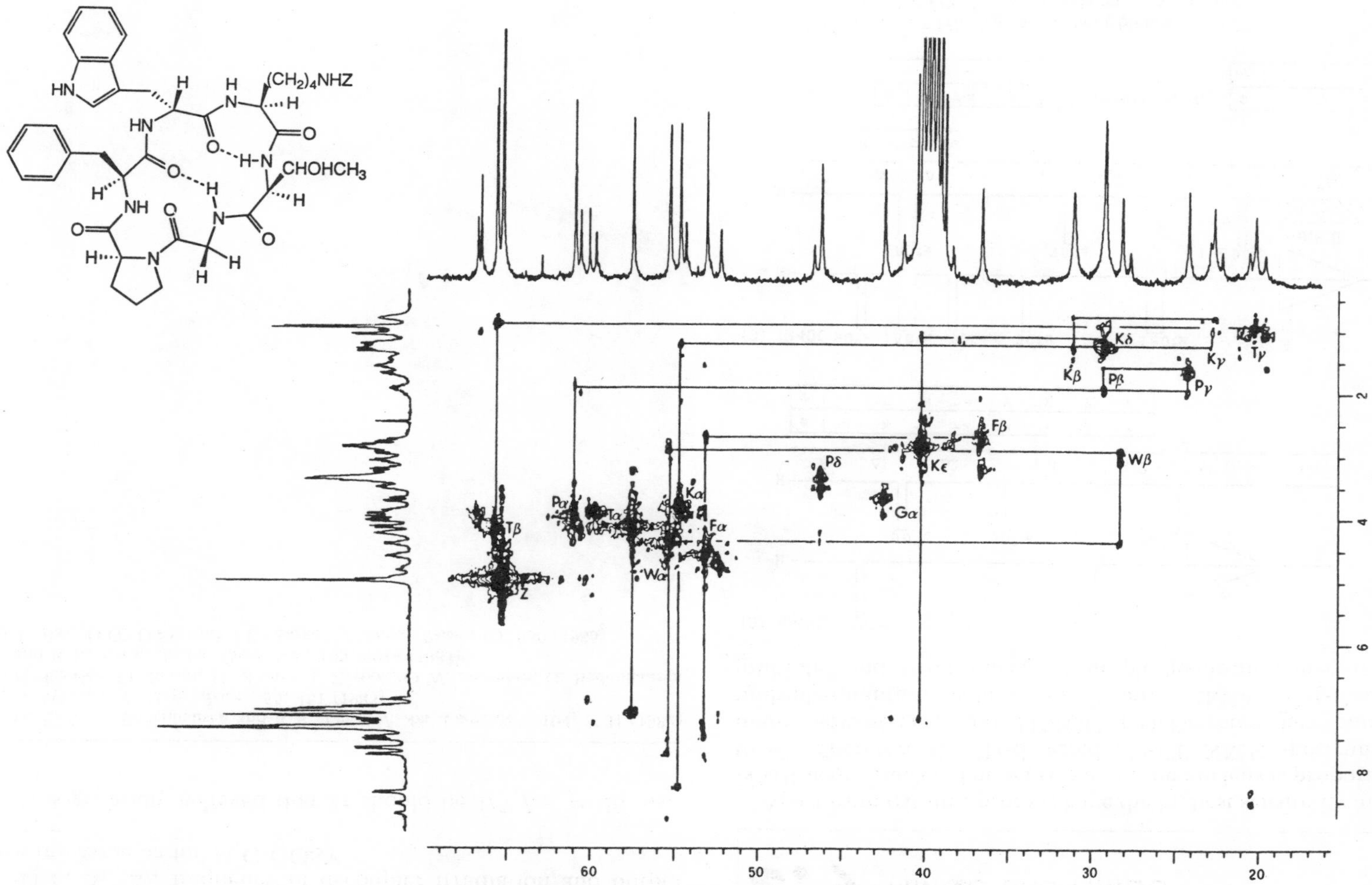

Fig. 2.29 Heteronuclear RELAY Experiment

(Reproduced by permission from Kessler, H., *J. Am. Chem. Soc.*, **105**, 6948 (1983)) thus clarifying the skeletal sequence of the molecule.

[Parameters Determined by the Spectroscopist]

1) t_1, Δ_1, Δ_2, frequency of decoupler irradiation and output are the same as for H,C-COSY.

2) τ

It is generally believed that 2τ should be $1/5\ J_{HH} \simeq 28$ ms.

---

1) G. Eich, G. Bodenhausen and R.R. Ernst, *J. Am. Chem. Soc.*, **104**, 3731 (1982)
2) G. Wagner, *J. Magn. Reson.*, **55**, 151 (1983)
3) H. Kessler, M. Bernd, H. Kogler, J. Zarbok, O.W. Sørensen, G. Bodenhausen and R.R. Ernst, *J. Am. Chem. Soc.*, **105**, 6944 (1983)
4) A. Bax, D.G. Davis and S.K. Sarker, *J. Magn. Reson.*, **63**, 230 (1985)

2.9 HMQC and HMBC

Apart from tritium, protons have the highest sensitivity among NMR active nuclei. The sensitivity of the nucleus is proportional to γ^3. Therefore the ^{1}H-detected 2D-FT NMR spectrum has many advantages, and HMQC (^{1}H-Detected heteronuclear multiple-quantum coherence)[1–3] and HMBC (^{1}H-Detected multiple-bond heteronuclear multiple-quantum coherence)[3, 4]

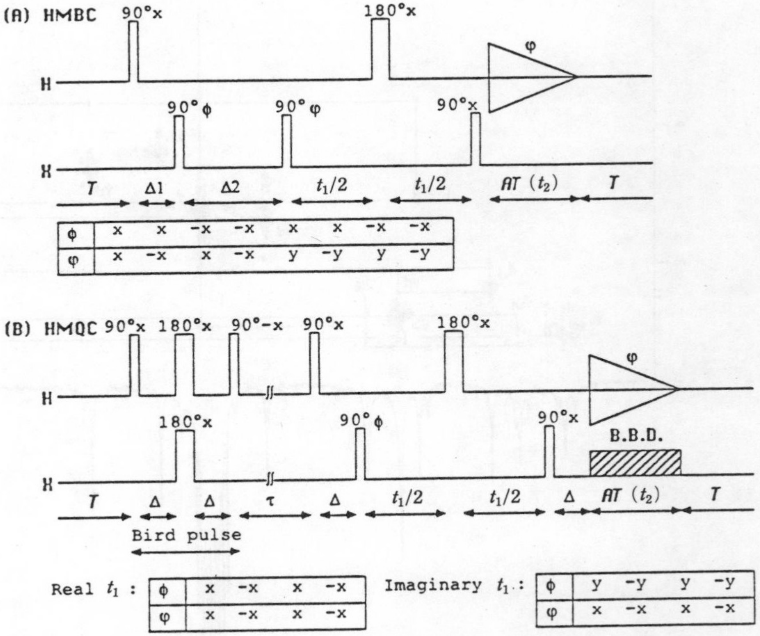

Fig. 2.31 (A) $\Delta_1 = 1/2\ ^1J_{XH}$, Δ_2 is about 60 ms.
(B) $\Delta = 1/2\ ^1J_{XH}$, τ depends on T_1 of proton.

techniques become increasingly important. HMQC is designed for the observation of the connectivity with direct coupling. HMBC can be used to observe long-range coupling. Usually, long-range H,C-COSY has low sensitivity so that a large amount of sample is required in comparison with the H,C-COSY technique. Because of its high sensitivity, HMBC is expected to be the most powerful method for the detection of long-range coupling.

Both methods must completely eliminate proton signals which are not coupled to the X-nucleus. Usual HMQC for small molecules has the BIRD pulse and the appropriate delay time τ in the beginning of the pulse sequence, which inverts protons not coupled to the X-nucleus, and attenuates the signals of these protons in the same manner as the WEFT method.

1) A. Bax, R.H. Griffey and B.L. Hawkins, *J. Magn. Reson.*, **55**, 301 (1983)
2) A. Bax and S.Subramanian, *J. Magn. Reson.*, **67**, 565 (1986)
3) M.F. Summers, L.G. Marzilli and A. Bax, *J. Am. Chem. Soc.*, **108**, 4285 (1986)
4) A. Bax and M.F. Summers, *J. Am. Chem. Soc.*, **108**, 2093 (1986)

2.10 2D-INADEQUATE

When Δ in the INADEQUATE experiment (Section 1.12) is replaced with t_1 and used as the evolution period (Fig. 2.32 (A)),[1–3] a 2D-FT NMR map is obtained in which the double quantum coherence frequency on the rotating frame (sum of offset frequencies) is correlated to each ^{13}C chemical shift. In this experiment, ^{13}C nuclei having the same double quantum coherence frequency are coupled with each other, and the connections in the carbon skeleton are found by tracing them. If the pulse sequence with echo type data collection as shown in Fig. 2.32 (B)

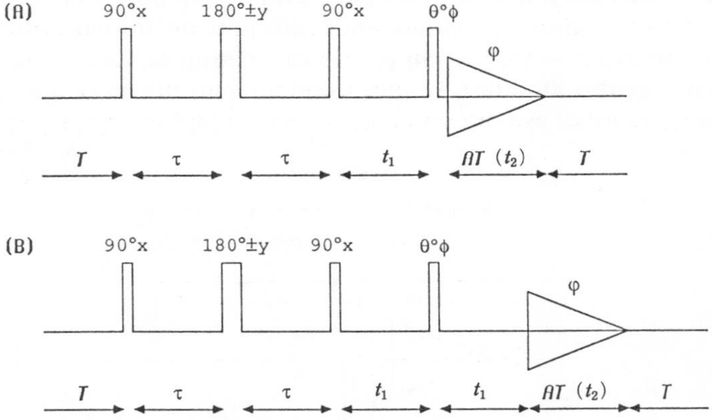

φ	x	y	−x	−y
φ	x	−y	−x	y

θ°=45° or 135°

Fig. 2.32 $\tau=(2n+1)/4J_{CC}$ ($n=0, 1, 2\cdots$)

is used,[4–6] then the double quantum coherence frequency is reduced to half and becomes the same as COSY.

The method is extremely attractive since it permits direct determination of the connections in the carbon skeleton, but the sensitivity is low. It remains, however, a most promising method anticipating improvement of ^{13}C probe.

[Parameters Determined by the Spectroscopist]

Delay time τ

The optimum conditions for obtaining double quantum coherence may be determined from equation (10).

$$\tau = (2n+1)/4\, J_{CC} \qquad (n = 0,\ 1,\ 2\ldots) \qquad (10)$$

τ is usually set at about 6 ms for spin-spin coupling when ^{13}C and ^{13}C are adjacent.

1) A. Bax, R. Freeman and T.A. Frenkiel, *J. Am. Chem. Soc.*, **103**, 2102 (1981)
2) A. Bax, T.A. Frenkiel, R. Freeman and M.H. Levitt, *J. Magn. Reson.*, **43**, 478 (1981)
3) T.H. Mareci and R. Freeman, *J. Magn. Reson.*, **48**, 158 (1982)
4) D.L. Turner, *J. Magn. Reson.*, **49**, 175 (1982)
5) D.L. Turner, *J. Magn. Reson.*, **53**, 259 (1983)
6) A. Bax and T.H. Mareci, *J. Magn. Reson.*, **53**, 360 (1983)

2.11 Multiple-quantum Filter

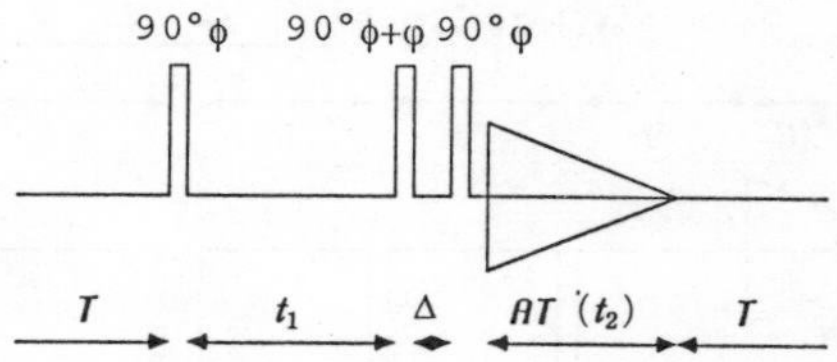

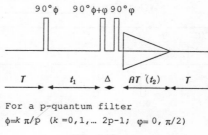

For a p-quantum filter
$\phi = k\,\pi/p$ ($k = 0, 1, \ldots$ 2p-1; $\varphi = 0,\ \pi/2$)

Fig. 2.33

In COSY, singlet peaks are often too intense so that S/N of the overall spectrum may be poor, and overlapping signals with the singlets may be difficult to see. What we are seeking in COSY is information on multiple spin-coupled hydrogen nuclei. The above objectionable feature can be eliminated if the singlet peaks could be filtered. The pulse sequence proposed by Piantini *et al.* (Fig. 2.33)[1] makes this possible through the use of a multi-quantum filter. The radiofrequency phase of each pulse is controlled as shown in Fig. 2.33. In the case of a double quantum filter, ϕ is varied in the order 0, $\pi/2$, $3\pi/2$ and singlet peaks which are incapable of double quantum transition are eliminated. By extending the design to triple and quadruple quantum filters, it should be possible to classify the spin systems.[2] (In the case of a triple or quadruple quantum filter, the apparatus must be able to control the radiofrequency phase of the pulse in 60° or 45° steps.)

Actual examples are shown in Figs. 2.34 and 2.35. Fig. 2.34 shows the double and triple quantum filters described by Piantini *et al.*.[1] In Fig. 2.34(b), singlets originating in dioxane and DMSO are removed by the double quantum filter and only the

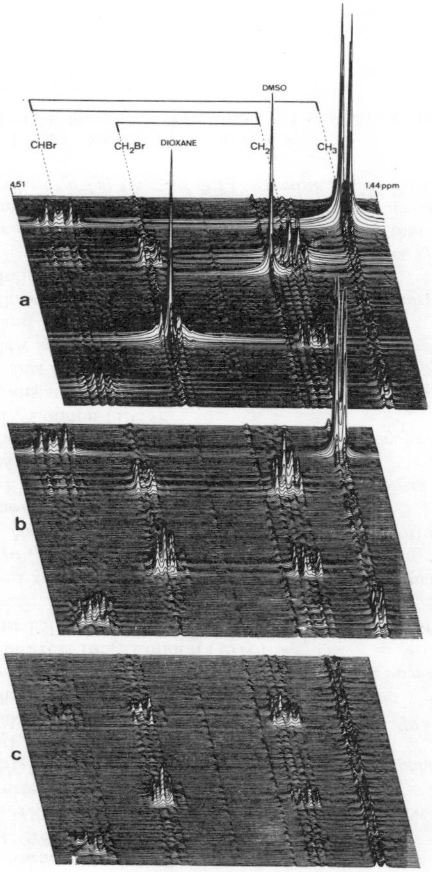

a : Usual COSY spectrum
b : The double quantum filter
c : The triple quantum filter[1)]

Fig. 2.34
(Reproduced by permission from Pianti, U, *et. al.*, *J. Am. Chem. Soc.*, **104**, 6800 (1982))

1,3-dibromobutane peak is seen. In Fig. 2.34(c), the cross peak of methyl, a doublet, is removed by the triple quantum filter.

Figure 2.35 is an example of the quadruple quantum filter described by Shaka and Freeman.[2)] 2,3-Dibromothiophene,

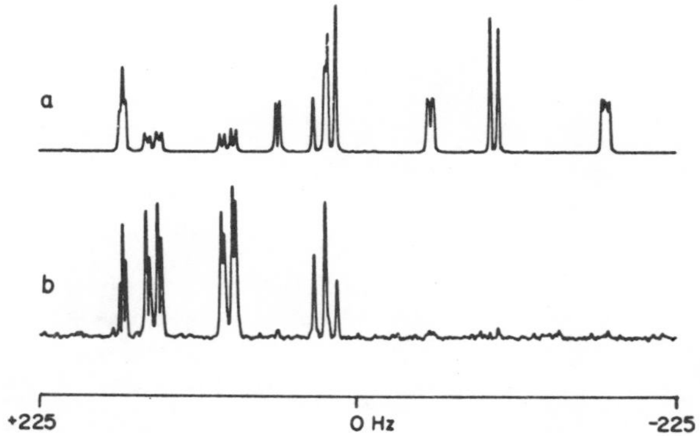

a : Normal spectrum
b : The quadruple quantum filter

Fig. 2.35 (Reproduced by permission from Shaka, A. J. *et al.*, *J. Magn. Reson.*, **51**, 169 (1983))

2-furoic acid and 1-bromo-3-nitrobenzene have 2-, 3- and 4-spin systems, respectively, but with the use of the quadruple quantum filter, only the signal of 1-bromo-3-nitrobenzene is left.

---

1) U. Piantini, O.W. Sørensen and R.R. Ernst, *J. Am. Chem. Soc.*, **104**, 6800 (1982)
2) A.J. Shaka and R. Freeman, *J. Magn. Reson.*, **51**, 169 (1983)

Bibliography

1) Y. Yukawa, (Ed.), *Lectures on Experimental Chemistry, Series 12. Nuclear Magnetic Resonance*, Maruzen Tokyo (1967) (*in Japanese*)
2) R.M. Lynden-Bell and R.K. Harris, *Nuclear Magnetic Resonance Spectroscopy*, Thomas Nelson & Sons Ltd., Middlesex, England (1969)
3) M. Nakasaki, *Nuclear Magnetic Resonance Analysis for Organic Chemists.*, Tokyo Kagaku Dojin (1976) (*in Japanese*)
4) K. Wüthrich, *NMR in Biological Research: Peptides and Proteins*, North-Holland Publishing Co., Amsterdam (1976)
5) E.D. Becker, *High Resolution NMR—Theory and Chemical Applications*, Academic Press, New York (1980)
6) K. Tori, T. Takeuchi and K. Furukawa, *Applied NMR—The usage of CW · FT NMR*, Kodansha, Tokyo (1984) (*in Japanese*)
7) K. Müllen and P.S. Pregosin, *Fourier Transform NMR Techniques—A Practical Approach*, Academic Press Inc., London (1976)
8) T.C. Farrar and E.D. Becker, *Pulse and Fourier Transform NMR—Introduction to Theory and Methods*, Academic Press, New York and London, (1971)
9) T. Miyazawa and Y. Arata, ed.: NMR—Review and Experimental Guide [I, II]. Nankodo, Tokyo (1983) (*in Japanese*)
10) C. Yamazaki, *Nuclear Magnetic Resonance Spectroscopy*, Kyoritsu Publ., Tokyo (1984) (*in Japanese*)
11) Y. Deguchi (Ed.), *New Lectures on Experimental Chemistry, Series 3, Basic Technique 2, Magnetism*, Maruzen, Tokyo (1976) (*in Japanese*)
12) R.K. Harris, *Nuclear Magnetic Resonance Spectroscopy—A Physicochemical View*, Pitman Books Ltd., London (1983)
13) H. Günther (translated by R.W. Gleason), *NMR Spectroscopy—An Introduction*, John Wiley & Sons Ltd., New York (1980)
14) J.W. Akitt, *NMR and Chemistry—An introduction to the Fourie transform multinuclear era*, Chapman and Hall Ltd., London (1983)
15) A. Bax, *Two-Dimensional Nuclear Magnetic Resonance in Liquids*, Delft University Press, Delft, Holland (1982)
16) R.R. Ernst, G. Bodenhausen and A. Wokaun, *Principles of Nuclear Magnetic Resonance in One and Two Dimensions*, Oxford University Press, Oxford (1987)
17) W.R. Croasmun and R.M.K. Carlson (Ed.), *Two-Dimensional NMR Spectroscopy—Applications For Chemists and Biochemists*, VCH Publishers, Inc., New York (1987)
18) A. Bax and L. Lerner, *Science*, **232**, 960 (1986)
19) H. Kessler, M. Gehrke and C. Griesinger, *Angew. Chem. Int. Ed. Engl.*, **27**, 490—536 (1988)

Index

A

absolute value 219
AB type signal 4
2-acetonaphthone 210
acetoxypachydiol dibenzoate 4
acquisition time 192
Actinomyces 52
Ad bax 210
adenanthin 94
adenine 132
adenosine 60
ailanthone 24, 120, 126, 128
Ailanthus altissima 24
alanine 172
albocycline 92
allylic coupling 82, 96
AMX 197
anomeric carbon 152
Aphanamixis grandifolia 96
aphanamol-I 96, 98, 102, 122, 146
APT 180
aspartic acid 30, 104, 106
attached proton test 180

B

bassianolide 26, 134
Bax 206, 217, 224
Beauveria bassiana Vaillant 26
BIRD pulse 227
Bloch 190
Bloch's equation 191, 200
Boltzmann distribution 22, 48, 190
Bombyx mori Linnaeus 26
brevetoxin A 114
brevetoxin B 186
broadband proton decoulping 40, 194, 214
broadband proton decoupling spectrum 32, 34, 36, 44, 46
1-bromo-3-nitrobenzene 229

C

CAMELSPIN 222, 223
CAMELSPIN spectrum 136
carrier frequency 191
Carr-Purecell-Meiboom-Gill method 201
Cartesian coordinate system 190
carvone 210
chemical exchange 197, 218
chemically exchangeable proton 62
chemically exchanging proton 134
chemical shift term 198
cholesteryl acetate 58, 180
cholic acid 12
chromazonarol 84, 86, 88, 90, 108, 110, 124
COLOC 215
COLOC spectrum 142, 164, 168, 170
compactin 52, 80, 82, 142, 144, 166
complete decoupling spectrum 32
contour level 88
contour plot 66, 72
correlation spectroscopy 211
correlation spectroscopy via long-range coupling 164
correlation time 126, 130
corss relaxation 220
COSY 76, 213, 214
COSY spectrum 78, 80, 82, 84, 86, 88, 90, 92, 94, 96, 102
COSY-45 211, 212
COSY-45 spectrum 100, 102, 106
CPMG 201
CPMG method 200
crenulacetal A 18
cross peak 76
cross-relaxation 197, 207, 218
cross-relaxation appropriate for minimolecules emulated by locked spin 136
cross relaxation term 130
cyanoginosin RR 170
cyanoviridin RR 170, 172
cytosine 132

D

DANTE 204
DCOSY 211, 213
DCOSY spectrum 98
decoupling 8
delay time 193
density matrix 208
deoxyribonucleotide 132
DEPT 34, 48, 50, 52, 54, 203
detection period 207
diagonal peak 76
1, 3-dibromobutane 229
2, 3-dibromothiophene 229
dictyodial 36
Dictyotaceae 4, 18, 36
Dictyota dichotoma 164
dictyotalide B 164
differential magnetization transfer 223
digital filter 193
digital resolution 84, 86, 90
dihedral angle 70
gem-dimethyl 94
gem-dimethyl system 86
2D-INADEQUATE 227
2D-INADEQUATE spectrum 178, 180, 182, 184,

186
dioxane 228
dipole-dipole interaction 130, 221
distance geometry method 132
distortionless enhancement by polarization transfer 50, 203
2D *J*-resolved spectroscopy 201
DMSO 228
DNA 12-mer 132
double exponential multiplication 193
double quantum coherence 205, 207
double quantum filter 104, 228
double quantum transition 130, 180
double quantum transition frequency 178
DQF-COSY 104, 108

E

E. COSY spectrum 106
Ernst 206, 220, 221
ethyl acetate 76
Eu $(fod)_3$ 6
evolution period 207
exponential filtering 193
exponential multiplication 193, 194
extreme narrowing condition 130

F

fast FT algorithm 193
FFT 193
FID 191
flip angle 192
fold-back 192
Fourier transform 191, 192
D-fucofranoside 152
2-furoic acid 229
free induction decay 191
Freeman 202, 229

G

gated decoupling 196
gated proton irradiation spectrum 38, 44
gated spin echo method 201
Gaussian line shape 193
Gaussian multiplication 193, 194
geminal coupling 100
generalized NOE 197
glutamic acid 172
gramicidin S 112, 130, 136
guanidine 136
guanine 132
Gymnodinium breve 114

H

Hahn's spin echo method 200, 201, 208
half-height width 94
H, C-COSY spectrum 138, 140, 142, 144, 146, 148, 150, 152, 154, 160, 168
H-detected heteronuclear multiple-quantum coherence 160, 226
H-detected multiple-bond heteronuclear multiple-quantum coherence 226
heteronuclear COSY 138, 156
heteronuclear *J*-resolved spectroscopy 209
heteronuclear *J*-resolved spectrum 74
heteronuclear multiple-bond connectivity 160
H, H-COSY 76, 152
HMBC 226, 227
HMBC spectrum 160, 168, 170, 172
HMQC 226
HMQC spectrum 160
HOHAHA 222, 223
HOHAHA spectrum 112, 114, 116, 170, 176
homoallylic coupling 96
homodecoupling 194, 195
homonuclear correlation spectroscopy 76
homonuclear COSY 138
homonuclear Hartmann-Hahn spectroscopy 223
homonuclear Hartmann-Hahn spectrum 112
Hund's rule 28
H, P-COSY spectrum 156
H, X-COSY 214
20-hydroxyecdysone 20

I

INADEQUATE 54, 205
incredible natural abundance double quantum transfer experiment spectroscopy 178
INDOR 197
INDOR spectrum 100
INEPT 34, 48, 50, 54, 202
INEPT-INADEQUATE 206
insensitive nuclear enhancement by polarization transfer 48, 202
inverse gated decoupling 196
inverse gated proton decoupling spectrum 40
inversion recovery 56, 58, 199
inversion recovery method 60
β-ionone 10, 16, 28, 32, 40, 42, 44, 48, 54, 56, 74, 78, 118, 138, 162
isobutylaldehyde 36

J

$^1J_{CH}$ 138, 162
$^2J_{CH}$ 162, 166
$^3J_{CH}$ 162, 166
Jeener 211
J modulation 201
JR 62, 204, 205
J-resolved spectrum 66, 68, 70, 72, 208
J-selective filter 215
jump and return 62, 204, 205

K

Karplus relationship 70
Kessler 224

L

laboratory frame 190
leucine 112, 136
light water 62
lipid A 156
longitudinal magnetization 207
longitudinal relaxation 200
longitudinal relaxation time 56, 199
longitudinal relaxtion time constant 190
long-range coupling 10, 14, 36, 38, 44, 46, 70, 78, 82, 86, 92, 94, 98, 162, 164, 172, 195, 209, 210, 212
long-range coupling constant 54
long-range H, C-COSY 215, 227
long-range H, C-COSY spectrum 162, 166
long-range selective proton decoupling 44, 195
Lorentz-Gauss fundtion 193
Lorentzian-Gaussian transformation 194
Lorentzian line shape 193
LSPD 38, 44, 46, 54, 195
LSPD spectrum 142

M

Macura 221
magnetic dipole moment 190
magnetogyric ratio 34, 190, 202, 221
maximum entropy method 168
L-menthol 140, 178
metasequoia tree 148
metasequoic acid A 148, 150
N-methyldehydroalanine 223
N-methylleucine 26
methyl-9, 12-octadecadienoate 6
methyl-2, 3, 5-tri-*O*-acetyl-β-methyl-2, 3, 5-tri-*O*-acetyl-β-D-fucofuranoside 154
microcystin RR 170
Microcystis vridis 170
mixing period 207
mixing time 118, 136, 218
MLEV-17 223
molecular motion 130
molecular movement 56, 58, 60
mosesin-2 14
Moses Sole 14
mugineic acid 68
multiple-quantum filter 228

N

negative NOE 22, 24, 26, 126, 128
net magnetization transfer 223
NOE 16, 32, 118, 126, 196, 202, 218, 220
NOE difference spectrum 16, 20, 196
NOEDS 16, 18, 22, 196
NOESY 207, 218
NOESY spectrum 118, 120, 124
noise decoupling 32
nuclear overhauser and exchange spectroscopy 118
nuclear overhauser effect 16

O

off-resonance decoupling 44,203
off-resonance proton decoupling 48, 194
off-resonance proton decoupling spectrum 34
ornithine 112, 136
oscillating magnetic fied 191

P

parthenolide 36
partially relaxed FT 200
particular problem 136
Pauli exclusion principle 28
PCOSY 211
phase cycling 192
phase-sensitive COSY 212
phase-sensitive double quantum filtered COSY 104
phase-sensitive DQF-COSY 104
phase-sensitive DQF-COSY spectrum 106, 108, 110
phase-sensitive H, C-COSY spectrum 158
phase-sensitive NOESY 219
phase-sensitive NOESY spectrum 126, 128, 130, 132, 134
phenylalanine 112, 136
Piantini 228
Planck's constant 190
polarization transfer 202, 203, 216
positive NOE 126
power spectrum 219
precession frequency 190
preparation period 207
presaturation 60
PRFT 199, 200
progressive relation 197, 198
proton coupled spectrum 38, 46
pseudo-echo-like FID 194
pseudo-INDOR 197
Ptychanthus striaitals 22
1 pulse experiment 192
1-1 pulse sequence 62
1-3-3-1 pulse 204, 205
1-3-3-1 pulse sequence 62
pulse rotation angle 192
purine 132
pyrimidine 132

Q

QPD 191, 192, 193
quadrature image 192
quadrature phase detection 192
quadruple quantum filter 229
quantum number 32

R

Rabdosia adenantha (Diels) Hara 94
radiofrequency phase 192
receiver phase 192
Redfield 214 204
regressive relation 197, 198
RELAY 224
relayed coherence transfer 174
relayed H, C-COSY 174
relayed H, C-COSY specturm 174, 176
relayed H, C-COSY/HOHAHA spectrum 176
residual coupling 34, 36
ROESY 222, 223
ROESY spectrum 136
rotating frame 191, 207, 222
rotating frame nuclear overhauser and exchange spectroscopy 136
rotenone 182

S

Salvia miltiorrhiza 46
salvilenone 46
α-santonin 2, 66, 174, 176
saturation 22, 60
saturation transfer 26, 62, 196, 197, 198
SDDS 8, 10, 12, 195
selective excitation 204
selective population transfer 8, 28, 30, 196, 197
selective proton decoupling 194
selective proton decoupling spectrum 42
semburin 70
Shaka 229
shark repellent 14
shift reagent 6
sine-bell filter 194
sine-bell function 193
single phase detection 192
single quantum transition 130
solvent effect 4
SPD 192
spin decoupling 8, 194
spin decoupling difference spectrum 8, 10, 195
spin echo 200, 203
spin-lock experiment 222

spin-locking 136
spin operator 208
spin quantum number 190
spin-spin coupling 207, 208, 209
spin-spin coupling constant term 198
spin tickling 8
SPT 28, 30, 196, 197
SPT spectrum 100, 106
stacked plot 66, 72
State 220
State's method 212, 220
Static magnetic field 190
steady state NOE 196
streptomyces 92
striatene 22
strychnine 158, 160, 168
sucrose 62
Swertia japonica makino 70
S/N 62
S/N ratio 193

T

T_1 32, 56, 58, 60, 199
T_1 measurement 199
T_2 200
T_2 measurement 200
TANGO 162
thymine 132
time proportional phase increment 220
TPPI 212, 220
transient NOE 197
transverse magnetization 207
transverse magnetization vector 191
transverse relaxation time 200
transverse relaxation time constant 190
trapezoidal multiplication 193, 194
trichilin-A 72
triple quantum filter 229
triple resonance 44

V

valine 112, 136
vicinal coupling 100

W

WALTZ-16 32
water eliminated FT 200
water eliminated FT-NMR 60
water signal 4, 60
WEFT 60, 199, 200, 227
W-type coupling 86, 90, 94, 98
Wüthrich 220

Z

zero quantum transition 130